BOOKS DO FURNISH A LIFE

BOOKS DO FURNISH A LIFE

READING AND WRITING SCIENCE

Richard Dawkins

EDITED BY GILLIAN SOMERSCALES

BANTAM PRESS

TRANSWORLD PUBLISHERS
Penguin Random House, One Embassy Gardens,
8 Viaduct Gardens, London SW11 7BW
www.penguin.co.uk

Transworld is part of the Penguin Random House group of companies
whose addresses can be found at global.penguinrandomhouse.com

First published in Great Britain in 2021 by Bantam Press
an imprint of Transworld Publishers

A CIP catalogue record for this book
is available from the British Library.

ISBNs 9781787633681 (hb)
9781787633698 (tpb)

Typeset in 12/15.5 pt Minion Pro by Jouve (UK), Milton Keynes
Printed and bound in Great Britain by Clays Ltd, Elcograf S.p.A.

The authorized representative in the EEA is Penguin Random House Ireland,
Morrison Chambers, 32 Nassau Street, Dublin D02 YH68.

Penguin Random House is committed to a sustainable
future for our business, our readers and our planet. This book
is made from Forest Stewardship Council® certified paper.

In memory of
Peter Medawar

CONTENTS

Editor's introduction   1

Author's introduction: The literature of science   4

I   TOOLS OF TWO TRADES: WRITING SCIENCE   19

In conversation with Neil deGrasse Tyson:
    On science and scientists, in public and private   21

The uncommon sense of science   38

Are we all related?   44

The timeless and the topical   47

Fighting on two fronts   54

Pornophilosophy   57

Determinism and dialectics: a tale of sound and fury   62

Tutorial-driven teaching   69

Life after light   76

A scientific education and the Deep Problems   79

Rationalist, iconoclast, Renaissance man   85

Revisiting *The Selfish Gene*   90

CONTENTS

**II WORLDS BEYOND WORDS:
CELEBRATING NATURE**   103

In conversation with Adam Hart-Davis:
     Evolution and plain writing in science   105

Close encounters with the truth   116

Conserving communities   120

Darwin on the slab   129

Life within life   132

Pure delight in a godless universe   138

Travelling with Darwin   143

Pictures of paradise   158

**III INSIDE THE SURVIVAL MACHINE:
EXPLORING HUMANITY**   161

Inconversation with Steven Pinker:
     Language, learning and debugging the brain   163

Old brain, new brain   176

Breaking the species barrier   183

Branching out   187

Darwinism and human purpose   192

Worlds in microcosm   199

Real genes and virtual worlds   221

Nice guys (still) finish first   229

Art, advertisement and attraction   234

From African Eve to the Banda strandlopers   240

We are stardust   253

The descent of Edward Wilson   255

IV THE MINER'S CANARY: SUPPORTING SCEPTICISM 267

In conversation with Christopher Hitchens:
Is America heading for theocracy? 269

Witness of internal delusion 279

Kicking the habit 284

The unburdening lightness of relief 287

A public and political atheist 290

The great escape 296

In His own words: a portrait of God 298

Liberation from theology 302

The God Temptation 309

The intellectual and moral courage of atheism 313

V COUNSEL FOR THE PROSECUTION: INTERROGATING FAITH 327

In conversation with Lawrence Krauss:
Should science speak to faith? 329

Defending the wall of separation 338

A moral and intellectual emergency 344

Unmasking the design illusion 347

'Nothing will come of nothing': why Lear was wrong 352

The fast food thesis: religion as evolutionary by-product 357

An ambitious banana skin 361

Heavenly twins 365

A tale of horror and heroism 368

CONTENTS

VI TENDING THE FLAME: EVANGELIZING EVOLUTION 373

In conversation with Matt Ridley:
From Darwin to DNA – and beyond 375
The 'little Penguin' relaunched 386
Foxes in the snow 393
Telling truth in a dark time 401
Irresponsible publishing? 407
Inferior design 414
The only kind of truth that works 418

Epilogue: To be read at my funeral 429

Sources and acknowledgements 431
Bibliography of works cited 437
Index 441

EDITOR'S INTRODUCTION

Never has the communication of science been more important than it is today. If, as Francis Bacon – literal and figurative Renaissance man – believed, knowledge itself is power, then never have humans had more power to act in ways beneficial to the future of the planet, its very fabric and its myriad inhabitants – and yet never, it seems, have they had less political will to make the necessary changes.

We live in a time when scientific knowledge and technological advance seem to be far outstripping the will to use them wisely. So a great burden falls on those with the knowledge that can inform human decision-making across the gamut of political, social, educational and commercial activity, and on those with the linguistic talent to command attention, to entice, to startle, above all to persuade. Those with both are the people who can, and who must, today speak truth to power if humanity is not to waste its potential in wasting the planet.

As I write this, humanity is hearing the loudest wake-up call that has rung out for many years, as the Covid-19 pandemic spreads across the world. Huge amounts of dedicated endeavour, political will, popular passion and rapid action have gone into efforts to combat this disease, its causes and its effects, all in a matter of months. One could forgive a cynic for noting the contrast with the laggardly response over decades to the long, slow burn of climate change – which is still going on while we all try to save our own lives and livelihoods. On both fronts, we rely on science to show us ways to cope, ways to survive, ways to improve; and we rely on scientists not only to do their complex, painstaking, demanding work, but to tell the rest of us what they are doing, and how, and the likely effects of their discoveries.

Never, then, has the communication of science been more important; and never has there been more pressure on the communicators. We live amid a multiplicity of media outlets, a barrage of multiway argument, revelation and contestation, a plethora of on- and offline academic publishing streams, constant quick-fire exchanges on social (and all too often anti-social) media. Where in this cacophony are we to find patient argument alongside passionate partisanship, the excitement of discovery alongside the determined discipline of interrogation, all in the cause of reason and science, respect and responsibility for this Earth we inhabit?

Well, we can read Richard Dawkins, for a start. Fortunately for the rest of us, there are across the world a great number of determined individuals – thinkers, researchers, speakers, writers – dedicated to the work of both doing and communicating science. They work alone and in teams, standing on the shoulders of giants, joining hands with their coevals, reaching out to inform and draw in those outside their own circles. Richard Dawkins is one of the most eminent of them – and, with characteristic energy and generosity of spirit, he is also one of the most active in promoting the efforts of others engaged in the same enterprise.

Hence this collection of short writings by an acknowledged master of the art of scientific communication. All are in one way or another connected with books, mostly science books – the books that have furnished Richard's life in science. The forewords, afterwords, introductions, reviews, essays reproduced here have all been composed to support, criticize or comment on work produced by others; to contribute to the vital task of spreading what we know through scientific method to be true, and defending it from those who would deny it, refuse it, misrepresent it.

In a volume all about communication, what better way to introduce each section than with a conversation? Each of the six parts of this collection begins with the edited transcript of a dialogue between Richard and another writer which reflects on its themes and relates them to the pressing issues of our time. And the collection as a whole is introduced by a new essay Richard has written specifically for this volume, in which he reflects upon 'The literature of science'.

The work of scientific communication is never-ending. We should all be grateful to those who have dedicated their lives to it – not only for the science they do, but for the words they write, both in and about the books with which we may furnish our own lives.

G.S.

AUTHOR'S INTRODUCTION

THE LITERATURE OF SCIENCE

Literature:

a   . . . that kind of written composition valued on account of its qualities of form or emotional effect.

b   The body of books and writings that treat of a particular subject.

(Shorter Oxford English Dictionary)

One of my teachers at Oxford encountered a junior colleague deep in the science branch of the Bodleian Library and stooped to murmur in the engrossed reader's ear. 'Ah, dear boy, I see you are consulting the literature. Don't. It will only confuse you.' 'Consulting' and 'the' give the game away. He was using 'literature' in the special way scientists do, a version of the *OED*'s '**b**' definition above. 'The' literature, for a scientist, is all those papers, often abstrusely and densely written, which pertain to a particular research topic. John Maynard Smith was once heard to say, 'There are those who read the literature. I prefer to write it.' His witticism didn't do himself justice, for he was a generous scholar who scrupulously read and credited the work of other scientists. But his quip again serves to illustrate the two meanings of 'literature'.

By 'the literature of science' in this essay, I mean something closer to the '**a**' definition from the *OED* above. I am talking about science as literature, good writing on the theme of science. This usually means books rather than scientific journals. As an aside, I think that's a pity. There's no obvious reason why a scientific paper shouldn't be gripping and entertaining. No reason why scientists shouldn't enjoy the articles it is their professional duty to read. During my spell as editor of the journal *Animal Behaviour* I tried to encourage authors to forsake not only the self-effacing scientific passive ('A different approach will be taken by the present author') but also the traditional and dreary 'Introduction, Methods, Results, Discussion' convention, in favour of telling a story. But now, to books.

I said 'good writing about science' and that may give the wrong impression. It doesn't have to mean 'fine' writing, certainly not if that conveys – as it can – a mood of pretentiousness, or *belles lettres*. I shall come on to Peter Medawar, the dedicatee of this book, in the second half of this essay for he was, in my opinion, the master of scientific literature in the sense that I intend. In his words, 'a scientist's fingers, unlike a historian's, must never stray toward the diapason'. Well, perhaps not quite never. The occasional purple passage is justified by the romance of science – the unimaginable scale of the expanding universe, the stately majesty of geological deep time, the complexity of a living cell, coral reef or tropical rainforest. The natural-history prose poetry of a Loren Eiseley or Lewis Thomas, the cosmic reveries of Carl Sagan, the prophetic sagacity of Jacob Bronowski? Medawar would not – certainly should not – censor them.

Science doesn't need languaging-up to make it poetic. The poetry is in the subject matter: reality. It needs only clarity and honesty to convey it to the reader and, with a little extra effort, to deliver that authentic tingling up the spine which is sometimes thought the prerogative of art, music, poetry, 'great' literature in the conventional sense. That sense is the one embraced by the awarders of the Nobel Prize in Literature. It explains why the prize is almost always given to a novelist or poet or playwright, occasionally to a philosopher, so far never to a real scientist. The only arguable exception is Henri Bergson who, if he could be called a scientist at all, sets a distinctly unfortunate precedent. I wonder whether it has simply never occurred to the Nobel committee that science – the poetry of reality – is a suitable vehicle for great literature. Here's Sir James Jeans, writing in 1930 of *The Mysterious Universe*:

> Standing on our microscopic fragment of a grain of sand, we attempt to discover the nature and purpose of the universe which surrounds our home in space and time. Our first impression is something akin to terror. We find the universe terrifying because of its vast meaningless distances, terrifying because of its inconceivably long vistas of time which dwarf human history to the twinkling of an eye, terrifying because of our extreme loneliness, and because of the material insignificance of our home in space – a millionth part of a grain of sand out of all the sea-sand in the world. But above all else, we find the universe terrifying because it appears to be indifferent to life like our own; emotion, ambition and achievement, art and religion all seem equally foreign to its plan.

Carl Sagan later said something similar in his famous *Pale Blue Dot* soliloquy:

> Look again at that dot. That's here. That's home. That's us. On it everyone you love, everyone you know, everyone you ever heard of, every human being who ever was, lived out their lives. The aggregate of our joy and suffering, thousands of confident religions, ideologies, and economic doctrines, every hunter and forager, every hero and coward, every creator and destroyer of civilization, every king and peasant, every young couple in love, every mother and father, hopeful child, inventor and explorer, every teacher of morals, every corrupt politician, every 'superstar,' every 'supreme

leader,' every saint and sinner in the history of our species lived there – on a mote of dust suspended in a sunbeam.

The Earth is a very small stage in a vast cosmic arena. Think of the rivers of blood spilled by all those generals and emperors so that, in glory and triumph, they could become the momentary masters of a fraction of a dot. Think of the endless cruelties visited by the inhabitants of one corner of this pixel on the scarcely distinguishable inhabitants of some other corner, how frequent their misunderstandings, how eager they are to kill one another, how fervent their hatreds.

Our posturings, our imagined self-importance, the delusion that we have some privileged position in the Universe, are challenged by this point of pale light. Our planet is a lonely speck in the great enveloping cosmic dark. In our obscurity, in all this vastness, there is no hint that help will come from elsewhere to save us from ourselves.

The Earth is the only world known so far to harbor life. There is nowhere else, at least in the near future, to which our species could migrate. Visit, yes. Settle, not yet. Like it or not, for the moment the Earth is where we make our stand.

It has been said that astronomy is a humbling and character-building experience. There is perhaps no better demonstration of the folly of human conceits than this distant image of our tiny world. To me, it underscores our responsibility to deal more kindly with one another, and to preserve and cherish the pale blue dot, the only home we've ever known.

In 2013 Carolyn Porco who, together with Neil deGrasse Tyson, is the nearest approach we have to a Carl Sagan *de nos jours*, initiated a beautiful commemoration by inviting the world's population to look up and smile at the camera, as her Cassini imaging team photographed us from Saturn, 'Pale Blue Dot' range, 898 million miles:

Look up, think about our cosmic place, think about our planet, how unusual it is, how lush and life-giving it is, think about your own existence, think about the magnitude of the accomplishment that this picture-taking session entails. We have a spacecraft at Saturn. We are truly interplanetary explorers. Think about all that, and smile.

For me, Carolyn had already earned her niche in the gallery of poetic scientists when she raced against time to have the ashes of her beloved mentor Eugene Shoemaker, narrowly denied in life his ambition to be the first geologist on the moon, included in the payload of a rocket that was about to go there, accompanied by these lines chosen by her from Shakespeare:

And when he shall die,
Take him and cut him out in little stars,
And he will make the face of heaven so fine
That all the world will be in love with night
And pay no worship to the garish sun.*

Peter Atkins, one of the finest English stylists among living scientists, takes the terror of the empty void and tames it with a blithe insouciance which some might condemn as scientistic but which I find magnificent:

In the beginning there was nothing. Absolute void, not merely empty space. There was no space; nor was there time, for this was before time. The universe was without form and void.

By chance there was a fluctuation, and a set of points, emerging from nothing and taking their existence from the pattern they formed, defined a time. The chance formation of a pattern resulted in the emergence of time from coalesced opposites, its emergence from nothing. From absolute nothing, absolutely without intervention, there came into being rudimentary existence. The emergence of

---

* *Romeo and Juliet*, Act III, scene 2.

the dust of points and their chance organization into time was the haphazard, unmotivated action that brought them into being. Opposites, extreme simplicities, emerged from nothing.

Lawrence Krauss develops a similar theme in *A Universe from Nothing*, for which I wrote the afterword (see pages 353–7 below). The book from which I took the Atkins quote, *The Creation*, ends with a fanfare of confidence in the power of science:

When we have dealt with the values of the fundamental constants by seeing that they are unavoidably so, and have dismissed them as irrelevant, we shall have arrived at complete understanding. Fundamental science then can rest. We are almost there. Complete knowledge is just within our grasp. Comprehension is moving across the face of the Earth, like the sunrise.

The chemist Peter Atkins could fairly be called a prose poet but he never lets style take precedence over clarity. Now, here is a biologist who sees the world through the eyes of a poet, but still decidedly scientific eyes, Loren Eiseley:

Since the first human eye saw a leaf in Devonian sandstone and a puzzled finger reached to touch it, sadness has lain over the heart of man. By this tenuous thread of living protoplasm, stretching backward into time, we are linked forever to lost beaches whose sands have long since hardened into stone. The stars that caught our blind amphibian stare have shifted far or vanished in their courses, but still that naked, glistening thread winds onward. No one knows the secret of its beginning or its end. Its forms are phantoms. The thread alone is real; the thread is life.*

---

* From *The Firmament of Time*.

The scientist and medical doctor Lewis Thomas writes the facts of reality and then strikes the imaginative spark that moves prose towards poetry:

> We live in a dancing matrix of viruses; they dart, rather like bees, from organism to organism, from plant to insect to mammal to me and back again, and into the sea, tugging along pieces of this genome, strings of genes from that, transplanting grafts of DNA, passing around heredity as though at a great party.[*]

You don't have to see viruses like that. The bare facts permit, but do not dictate, the literary image. Yet once it has been added, the reader better sees the point. Here's Lewis Thomas (in *The Lives of a Cell* again) on mitochondria:

> We are not made up, as we had always supposed, of successively enriched packets of our own parts. We are shared, rented, occupied. At the interior of our cells, driving them, providing the oxidative energy that sends us out for the improvement of each shining day, are the mitochondria, and in a strict sense they are not ours.

My former Oxford colleague Sir David Smith found an apt literary allusion in making a related point. Mitochondria have become so thoroughly integrated into the host cell that their origin as invading bacteria was only recently understood.

> In the cell habitat, an invading organism can progressively lose pieces of itself, slowly blending into the general background, its former existence betrayed only by some relic. Indeed, one is reminded of Alice in Wonderland's encounter with the Cheshire

---

[*] From *The Lives of a Cell*.

Cat. As she watched it, 'it vanished quite slowly, beginning with the tail, and ending with the grin, which remained some time after the rest of it had gone'.*

Peter Medawar, the Nobel Prize-winning zoologist and immunologist to whom this book is dedicated, was, I think, the greatest literary stylist among the scientists of the twentieth century and I shall use him as my exemplar for the next part of my essay. He was certainly the wittiest scientist I ever met. Indeed, if I were asked for a definition of 'wit' as opposed to simply telling jokes, I might define it ostensively as 'just about anything Peter Medawar ever wrote for the general public'. Listen to this, the opening words of his 1968 Romanes Lecture in Oxford, on 'Science and Literature':

I hope I shall not be thought ungracious if I say at the outset that nothing on earth would have induced me to attend the kind of lecture you may think I am about to give.

That wonderful Medawarian sally prompted a literary scholar, in his critical reply, to remark, 'This lecturer has never been thought ungracious in his life.'

Medawar could be cutting, he could ridicule, he had no patience with pretentious cant. But he never descended to vulgar abuse. His lampooning of what we might call 'francophoneyism' (I like to think Peter would have enjoyed the word) was merciless: his wit, that insouciant patrician mastery, the kind of thing that makes you want to rush out into the street when you read it, to show somebody – anybody. As a true stylist who used style in the service of clarity, he made short work of those self-regarding 'intellectuals' who raised style above content:

---

* In chapter 1 of *The Cell as a Habitat*, co-authored with M. H. Richmond.

Style has become an object of first importance, and what a style it is! For me it has a prancing, high-stepping quality, full of self-importance; elevated indeed, but in the balletic manner, and stopping from time to time in studied attitudes, as if awaiting an outburst of applause. It has had a deplorable influence on the quality of modern thought . . .

Returning to attack the same targets from another angle, Medawar wrote:

I could quote evidence of the beginnings of a whispering campaign against the virtues of clarity. A writer on structuralism in the *Times Literary Supplement* has suggested that thoughts which are confused and tortuous by reason of their profundity are most appropriately expressed in prose that is deliberately unclear. What a preposterously silly idea! I am reminded of an air-raid warden in wartime Oxford who, when bright moonlight seemed to be defeating the spirit of the blackout, exhorted us to wear dark glasses. He, however, was being funny on purpose.

Medawar ended this lecture on science and literature with a resounding declaration of his own:

In all territories of thought which science or philosophy can lay claim to, including those upon which literature has also a proper claim, no one who has something original or important to say will willingly run the risk of being misunderstood; people who write obscurely are either unskilled in writing or up to mischief. The writers I am speaking of are, however, in a purely literary sense, extremely skilled.

The obscurantism in question positively pleads to be satirized, and the physicist Alan Sokal rose to the occasion. His 'Transgressing the boundaries: towards a transformative hermeneutics of quantum gravity' is a sublime masterpiece: utter nonsense from start to finish, yet accepted for publication in a pretentious journal of literary culture, no doubt because meaningless gibberish was precisely what the journal existed to publish. More recently, Peter Boghossian, James Lindsay and Helen Pluckrose fooled journal editors into publishing a flurry of similar lampoons, this time satirizing what they called 'Grievance Studies': the politically influential genre of 'More Victim than Thou' self-pity. How Peter Medawar would have relished these hoaxes. Also the 'Postmodernism Generator', a computer program designed to churn out an indefinite number of spoof articles, indistinguishable from the senseless 'postmodern' real thing.

If you should feel there's something genially anti-French in my quotations from Medawar himself, your suspicions would not be allayed by the following, from his review of Pierre Teilhard de Chardin's *The Phenomenon of Man* – a candidate for the finest negative book review ever written.* But his targets are not limited to French intellectuals and their fellow-travellers. The contrast he makes with a once dominant school of German thought – 'these tuba notes from the depths of the Rhine' (what an image, how exquisitely Medawar!) – shows him to have been an equal opportunity pretension-puncturer:

> *The Phenomenon of Man* stands square in the tradition of *Natur-philosophie*, a philosophical indoor pastime of German origin which does not seem even by accident (though there is a great deal of it) to have contributed anything of permanent value to the storehouse of human thought. French is not a language that lends itself naturally to the opaque and ponderous idiom of nature-philosophy, and Teilhard has accordingly resorted to the use of

---

* Published in *Mind* (1961) and reprinted in *Pluto's Republic*.

that tipsy, euphoristic prose-poetry which is one of the more tire-some manifestations of the French spirit.

The affable good humour blunts the barbs, so that offence can hardly be taken. Once again, 'This lecturer has never been thought ungra-cious in his life.' Yet at the same time, though superficially blunted, the barbs somehow contrive to remain as sharp and penetrating as ever. Such a contrast to Samuel Johnson, in Peter's own splendid characterization, 'wielding, as ever, the butt end of his pistol'.

The conclusion of the Teilhard review takes on not just Teilhard himself but an entire underworld (he chose the word advisedly, as we shall see) of out-of-its-depth literary culture:

> How have people come to be taken in by *The Phenomenon of Man*? We must not underestimate the size of the market for works of this kind, for philosophy-fiction. Just as compulsory primary educa-tion created a market catered for by cheap dailies and weeklies, so the spread of secondary and latterly tertiary education has created a large population of people, often with well-developed literary and scholarly tastes, who have been educated far beyond their capacity to undertake analytical thought. It is through their eyes that we must attempt to see the attractions of Teilhard.

'Educated far beyond their capacity to undertake analytical thought.' Isn't that delicious beyond words?

Did Peter sometimes go too far? Women friends of mine have bridled at his explanation for the title of one of his books, *Pluto's Republic*:

> A good many years ago a neighbour whose sex chivalry forbids me to disclose exclaimed upon learning of my interest in philoso-phy: 'Don't you just adore Pluto's *Republic*?'

He certainly had a highly developed sense of mischief. When, at an unusually young age and still not well known, he was newly elected Jodrell Professor at University College, London, John Maynard Smith asked J. B. S. Haldane what this chap Medawar was like. Haldane tersely paraphrased Shakespeare: 'He smiles and smiles, and is a villain.' Do *we* conceal a guilty, vicarious mischief ourselves when we hug ourselves with delight at the following, in 'Lucky Jim', Medawar's review of James Watson's *The Double Helix*:

> It just so happens that during the 1950s, the first great age of molecular biology, the English schools of Oxford and particularly of Cambridge produced more than a score of graduates of quite outstanding ability – much more brilliant, inventive, articulate and dialectically skilful than most young scientists; right up in the Jim Watson class. But Watson had one towering advantage over all of them: in addition to being extremely clever he had something important to be clever about.*

I think a certain amount of offence *was* taken at that, for which Peter, gracious as ever, offered a kind of half apology. And he had the enormous advantage of being – at a level possibly unprecedented in the annals of Nobel scientists – deeply read and cultivated in literature. A real polymath – to use that overworked word in its true sense – who could hold his own against scholars in almost any field as well as his own.

His polymathy calls to mind his generous, and beautifully written, pen portrait of a genuine hero, D'Arcy Thompson:†

> an aristocrat of learning whose intellectual endowments are not likely ever again to be combined within one man. He was a classicist of sufficient distinction to have become President of the Classical

---

* Published in the *New York Review of Books*, 28 March 1968.
† Published in *Pluto's Republic*.

Associations of England and Wales and of Scotland; a mathematician good enough to have had an entirely mathematical paper accepted for publication by the Royal Society; and a naturalist who held important chairs for sixty-four years . . . He was a famous conversationalist and lecturer (the two are often thought to go together, but seldom do), and the author of a work which, considered as literature, is the equal of anything of Pater's or Logan Pearsall Smith's in its complete mastery of the bel canto style. Add to this that he was over six feet tall, with the build and carriage of a Viking and with the pride of bearing that comes from good looks known to be possessed.

Did Peter know how much of himself was in that description? I doubt it. If I am asked to name a single scientist whom I regard as a role model, and more particularly a single author whose writing style inspired me more than any other, I would nominate that other aristocrat of learning, Peter Medawar.

Without being in the Medawar class, a scientist can deeply love literature and many do. Books have always been an important part of my life. Unlike many biologists I came to my subject not through a love of birds or natural history in the wild (that came later) but through books and a preoccupation with the deep philosophical questions of existence. My childhood devotion to Doctor Dolittle kindled my love of animals and my moral concern for their welfare. In *Science in the Soul* I even compared Hugh Lofting's protagonist to my own gentle hero, Charles Darwin, the young 'Philos' of the *Beagle*. I progressed to the *Eagle* comic and 'Dan Dare, Pilot of the Future'. The romance of space travel gripped me, even though the science was unnecessarily slipshod (you don't seize the joystick and zoom off in the general direction of Venus; you have to compute orbits, schedule slingshots and harness the calculated power of gravity). Later, the novels of Arthur C. Clarke put me straight on such solecisms and led me to a love of science fiction.

My schoolfriends and I avidly discussed the moral and political implications of *Brave New World* and moved on to Aldous Huxley's other novels which, though not science fiction in themselves, are manifestly written by someone deeply read in science and with an inside track on the minds and emotions of those who practise it. Scientists such as the dreamy Lord Tantamount in *Point Counterpoint* are among Huxley's most sympathetic characters. His *After Many a Summer** is surely influenced by his scientific reading, including his brother Julian's research on axolotls, injecting them with thyroid hormone and turning them into salamanders never before seen. We are juvenile apes; so, if we lived two hundred years, would we turn into hairy quadrumana like Huxley's fictional Earl of Gonister?

I have learned some of my science from reading science fiction (see for instance my foreword to Fred Hoyle's *The Black Cloud* on page 79), and this leads me to wonder why it might sometimes take fiction to teach such lessons. Why do we enjoy fiction? What is the fascination of stories about non-existent people and things that never happened, and why do we turn to them for light relief after reading about things that did? Hardly a good example of light relief, but why did William Golding write *Lord of the Flies* as fiction, when he could have penned a prophetic treatise on human psychology with serious predictions of what would happen if a group of schoolboys should be marooned on an island with no adults? H. G. Wells did it both ways: prophetic fiction such as *The Time Machine*; and non-fictional speculation in his remarkable (and to modern readers deplorable in parts) *Anticipations of the Reaction of Mechanical and Scientific Progress upon Human Life and Thought*. But what is it that makes fiction so palatable? I think I dimly understand some kind of answer, but I am no literary scholar and it would be presumptuous of me to rush in where Henry James, E. M. Forster and Milan Kundera have trod.

---

* In some US editions the Tennyson quotation is completed: 'dies the swan'.

The title I have chosen for the present volume – *Books Do Furnish a Life* – reflects the love of books that has led me, over many years, to accept invitations to write forewords, introductions or afterwords to books that I admire, to contribute to collections of essays, and to write reviews of books that I might (or, less often, might not) admire. There follows a selection of these pieces, put together in collaboration once again with Gillian Somerscales (for whose literate conscientiousness I am, as ever, immensely grateful). Any of them could have featured in our previous anthology, *Science in the Soul*, but it made excellent sense to keep them back for a separate volume of book-related writings – a collection that I hope reflects something of the range and quality of the literature of science.

I

TOOLS OF TWO TRADES: WRITING SCIENCE

IN CONVERSATION WITH
NEIL DEGRASSE TYSON

ON SCIENCE AND SCIENTISTS,
IN PUBLIC AND PRIVATE

In April 2015 I met Neil deGrasse Tyson, Director of the Hayden Planetarium, at his office in New York. We talked for nearly an hour and a half, ranging over many topics of passionate concern to us both, and our conversation was filmed and recorded for Neil's radio show *StarTalk*. The following is an abbreviated transcript of parts of that conversation.

Are we just not wired up to think logically? Would you trust a doctor who you know holds unscientific beliefs? How do you get people to change their minds? How many scientists are religious – and does it matter?

NDT: So I've got here, live in the flesh, the one, the only, the inimitable Richard Dawkins. Richard, thanks for coming.

RD: Thank you very much.

NDT: I want to talk to you about the human mind's capacity to know, and to think, and to believe. You know, I look at how much trouble people have with mathematics, typically. If there's any one subject that the most number of people say 'I was never good at—', it's going to be math. And so I say to myself, 'If our brain were wired for logical thinking, then math would be everyone's easiest subject.' Everything else would be harder. So I'm kind of forced to conclude that our brain is not wired for logic.

RD: It's a very good point. And it's more than just that. I think there's also a kind of unwarranted pride in being bad at mathematics. You will never hear anybody saying how proud they are at being ignorant of Shakespeare, but plenty of people will say they are proud of being ignorant of mathematics.

NDT: Or if they don't use the word 'proud' they'll say, 'I was never good at math, ha ha ha' – they'll chuckle about it. Like it's a joke.

RD: Yes. There was a piece in one of the British newspapers where a science writer – I think a science journalist – was lamenting the fact that many people in Britain think it takes one month for the Earth to orbit the sun, and the editor inserted there: 'Doesn't it? – Ed'. So he was, as it were, saying, 'I'm the editor of a national newspaper, and of course I don't think it really takes a month – but nevertheless it's OK to make a joke about being ignorant of this elementary point of astronomy,' which you would never, ever do about confusing Byron with Virgil or something like that—

NDT: Or ever be proud of such a thing. So, then you must admit that we as a human organism must have a great challenge before us to think rationally, logically, scientifically.

RD: Yes. You made the very interesting point that maybe we are not wired to be good at logic – well, you generalized from mathematics to logic – but I think it's an interesting point that our wild ancestors, needing to survive in the presence of lions and drought and famine and things . . . you'd think logic would be pretty important for survival – if not mathematics.

NDT: Well, maybe there were early people who said, 'Oh, there's a creature there with big teeth. Let me investigate it further . . .'

RD: Yes. In a way, that's right: to be curious might be a bad thing.

NDT: Curiosity doesn't always work.

RD: I had a cousin who, as a little boy, put his finger in the mains and got a shock, so he did it again just to make sure! So he was a real scientist, but that's not very good for survival.

NDT: Right! So perhaps the gut reaction to run, or to be scared, or to . . . chant . . . I guess what I'm getting at is there's so much of human civilization that derives, not from logical thinking, but from what we might simply call illogical thinking. Take illogical thinking and, say, art: I've got Van Gogh on the wall but no one's going to quiz him and say, 'How logical were you when you painted the starry night?' So what does it mean to object, then, to people who feel this way? Because I detach myself more from that battle than you do. You are on the front lines and I'm way back watching you do this, and I'm saying sometimes people just want to – feel, rather than think.

RD: Yes. I keep pushing back to the evolutionary origins of this. When you have to survive in a hostile environment, it may be that you do need a certain amount of illogical—

NDT: Gut feeling.

RD: Yes. It may be that you need to fear things which logic tells you you needn't . . . but maybe it's a matter of the odds that something is actually dangerous.

NDT: Or the cost to you if it is.

RD: The cost to you. If you see a sort of rustling in the trees, it could be a leopard about to jump on you but it's much more likely to be the wind, and the logical, rational explanation is probably it's the wind.

But when your survival depends upon the remote possibility – well, in fact not remote, the rather lower probability that it might be a leopard, the prudent thing is to be more risk averse than—

NDT: Than the statistics justify.

RD: Exactly, yes.

NDT: OK. So now we have a world where . . . we're prisoners of this genetic moulding that has occurred; and I guess my point is I don't object as much to that as you do.

RD: Yes, OK.

NDT: And . . . what's the phrase, it sticks in your craw? So, OK . . . bring it on!

RD: A former professor of astronomy at Oxford told me a story of an American astrophysicist who writes learned articles in astronomical journals, mathematical papers, and the mathematics is premised on the belief that the universe is between 13 and 14 billion years old. And this man writes his papers and he does his mathematics and everything – and yet he privately believes the world is only six thousand years old. Well, you may be tolerant of that because you may say, 'Well, as long as he gets his sums right, as long as his paper is well-researched . . .'

NDT: A well-researched paper, yes.

RD: I would say that man should be fired. He should not be a professor of astrophysics in an American university. And we might differ about that, because you might say his private beliefs are private, they're nothing to do with me, if he does his astronomy right then that's OK.

NDT: Yes, I agree with you that that's how I would react. What he does at home, on Sunday, that's his own thing – if it doesn't enter the science classroom then I don't care how he thinks.

RD: OK, then let me take an even more extreme example. It's fictitious in this case. Imagine you were going to consult a doctor, and I'll make him an eye doctor because they're sort of . . . above the waist, but you happen to know that he privately doesn't believe in the sex theory of reproduction. He believes that babies come from storks.

NDT: OK. I wouldn't go to that doctor.

RD: You wouldn't go to that doctor, but I've met plenty of people – especially in America – who say, 'It's not any of your business what he believes below the waist . . . he's an eye doctor. Is he competent? Can he repair your cataracts?' I don't think he should be employed in a hospital because what you're saying about that man is that he's got the kind of mind which is so adrift from reality that even if he's a competent eye surgeon, I don't think he could be trusted.

NDT: OK, so – interestingly – you're reacting in the way our ancestors hearing the rustle in the bushes are reacting, because most of the time it's wind, some of the time it's a leopard and that creates a fear factor that overrides everything else. He's a good eye surgeon, he or she is a good eye surgeon, but there's that lingering risk that the stork theory of reproduction might somehow affect the scalpel. So you fear that risk.

RD: I'm not sure it needs to affect the scalpel; I think it's something—

NDT: OK, so then you object on principle.

RD: I think so, yes.

NDT: Yes, not on practice. It's a principle thing.

RD: Take a professor of geography who believes in the flat Earth but—

NDT: —but otherwise makes perfect globes.

RD: Yes, quite. Yes, exactly. There are such people . . .

NDT: OK, so you're a principle person. You want the whole package to be consistent.

RD: I think so.

NDT: OK, so now, given that, what do you do about it? Because I don't really do anything. You want to change that; and we just admitted together that we are prisoners of mystical, magical ways of thinking – or illogical ways of thinking – and so you want to change the biological directive of the human mind. How do you do that?

RD: I like to use the phrase 'consciousness-raising'. I don't want to be dictatorial and say there should be a law against illogical thinking. I'm not that fascist! But—

NDT: You know what would happen, if you . . . let's imagine a future where all illogical people had to move to one particular state. That would be the state where all the music and art would come from, right? All the truly creative people are some of the least logical people I've ever met, yet they create and they make the world a little more interesting. But that's a different issue. OK, so what do you do? You want to consciousness-raise. Do you have tactics? Because I want to consciousness-raise too. So let's compare.

RD: OK. I suspect your tactics may be better than mine because your tactics, I think, are to lead by example.

NDT: Yes.

RD: Well, mine are to practise logic, practise science, expose the wonder of science. I like to do all that as well.

NDT: In fact, your book has the word 'wonder' in it. Your memoir – *An Appetite for Wonder*, which any scientist has and most people have, I think.

RD: Yes, and it's actually the subtitle of another of my books, *Unweaving the Rainbow* – the subtitle is *science, delusion and the appetite for wonder.* And that book, by the way, *Unweaving the Rainbow*, is my attempt to join poetry to science. The phrase, the title, comes from Keats's attack on Newton for unweaving the rainbow. Keats thought that Newton was destroying the poetry of the rainbow by explaining the spectrum. And the message of my book is that you don't: that by destroying the mystery you increase the poetry, you don't decrease it.

NDT: And I try to go there in all of my work. Whether or not I succeed, that's my intent. So where do you differ from this? Or where else do you go?

RD: I certainly want to go all the way with that. I'm with Richard Feynman, who said, 'When I look at a rose, I see the same beauty as a poet or a painter sees in the rose, but I also get poetic inspiration from the fact that I know that the colour is to attract insects and that's come about by natural selection.'

NDT: I feel the same way about beautiful sunsets. I think there's no more reproduced image when people want you to think of God, than a sunset with beams of light coming out.

RD: Impurities in the—

NDT: —in the atmosphere! So I too ... I deeply appreciate the splendour of a magnificent sunset with a curtain of twilight colours going from, you know, deep blue to sky blue and the red sun. But I also know that the surface of the sun is six thousand degrees and is really scattering the atmosphere, you have water droplets condensing to make clouds and so I agree with the Feynman approach to that. But where else do you go, where else do you take this?

RD: Maybe I go a little bit further in the direction of good-natured ridicule, of absurd ideas like astrology, like homeopathy.

NDT: You're saying it's good-natured, but the people who are on the other side of your wit and intelligence ... are they saying you're being good-natured?

RD: Possibly not. I don't really care about that. I have an eye to – not just to the astrologers I'm talking to, but for example the radio audience or whoever it is who are listening in.

NDT: Yes, the larger . . . because, right, you have visible platforms where you share this.

RD: We both do. And the point has often been made to me that if you call somebody an idiot you're not going to change his mind, and that's possibly true, but you may change the minds of a thousand people listening in and so I'm less inhibited about calling him an idiot.

NDT: OK, so when you have the conversation with the individual, knowing you have a platform, even if the individual is insulted or feels bad or feels stupid, you're relying on the fact that there are some other people who are perhaps on the fence who could be swayed by your arguments? So for me, I'm more for the one-on-one, I guess. I want to have the one-on-one conversation and the eavesdroppers are perhaps imagining themselves there. And that's my tactic, if I can call it that – I'm feeling the one-on-one more than I'm feeling the audience.

RD: You have a huge audience.

NDT: Can I give an example? How would you have handled this case? I have a relative, right, whose father died. He's my cousin, she's my niece, once removed. She's alone in the room with her father. The father is dead in a half-open casket. She reports to me – this is weeks later – and by the way, she's a real estate agent, and majored in accounting. She said her father sat up, and she had a conversation with him. And I said, 'What transpired?' and she said, 'He said, "Don't worry, I'm in a better place,"' and she said, "I'm glad, we're sad you're gone but I'm glad to hear."' And so that was the conversation. So I said, 'OK, how am I going to deal with this? This is family, how am I going to handle this?' Here's what I did. I said, 'Next time this happens, ask him questions that could be really useful on this side of that barrier, like "Where are you? Are you wearing clothes? Where did you get the clothes? Is there money where you are? What's the weather like? Who else is there? How old are you there? In your mind's eye are you young, or are you old? Is grandma there? How old is she? If grandma's where she is, would she make herself old? Or would she be young again?" Ask questions.'

RD: I think that's a terrific answer.*

NDT: Where you get information. And so now she's on notice. Every time I see her, she says, 'I got it, we're going to go there.' And so now she's got her own little experiment that she's going to do, next time dead people sit up and talk to her. So what would you – if I could just ask – what would you have done?

RD: I wouldn't have thought of that – I wish I had. I am genuinely curious – I mean, do you think she was lying? Do you think she had a hallucination?

NDT: Oh, I don't know. I mean, I'm a trained astrophysicist, so my first explanation is that it was a hallucination, because everything we know about dead people tells us they don't sit up and have a conversation, and of course there are no witnesses or video. But I didn't care whether – how real she thought it was. What I cared was that I gave her tools, I think; I gave her tools so that next time this happens, she can separate an objective reality from what might have been in her mind. And then she arrives at that conclusion by herself, not by me telling her she's hallucinating.

RD: Yes, I think that's terrific. I mean if she was still . . . grieving, and still mourning, then I would have been inhibited in saying, 'You were hallucinating.'

---

* Carl Sagan made a similar point about interrogating alleged aliens when they allegedly capture you: see p. 118.

NDT: Out of sensitivity.

RD: Yes, out of sensitivity.

NDT: So there is a soft side to Richard Dawkins.

RD: Oh, surely.

NDT: Let the record show – Richard – he can be a puppy!

RD: But setting that aside, I think I would say something like, 'I think you probably drifted off to sleep and had a dream,' something like that.

NDT: Because these are solid moments, right?

RD: I mean, we all dream, every night, and we have experiences in our dreams which are utterly unreal, utterly surreal, and most of us don't know we're dreaming. It gives you an insight into what it would be like to be insane, actually. Every single night I go insane because of the dreams that I have. Any rational person could immediately say, 'This is not real.' So I think that would be another way, a reasonably sympathetic way of saying it – but I like your way better.

NDT: I'm just saying that's kind of my MO, if you will. This is my method of interaction. And I can tell you that – just in my own life – there's been a sort of 'land grab' of me by atheists, to claim me as an atheist, and my objection is that I just don't want a title. I don't want to be labelled, on the grounds that if someone comes to me expecting that I fulfil a title, or some label indicates what I should say, then they will presuppose they already know my arguments in advance. And I'd rather they hear me from scratch, and hear me build an argument and build a conversation, and then we build it together. A very subtle point I'll just insert here: when I taught in college, it was at a time when you had those transparencies that you could write on, or you could pre-prepare them, and some professors would just slap down a fully prepared transparency and there were all the notes, and they would speak to the notes.

RD: As opposed to building it up.

NDT: As opposed to building it up. And I'd say, 'No! If you do that, they'll just copy!' Whereas if you draw the first part of the diagram – here's an axis of temperature, and here's time, and here's – and then you assemble the ideas together and you get a far deeper under-standing of what's going on.

RD: That's a very good didactic point for actually teaching. I'm not sure it applies to what you were saying before about being labelled. I mean you wouldn't—

NDT: No, it does in a conversation. If you don't know anything about me, you have to learn it. From scratch.

RD: But if you were known to be a rationalist, for example. If you were known to be a realist, if you were known to be somebody who bases his conclusions on evidence. Would you feel the need to hide that label, on the same grounds?

NDT: In the sense that someone coming into the conversation may be defensive in advance, they may have a posture in advance, they mvight try to line up some arguments in advance, and it denies the purity of a conversation that could have happened. The sincerity of a conversation that I value in a one-on-one encounter.

RD: Well, I think I would be the same. I mean, if I were going to have dinner with somebody, and I wanted to persuade her of my point of view, I don't think I'd say, 'Right, I'm an atheist!' I think I would develop the thing step by step. Develop the transparency one step at a time. But if I were living in a country – like the United States – where it was impossible to get elected if you had that label, I think it's a little bit like Gay Pride – it's like standing out and saying, 'I'm gay,' or 'I'm not gay, but I believe in gay marriage,' or something.

NDT: But I think part of the difference is that in the secular move-ment, there's an urge to get more people to think that way, or to be that way, on the grounds that you have a better society for having done so. Or a more rationally deciding society. I don't know a single gay person who has an objective to turn everybody gay. They just want themselves to be respected for what they are, but they're not trying to make everybody else gay. And that's different from, say, your book *The God Delusion*. There's no book entitled *The Straight Delusion* saying, 'We're the right way to be.' So to me there's a difference in objective between the gay movement and the secular movement, if you will.

RD: I think you're exaggerating the desire of the secular movement to convert everybody to our point of view. It's more, 'We want to convert you, not to atheism, but to the view that atheists should not be discriminated against.'

NDT: So, are we living in different times from a century ago? Did Darwin create a deeper divide? Was there a community of religious people who dug their heels in more strongly after Darwin? Because I don't remember this level of conflict. Maybe I just was unaware. I don't claim perfect knowledge of social and cultural mores around the world, but I remember days when religious people went to church, or synagogue, or mosque, on the weekend, and during the week you went to school and learned science.

RD: It does appear to be a real phenomenon that people who are not religious in America today are in danger of being ostracized. Though this is not true in places like Silicon Valley, where I've just been. I kept meeting people in Silicon Valley who said, 'Well, what's the problem? I'm an atheist, everybody knows I'm an atheist' – but they live in Silicon Valley!

NDT: How about the UK, or Europe in general? It's very atheistic, isn't that correct?

RD: Yes, and paradoxically, many European countries have an established church and that may be no accident. It may be that the established church makes religion kind of boring, whereas in America, religion is—

NDT: We have choice of religion.

RD: Free enterprise. Choice.

NDT: Yes.

RD: You advertise your megachurch.

NDT: And I can follow the preachers I like and then—

RD: And then you go to this church and this church, rather than that church and—

NDT: That's fascinating.

RD: In Britain people don't go to church except to be married and buried.

NDT: The brief time I spent there while we were filming *Cosmos*, that's when I learned. You know, there's this thing called the Anglican Church, but it's an administrative entity in practice, and beyond that nobody goes to church.

RD: Yes. It crowns the monarch—

NDT: An7.25 ptd then it's done!

RD: Yes.

NDT: And the rest of Europe is . . . ?

RD: I think Europe is very variable. I think in France and the traditionally Catholic countries – France, Italy, Spain – there's a very strong anti-clericalism, but the United States stands out like a sore thumb for its religiosity. You have to get, more or less, right over Europe to the Middle East before you start getting a similar preoccupation with religion.

NDT: Now, if we go back in time, essentially every famous scientist pre-twentieth century is religious. Galileo is religious, he's a devout Catholic. Newton was Anglican, but he objected to the Trinity; he had some issues. And it's not uncommon to have strongly religious people in modern times cite the religiosity of scientists of the past, and if you look at the numbers today – I haven't checked the very latest ones, but when I did check, in the United States, as many as a third of practising, publishing scientists would claim to be religious in the unambiguous way where you ask, 'Do you pray? Do you have an all-powerful being interceding in your daily affairs?' And they'd say 'Yes'. So you can't then say being religious is in and of itself the problem; you would have to modify that argument to say it's when you want to do *this* with your religion, then it's a problem. But otherwise, for all these other people, it's just fine.

RD: OK, let me take that point. First, I think we ought to make a big distinction between the historical point and the present day.

NDT: Sure, I conflated them. I'm sorry.

RD: Newton, Galileo . . . pre-Darwin. You couldn't *not* be religious pre-Darwin – at least, you could, but you would have to be very – very stalwart in your scepticism. Because when I look around the world, it kind of looks almost obvious that there had to be a designer until Darwin came along. Who can blame Newton and Galileo? So I'm deeply unimpressed by that argument. As for the one-third of scientists in America – that's approximately correct by the polls that I've seen. But if you move away from scientists generally to the elite scientists – studies have been done of both the American National Academy and the British Commonwealth Royal Society, the corresponding elite academies of science, and there it's about 10 per cent. Now, I've seen you make the point that we still have to worry about 10 per cent—

NDT: Well, I don't know if you can use the word 'worry': that was how people wanted to characterize me, but my actual point was that we have people such as yourself out there, you know, making the case to the public – but I don't see you making the case to the third of scientists, our professional brethren; and what hope do you have of converting the public, leading them to more rational ways, when our own scientific community is representing in just that way to the level of a third – and even in the elite group, that 10 per cent is not 0 per cent.

RD: That's right. But you have to put a little bit of caution on that. If you ask the scientists what they actually believe, they may say they're religious, they may say 'I'm Jewish' or 'I'm Christian'. If you actually ask them, the one-third, and perhaps more particularly the 10 per cent, what they believe, they will talk about the mystery of the universe – they have a sort of reverent attitude, which I have as well and I think you have. But then if you say, 'Do you actually believe in anything supernatural? I know you call yourself Christian, but do you believe that Jesus was born of a virgin and rose from the dead?', of course they don't. And so, you've got to subtract them off, I suspect; you subtract off the Einsteinians—

NDT: So the Einsteinian God is Spinoza's God, a God of the universe that is responsible for laws and things and responsible for the universe that science observes. It's just kind of untestable, that's all.

RD: I don't think it would even be 'responsible for the universe'; I think it's just God *is* the universe. Which is a bit different from thinking there's an intelligence that started it all. So I think you want to subtract them off. And then you are left with a few who actually do believe in the virgin birth, and I don't know what to make of them. I think that they are, as it were, traitors to science.

NDT: But they still do science! So you object philosophically.

RD: Like the astrophysicist I told you about. Well, we've been there already. But there's something else perhaps I'll tell you. My British foundation did a survey – we commissioned a public opinion poll – and we chose the very week of the census which took place in 2011, and the census in Britain actually asks what your religion is and you have to tick a box that says Christian, Jewish, Muslim etc. or none.

So we commissioned a professional polling organization to sample those who ticked the Christian box to find out what they really believed. Obviously it was only a sample, it was a couple of thousand, but it was done professionally. And we asked them questions like 'OK, you ticked the Christian box, do you believe Jesus is your lord and saviour?' No. 'Do you believe Jesus was born of a virgin?' No. 'Do you believe Jesus rose from the dead?' No. 'Then why do you call yourself a Christian?' Oh, because I like to think of myself as a good person. So that's the kind of level that people will sink to in agreeing to tick the Christian box to get the label, to accept the label of Christian. We then said well, OK, so you like to think of yourself as a good person – it wasn't sequential, they're all separate questions – but you like to think of yourself as a good person; so when you're faced with a moral dilemma in your own life, do you turn to your religion, or do you turn to your friends? Do you turn to your cultural background?

NDT: It's an excellent question. Beautiful question. I want to comment on that, but go on.

RD: And I think it was only about 9 per cent of people who ticked the Christian box who said they turn to their religion, although a majority said that they ticked the Christian box because they like to think of themselves as a good person. So all this is showing really is: be sceptical when people tell you that they are religious. Be sceptical when people tell you 'I am a Christian' or 'I am a Jew', especially if they say 'I'm a Jew'. That probably means that they are loyal to Jewish traditions and—

NDT: In America, generally, it means just that. Unless they're full-out Hasidic, practising. Judaism is a culture, more than it is a religion, here in the United States.

RD: Which is fine.

NDT: I was interviewed for the *New Yorker* magazine and at some point the interviewer asked, was I raised in any religion? I said yes,

I was raised Catholic, and that was actually the first time I had ever said that publicly. I never tried to hide it, it was just that no one ever asked. And I said, but it was kind of like we used to go to church weekly and then it kind of faded to once a month, then we became 'Ashes and Palms' Catholics where you just go on the holidays – and of course we celebrated Christmas. And the real point I wanted to make in this article was that it did not influence, in any obvious way, any decisions we made; my mother never came to us and said, 'You shouldn't do this because Jesus is watching.' There was no such interaction in the household. But there was the urge to say, in an article on Neil deGrasse Tyson, 'He was Catholic but now he's a scientist and lost his Catholic ways,' as though there was some big transition that happened.

RD: But there was no transition.

NDT: There was no transition! I see the urge of people to want to make these associations, but in our household there was never the thought, 'What would Jesus do?' Just: 'What would a rational, thinking person do in this situation?' And that's how my whole life unfolded.

I still would rather just have no label at all. The only 'ist' I am, as I've said, is a scientist, and beyond that – have a conversation with me, as we just did!

So Richard, thanks for coming through town. This is a long overdue conversation.

RD: Yes.

NDT: Every time I see you I say, you know, 'I want to tell him this', and 'I want to think about that', and get his take on this – so it was great to have you here. So thanks again.

RD: And thank you!

THE UNCOMMON
SENSE OF SCIENCE

This review of *The Unnatural Nature of Science* by Lewis Wolpert appeared in the *Sunday Times* in 1992. An eminent British embryologist born in South Africa in 1929, Dr Wolpert has a reputation as an outspoken, some would say scientistic (a compliment in my view and probably his, though not in theirs) champion of science. In mischievously humorous voice he is not reticent in expressing his doubts as to the usefulness of philosophy, especially certain fashionable schools of philosophy of science. Normally a cheerful and amusing companion, he has at times suffered bouts of severe depression, eloquently and movingly described in *Malignant Sadness*.

Pour one glass of water into the sea. Allow time for it to be thoroughly dispersed throughout the oceans of the world. Then scoop another glassful out of the sea, anywhere. Almost certainly you will retrieve at least one molecule of water from the original glass. This is because, as Lewis Wolpert says, 'there are many more molecules in a glass of water than there are glasses of water in the sea'. That simple statement has stunning implications. The cup of coffee that I am about to drink contains atoms that passed through Oliver Cromwell's bladder; and through yours, and the Pope's. It seems to unite us all in one big happy family. Wolpert, however, makes a less soppy and more interesting point: that science is far removed from common sense.

T. H. Huxley famously said: 'Science is nothing but trained and organised common sense . . . and its methods differ from those of common sense only as far as the guardsman's cut and thrust differ from the manner in which a savage wields his club.' Wolpert, I dare say, would underwrite that as far as certain aspects of scientific *method* are concerned, and provided that 'common sense' means robust good sense rather than vulgar folk wisdom. Vulgar folk wisdom, when faced with an uncanny coincidence, resorts eagerly to the supernatural. Robust good sense knows that the coincidence might have happened anyway. Scientists are statistical guardsmen trained to calculate its probability.

But that is scientific *method*. Wolpert would take his stand, I suspect, not so much on methods as on results. Science notoriously departs from common sense in the brain-tormenting upper reaches of quantum and relativity theory, but even classical Newtonian mechanics is no pushover for our wretched intuition. Who would have imagined that, if you drop one bullet as you fire another horizontally from a rifle, the two bullets will hit the ground simultaneously?

Wolpert mischievously flirts with the suggestion that 'if something fits in with common sense it almost certainly isn't science . . . Our brains – and hence our behaviour – have, in evolution, been selected for dealing with the immediate world around us.' I can testify that evolution itself, though it is childishly easy to understand when compared with, say, black hole singularities, falls foul of a doggedly obtuse common sense built to apprehend a human timescale spanning seconds to centuries, which is dazed into foolishness by the slow grind of geology's million-century mills.

Lewis Wolpert is a distinguished embryologist, a Fellow of the Royal Society who manages simultaneously to be a successful popularizer (it is easier to get away with the feat if you are an unsuccessful popularizer). His name was prominent in literary pages earlier this year as an articulate spokesman for science against an overpublicized whinnying chorus of lady novelists,* quality journalists and third-rate philosophers who complained that science had robbed humanity of its 'soul'. Wolpert saw them all off with authority and dispatch. He rightly doesn't mention them in his book, although he does epitomize their views with an equally silly quotation from a better writer, D. H. Lawrence: 'Knowledge has killed the sun, making it a ball with spots.'

This book is Wolpert's collected thoughts on science, its importance, how it is done and how it relates to other fields. After introducing his main thesis that science affronts common sense, Wolpert makes the interesting if not totally surprising observation that technology is quite separate from science and has sprung up in history far more frequently. Then follows the obligatory 'it all began with the Greeks' chapter. I've never understood why we have to be impressed by that fire, earth and water stuff, but nobody who writes this sort of book seems able to leave it out, and Wolpert even manages to sound enthusiastic.

We scientists have our ambitions and our human frailties, and Lewis Wolpert exposes them frankly. Some worry jealously about priority, or at least crave admiration from peers. J. B. S. Haldane, whose centenary falls on the day that I write, emerges as the honourable exception whose 'great pleasure was to see his ideas widely used even though he was not credited with their discovery'. A few scientists cheat, changing numbers or inventing experiments that never happened, in the service of a cherished hypothesis. What is significant about these cases is not their existence but the unmitigated horror with which they are regarded by the scientific community.

If a gardener asks us to pay him in used fivers, we wink knowingly

---

* Perish the thought that I should attack female novelists as a class. At the time this review was published, readers would have had no difficulty in identifying the particular 'lady' I had in mind. To name her now, so long afterwards when she might well have changed her mind, would be uncharitable.

and don't tell the tax man. If a friend Joads the railway,* travelling without a ticket, we are less indulgent but still don't expose him. But a scientist proved to have fabricated a data point would be banished unlamented from his profession without a second chance. Admittedly, precisely because scientific cheating is such a heinous disgrace, professors tend to close ranks about a colleague so accused, and put would-be whistle-blowers through distressing hoops to substantiate their case. But all scientists at least pay lip service to the view that a proven cheat must not practise as a scientist ever again and must presumably retrain in the law or some other profession more suited to his talents.

Nobody denies that scientists sometimes violate their standards, falsifying, or at least massaging, evidence to support a case. What is impressive about science is that it has such high standards to violate. Barristers are paid (putting it mildly) to do little else than massage evidence to support a case. Politicians and journalists are respected for doing the same in the service of a 'policy' and an 'angle' respectively. The reason for the peculiar fastidiousness of scientists is not far to seek. In everyday life it is accepted that cheating is prevented only by constant policing. Receipts, countersignatures and evidences of identity are routinely demanded with no offence taken. But in science (and certain other fields of scholarship) the entire enterprise depends upon trust, with no checks or policing. If a scientist, working alone without witnesses, reports that he did X and observed Y, there is no time for his colleagues to do anything other than assume that X was indeed what he did and Y was indeed what he observed. If cheating became widespread the whole enterprise would crumble. This is why cheating in science is such an unforgiven sin. If professional standards rubbed off on personal behaviour, scientists would surely

---

* The philosopher C. E. M. Joad (1891–1953) was a popular English broadcaster and public intellectual until his downfall when his chronic habit of travelling on trains without a ticket was very publicly exposed. He can't have needed the money: it must have been a kind of game for him, a battle of wits, albeit a foolish one.

be the most moral class of persons in the world. They are also naturally trusting, which is why, as Wolpert notes, it is a good idea – on the
'thief to catch a thief' principle – to co-opt professional conjurors*
into the unmasking of psychics, mediums and other charlatans.

---

* A notorious example was the recruiting of James 'The Amazing' Randi by
the then editor of *Nature*, John Maddox, to investigate the claim by
Jacques Benveniste and others that 'water has a memory'. Such a claim is
central to the paradoxical homeopathic doctrine that an almost infinite
degree of dilution does not weaken – indeed strengthens – the medicinal
power of an ingredient. This could only be true if there existed some
hitherto unknown and utterly revolutionary principle of physics:
something about the molecular constitution of water would have to
enable it to retain a 'memory', a mysterious imprint of having once been
in contact with the now missing ingredient. The scientist who could
demonstrate such a remarkable conclusion would win Nobel Prizes in
both Physics and Medicine, and I have often waxed sarcastic about the
failure of homeopaths even to attempt a serious investigation of it. Instead
they are content to allege apparent cures of patients that could easily
result from the well-established placebo effect. The Benveniste team were
an exception. They appeared to have done the research properly, and
they submitted a paper to *Nature*. John Maddox took the courageous
decision to publish it. However, in view of the extreme importance and
surprisingness of the result, he stated a condition. This was that the
experiment should be repeated, in Benveniste's lab, under the supervision
of himself and two colleagues, one of them being James Randi. Randi
habitually deployed his world-class knowledge of the conjuror's art in
sceptical investigation of paranormal claims. He unmasked spiritualist
mediums, publicity-seeking spoon-benders and other such people on
many occasions. The conclusion from this and other attempts at
replication seems to be that, under controlled double-blind conditions,
the Benveniste Effect occasionally seems to show itself – but only when
a particular member of Benveniste's team is running the experiment. Go
figure, as the Americans say.

Wolpert has some pithy things to say about the role of philosophers in science. He concludes that they are mostly harmless, but goes on to single out one pernicious influence from social science, so-called 'cultural relativism'. 'Even statements like 2 + 2 = 4 are treated as legitimate targets for sociological questioning, and so too are logic and rationality.' I'd have assumed that Wolpert was exaggerating had I not met certain sociologists, holding down salaries in universities and influencing students. Wolpert provocatively declares that 'Scientists can be very proud to be naive realists.' Sociologists can be grateful for at least a measure of scientific realism every time they board a jet aeroplane rather than a magic carpet or a reindeer-powered aerial sledge.

Aeroplanes fly because engineers assume things like 2 + 2 = 4. I wrote that, not Lewis Wolpert. Nevertheless it is fair to say that, if you find it facile, simplistic, 'reductionist' or naive, you probably won't like Wolpert's book. Nor will you like it if you think that scientific truth is ultimately based on faith and has no special status compared with astrology, religion, tribal mythology or Freud. Nor if you think that today's scientific world-view is no advance over the world-views of earlier eras. Nor if you think that science has killed the human soul. If you subscribe to any of these persuasions you'll probably find the book irksome. If you have more sense, you'll enjoy it. In deference to the book's central theme, I suppose I should add that if you have too much common sense you'll find it salutary. Whichever is the case, read it.

ARE WE ALL RELATED?

A former head of my Oxford college was once heard to say, 'When I began my career as a young lecturer, I was told to make only one point per lecture. And nowadays I am told that even that is considered excessive.' This little piece makes only one point but a counter-intuitive and therefore worthwhile one. It is correspondingly brief, and indeed it was commissioned to be brief as befits a contribution to an anthology for children, published in 2012 under the title *Big Questions from Little People*.

Yes, we are all related. You are a (probably distant) cousin of the Queen, and of the President of the United States, and of me. You and I are cousins of each other. You can prove it to yourself.

Everybody has two parents. That means, since each parent had two parents of their own, that we all have four grandparents. Then, since each grandparent had to have two parents, everyone has eight great-grandparents, and sixteen great-great-grandparents and thirty-two great-great-great-grandparents and so on.

You can go back any number of generations and work out the number of ancestors you must have had that same number of generations ago. All you have to do is multiply two by itself that number of times.

Suppose we go back ten centuries, that is to Anglo-Saxon times in England, just before the Norman Conquest, and work out how many ancestors you must have had alive at that time. If we allow four generations per century, that's about forty generations ago.

Two multiplied by itself forty times comes to more than a thousand trillion. Yet the total population of the world at that time was only around three hundred million. Even today the population is seven billion, yet we have just worked out that a thousand years ago your ancestors alone were more than 150 times as numerous. And we've so far dealt only with your ancestors. What about my ancestors, and the Queen's and the President's? What about the ancestors of every one of those seven billion people alive today? Does each one of those seven billion people have their own thousand trillion ancestors?

To make matters worse, we've so far only gone back ten centuries. Suppose we go back to the time of Julius Caesar: that's about eighty generations. Two multiplied by itself eighty times comes to more than a thousand trillion trillion. That's more than a billion people packed into every square yard of the Earth's land area. They'd be standing on top of each other, hundreds of millions deep!

Obviously we must have done our sums wrong. Were we wrong to say that everybody has two parents? No, that is definitely right. So, does it follow that everyone has four grandparents? Well, sort of yes, but not four *separate* grandparents. And that is exactly the point. First cousins sometimes marry. Their children have four grandparents, but instead of eight great-grandparents they only have six (because two great-grandparents are shared).

Cousin marriage cuts down the number of ancestors in our calculation. First-cousin marriages are not particularly common. But the same idea of cutting down the number of ancestors applies to marriages between distant cousins. And that is the answer to the riddle of the very big numbers that we calculated: we are all cousins. The real population of the world at the time of Julius Caesar was only a few million, and all of us, all seven billion of us, are descended from them. We are indeed all related. Every marriage is between more or less distant cousins, who already share lots and lots of ancestors before they have children of their own.

By the same kind of argument, we are distant cousins not only of all human beings but of all animals and plants. You are a cousin of my dog and of the lettuce you had for lunch, and of the next bird that you see fly past the window. You and I share ancestors with all of them. But that is another story.

THE TIMELESS AND
THE TOPICAL

In 2000 Houghton Mifflin launched an annual anthology entitled *The Best American Science and Nature Writing*. I was invited by Tim Folger, the series editor, to be guest editor of the 2003 edition; this is an edited version of my introduction to that year's collection.

Carl Sagan gave one of his last books the characteristically memorable subtitle *Science as a candle in the dark*. Its equally memorable main title was *The Demon-Haunted World*,* and its theme was the darkness of ignorance – and the fear that darkness breeds. In the words of a prayer which I early learned from my Cornish grandmother,

> From ghoulies and ghosties and long-leggety beasties
> And things that go bump in the night
> Good Lord deliver us.

Some say it is Scottish, not Cornish, but wherever it comes from the sentiments are shared worldwide. People are afraid of the dark.

---

* My review of Sagan's book for *The Times* is reproduced below (see pp. 116–19).

Science, as Sagan argued and personally exemplified, has the power to reduce ignorance and dispel fear. We should all read science and learn to think like scientists, not because science is useful (though it is), but because the light of knowledge is wonderful and banishes the debilitating and time-wasting fear of the dark.

Unfortunately, science arouses fears of its own, usually because of a confusion with technology. Even technology is not inherently frightening, but it can, of course, do bad things as well as good. If you want to do good, or if you want to do bad, science will provide the most effective way in either case. The trick is to choose the good rather than the bad, and what I fear is the judgement of those to whom society delegates that choice.

Science is the systematic method by which we apprehend what is true about the real world in which we live. If you want consolation, or an ethical guide to the good life, you can look elsewhere (and may be disappointed). But if you want to know what is true about reality, science is the only way. If there were a better way, science would embrace it.

Science can be seen as a sophisticated extension of the sense organs nature gave us. Properly used, the worldwide cooperative enterprise of science works like a telescope pointing towards reality; or, turned around, a microscope to dissect details and analyse causes. So understood, science is fundamentally a benign force, even though the technology that it spawns is powerful enough to be dangerous when abused. Ignorance of science can never be a good thing, and scientists have a paramount duty to explain their subject and make it as simple as possible (though no simpler, as Einstein rightly insisted).

Ignorance is usually a passive state, seldom deliberately sought or intrinsically blameworthy. Unfortunately, there do seem to be some people who positively prefer ignorance and resent being told the truth. Michael Shermer, debonair editor and proprietor of *Skeptic* magazine, tells of the audience reaction when he unmasked a professional charlatan on stage. Far from showing Shermer the gratitude he deserved for exposing a fake who was conning them, the audience was hostile. 'One woman glared at me and told me it was "inappro-priate" to destroy these people's hopes during their time of grief.'

Admittedly, this particular phoney's claim was to communicate with the dead, so the bereaved may have had special reasons for resenting a scientific debunker. But Shermer's experience is typical of a more general mood of protective affection for ignorance. Far from being seen as a candle in the dark, or as a wonderful source of poetic inspiration, science is too often decried as poetry's spoilsport.*

A more snobbish denigration of science can be found in some, but by no means all, literary circles. 'Scientism' is as dirty a word as any in today's intellectual lexicon. Scientific explanations that have the virtue of simplicity are derided as 'simplistic'. Obscurity is often mistaken for profundity; simple clarity can be taken for arrogance. Analytical minds are denigrated as 'reductionist' – as with 'sin', we may not know what it means, but we do know that we are against it. The Nobel Prize-winning immunologist and polymath Peter Meda-war, not a man to suffer fools gladly, remarked that 'reductive analysis is the most successful research stratagem ever devised', and continued: 'Some resent the whole idea of elucidating any entity or state of affairs that would otherwise have continued to languish in a familiar and nonthreatening squalor of incomprehension.'†

Non-scientific ways of thinking – intuitive, sensitive, imaginative (as if science were *not* imaginative!) – are thought by some to have a built-in superiority over cold, austere, scientific 'reason'. Here's Medawar again, this time in his celebrated lecture 'Science and Literature'.‡ 'The official Romantic view is that Reason and the Imag-ination are antithetical, or at best that they provide alternative pathways leading to the truth, the pathway of Reason being long and

---

* Keats's decrying of Newton for that very thing handed me the title, and theme, of my book *Unweaving the Rainbow*.

† From *Aristotle to Zoos*.

‡ Given in Oxford in 1968. For more on this lecture, and on Peter Medawar, see my essay on 'The literature of science', pp. 4–18 above.

winding and stopping short of the summit, so that while Reason is breathing heavily there is Imagination capering lightly up the hill.'

Medawar goes on to point out that this view was even once supported by scientists themselves. Newton claimed to make no hypotheses, and scientists generally were supposed to employ 'a calculus of discovery, a formulary of intellectual behaviour which could be relied upon to conduct the scientist towards the truth, and this new calculus was thought of almost as an antidote to the imagination'.

Medawar's own view, inherited from his 'personal guru' Karl Popper and shared by most scientists today, was that imagination is seminal to all science but is tempered by critical testing against the real world. Creative imagination and critical rigour are both to be found in this collection of contemporary American scientific literature.

For a non-American to be invited by a leading American publisher to anthologize American writings about science is an honour, the more so because American science is, by almost any index one could conjure, pre-eminent in the world. Whether we measure the money spent on research or count the numbers of active scientists working, of books and journal articles published, or of major prizes won, the United States leads the rest of the world by a convincing margin. My admiration for American science is so enthusiastic, so downright *grateful*, that I hope I may not be thought presumptuous if I sound a note of discordant warning. American science leads the world, but so does American *anti*-science. Nowhere is this more clearly seen than in my own field of evolution.

Evolution is one of the most securely established facts in all science. The knowledge that we are cousins to apes, kangaroos, and bacteria is beyond all educated doubt: as certain as our (once doubted) knowledge that the planets orbit the sun, and that South America was once joined to Africa, and India distant from Asia. Particularly secure is the fact that life's evolution began a matter of billions of years ago. And yet, if polls are to be believed, approximately 45 per cent of the population of the United States firmly believes, to the contrary, an elementary falsehood: that all species separately owe their existence to 'intelligent design' less than ten thousand years ago.* Worse, the nature of American democratic institutions is such that this perversely ignorant half of the population (which does *not*, I hasten to add, include leading churchmen or leading scholars in any discipline) is in many districts strongly placed to influence local educational policy. I have met biology teachers in various states who feel physically intimidated from teaching the central theorem of their subject. Even reputable publishers have felt sufficiently threatened to censor school textbooks of biology.

That 45 per cent figure really is something of a national educational disgrace. You'd have to travel right past Europe to the theocratic societies around the Middle East before you hit a comparable level of anti-scientific miseducation. It is bafflingly paradoxical that the United States is by far the world's leading scientific nation while *simultaneously* housing the most scientifically illiterate populace outside the Third World.

Sputnik, the Russian satellite launched in 1957, was widely seen as a salutary lesson, spurring the United States out of complacency and into redoubled educational efforts in science. Those efforts

---

* In the intervening years the figure has dropped, I'm glad to say. In 2019 it stood at 38 per cent. Along with religious indoctrination, part of the problem doubtless lies in education. Middle-school science teachers in America are ill-equipped to withstand the hostile onslaught of religiously inspired attacks, from pupils, ignorant parents and school boards, as soon as they reach the evolution section of the curriculum. I'm proud that my charitable foundation, the Richard Dawkins Foundation for Reason and Science (RDFRS) is trying to do something about this sorry state of affairs. We're doing it through TIES, the Teacher Institute for Evolutionary Science. Under its Director Bertha Vasquez, herself a charismatic teacher, TIES runs workshops to train teachers in how to teach evolution. Using resources such as PowerPoint presentations created by Bertha, TIES workshops have now been run in all fifty states, with highly successful results.

paid off spectacularly, for example in the dazzling successes of the space programme and the Human Genome Project. But more than forty years have passed since Sputnik, and I am not the only Americophile to suggest that another such fright may be needed. Short of that – well, in any case – we need excellent scientific writing for a general audience. Fortunately that high-quality commodity is in abundant supply in America, which has made the compiling of this anthology both easy and a pleasure. The only difficulty, indeed the only pain, has been in deciding what to leave out.

Should a collection such as this be timely or timeless? Topical and of-the-moment? Or *sub specie aeternitatis*? I think both. On the one hand, the volume is one of a series, tied to a particular year, sandwiched between predecessors and successors. That nudges us in the direction of topicality: what are the hot scientific subjects of 2003; what are the current political and social issues that scientific writings of the previous year might illuminate? On the other hand, science's ambitions – more so, I venture, than any other discipline's – approach the timeless, even the eternal. Laws of nature that changed from year to year, or even from eon to eon, would seem too parochial to deserve the name. Of course our *understanding* of natural law changes – for the better – from decade to decade, but that is another matter. And, within the unchanging laws of the universe, their physical manifestations change, on timescales from gigayears to femtoseconds.

Biology, like physics, anchors itself in uniformitarianism. Its defining engine – evolution – is change, change par excellence. But evolution is the same *kind* of change now as it was in the Cretaceous, and as it will be in all futures we can imagine. The play's the same, though the players that walk the stage are different. Their costumes are similar enough to connect, say, triceratops with rhinoceros, or allosaurus with tiger, in ecological continuity. If an ecologist, a physiologist, a biochemist and a geneticist were to mount an expedition to the Cretaceous or the Carboniferous, their 2003-vintage skills and education would serve them almost as well as if they were going to, say, Madagascar today. DNA is DNA, proteins are proteins. They and their interactions change only trivially. The principles of Darwinian natural selection, of Mendelian and molecular genetics, of physiology and ecology, the laws of island biogeography, all these surely applied to dinosaurs, and before them to mammal-like reptiles, just as they apply now to birds and modern mammals. They will still apply in a hundred million years' time, when we are extinct and new faunistic players have taken the stage. The leg muscles of a tyrannosaur in hot-breathed pursuit were fuelled by ATP such as any modern biochemist would recognize, charged up by Krebs cycles indistinguishable from the Krebs cycles of today. The science of life doesn't change from eon to eon, even if life itself does.

So far, so timeless. But we live in 2003. Our lives are measured in decades and our psychological horizons crammed somewhere between seconds and centuries, seldom reaching further. Science's laws and principles may be timeless, but science bears mightily upon our fleeting selves. The science and nature writing of 2002 is not the same as that of ten years ago, partly because we now know more about what is eternally true, but also because the world in which we live changes, and so does science's impact upon it. Some of the essays and articles in this book are firmly date-stamped; some are timeless. We need both.

FIGHTING ON TWO FRONTS

In 2013, the literary agent and science impresario John Brockman convened a small meeting of colleagues to honour the great anthropologist and field ethnographer Napoleon Chagnon (https://www.edge.org/conversation/napoleon-chagnon-blood-is-their-argument). Along with Chagnon himself, those present comprised the philosopher Daniel Dennett, the psychologist Steven Pinker, the evolutionary theorist David Haig and the primatologist Richard Wrangham. The proceedings of the meeting were published on John Brockman's *Edge* website. I wasn't able to attend but John invited me to write the introduction. When Chagnon died in 2019, the proceedings of the meeting were re-posted on the *Edge* website – including my introduction, which is reproduced here, slightly abbreviated.

I first met Nap Chagnon at a conference in Paris in the late 1970s, convened by Robin Fox and others, the purpose of which was to foster a dialogue between anthropologists and evolutionary biologists. Such luminaries as Robert Trivers, John Maynard Smith and Richard Alexander were there. Some of the invited anthropologists pointedly stayed away, one of them saying: 'Why do we need to meet with biologists? Let's just send them a reading list.' Such arrogance perhaps sheds a little light on the otherwise baffling vilification by career social anthropologists of Napoleon Chagnon, that outstanding example of an anthropologist who took the trouble to read up on evolutionary biology and apply it to his research.

From that Paris conference, I treasure the memory of Chagnon's performance, late one evening in the bar, of a Yanomamö war dance. Doubtless pompously humourless pedants of today would condemn it as 'cultural appropriation'.

Though he was a strong and robust personality, it is doubtful whether this distinguished scholar and field researcher ever fully recovered from the personal wounds inflicted by sections of the academic anthropology community, egged on by an infamously ill-informed and malicious book, now thankfully sunk without trace along with its author.

Napoleon Chagnon is a Living World Treasure. Arguably our greatest anthropologist, he is brave on two fronts. As a field worker in the Amazon forest he has lived, intimately and under conditions of great privation, with 'The Fierce People' at considerable physical danger to himself. But the wooden clubs and poison-tipped arrows of the Yanomamö were matched by the verbal clubs and toxic barbs of his anthropologist colleagues in the journal pages and conference halls of the United States. And it is not hard to guess which armamentarium was the more disagreeable to him.

Chagnon committed the unforgivable sin, cardinal heresy in the eyes of a certain kind of social scientist: he took Darwin seriously. Along with a few friends and colleagues, Chagnon studied the up-to-date literature on natural selection theory, and with brilliant success he applied the ideas of Fisher, Hamilton, Trivers and other heirs of Darwin to a human tribe which probably ran as close to the cutting edge of natural selection as any in the world. It is sobering to reflect on how unconventional a step this was: science bursting into the quasi-literary world of the anthropology in which the young Chagnon was trained. Still today, in many American departments of social science, for a young researcher to announce a serious interest in Darwin's dangerous idea – even an inclination towards scientific thinking at all – can come close to career suicide.

In Chagnon's case the animosity spilled over from mere academic disagreement to personal slander, which was not only untrue but diametrically opposite to the truth about this ethnographer and his decent and humane relationship with his subjects and friends. The episode serves as a dark lesson in what can happen when ideology is allowed to poison the well of academic study. I read up on the whole affair from a safe distance at the time, and was sufficiently shocked to break off friendly relations with my then publisher, over his decision to promote the now discredited book that spearheaded the attack. The whole episode is thankfully in the past, but it blighted Chagnon's career, and I don't know whether the lesson for social science has been adequately learned.

Chagnon came along at just the right time for the Yanomamö and for scientific anthropology. Encroaching civilization was about to close the last window on a tribal world that embodied vanishing clues to our own prehistory: a world of forest 'gardens', of kin groups fissioning into genetically salient sub-groups, of male combat over women and transgenerational revenge, complex alliances and enmities; webs of calculated obligation, debt, grudge and gratitude that might underlie much of our social psychology and even law, ethics and economics. Chagnon's extraordinary body of work will long be mined, not just by anthropologists but by psychologists, humanists, litterateurs, scientists of all kinds: mined for . . . who knows what insights into the deep roots of our humanity?

PORNOPHILOSOPHY

This review of *Mystery Dance: on the evolution of human sexuality* by Lynn Margulis and Dorion Sagan, published in *Nature* in 1991, is perhaps the cruellest book review I have ever written. I debated with myself whether it should be included here, but pretentious obscurantism gets my goat every time – especially the lamentably influential school of francophoney obscurantism which seems to be largely to blame for the nonsense in *Mystery Dance*. See, for instance, the review entitled 'Postmodernism disrobed', collected in my earlier anthology, *A Devil's Chaplain*. Such pretension has chronically infected the social science departments of many universities – and has been justly satirized in pitch-perfect hoax articles by a number of writers, beginning with Alan Sokal's splendid 'Transgressing the boundaries: towards a transformative hermeneutics of quantum gravity'. Bizarrely, one of the more recent hoaxers, Peter Boghossian, found himself on the receiving end of a disciplinary hearing at his university as a result. His satire, perhaps, cut too near the bone for comfort.

Lynn Margulis is a distinguished biologist: a courageous outsider whose eventual hard-won triumph constitutes one of the important scientific revolutions of our age, the startling realization that our cells are nothing less than symbiotic colonies of bacteria. It would have been good to have heard her views 'on the evolution of human sexuality', the subtitle of this book.

As for what we actually get, I imagine that she is less to blame than her co-author (and son) Dorion Sagan. Indeed, other evidence implicates him as a sucker for those trendy French 'philosophers' whose influence liberally, irrelevantly and distractingly pervades – indeed, ruins – the book. Their 'philosophy' avowedly consists in playing games, which might seem harmless enough. But these charlatans want it both ways – to enjoy a self-indulgent frivolity while simultaneously being thought profound.

Had you noticed that the French word *lit* means both 'read' and 'bed', *hein? Tiens, quel joli* joke. Even better, 'semantics and semiotics bear an evocative resemblance to the sexual word semen'. Oh, such formidable deconstruction, is it not? And 'the English verb mean shares roots with moan' (nudge nudge). But attend, I can cap (wink wink) that trope there. At my school the slang word for erection was – but no I cannot bare it, it has such a droll signification – 'root'.

What on earth have these puerile self-confessed games to do with anything? There are several thousand languages in the world, a rich statistical sample. Has the French 'philosopher' done a systematic survey of them to test his hypothesis that words about sex and about meaning are connected? Margulis the tough-minded scientist would demand no less when refereeing a scientific paper. But chic savants can apparently indulge in idle tossing off and get away with it. I mustn't be too high and mighty, though: I recently heard a lecture by an Oxford disciple of this school of modish French dippiness, and it took me a full five minutes to see through him (it was when he made his big point about Jesus' name and Jacques Derrida's both beginning with J).

The philosophical priorities in this book can be gauged from the authors' bizarre estimation of Heidegger as 'perhaps the most influential philosopher of the twentieth century'. He was the old Nazi, you remember, who single-handedly wrestled with the problem of 'being'. Without Heidegger it would have escaped us that 'the nothing annihilates itself'. Even Lacan (a favourite among Margulis and Sagan's French mentors) refers to Heidegger's 'dustbin style in which currently, by use of his ready-made mental jetsam, one excuses one-self from any real thought'.

The book is built around a recurring theme of striptease. The androgynous stripper peels off garment after garment, metaphorically revealing our evolutionary past. S/he is forever twiddling round and changing sex – sorry, 'gender' – on successive gyrations of the mystery dance. At times the language suggests a cack-handed attempt at eroticism. Or it may seek to embarrass, in which case it succeeds:

> As she climaxes a man briefly appears beneath her, existing only long enough to ejaculate before disappearing again into her shuddering loins [apologies to W. B. Yeats not offered]. In retrospect the audience realized that they saw her give ejaculatory virtual birth to a full-grown male. The body turned and below the rippling abdomen of the seven-veiled striptease artist was an obscure pubis, dark and hairy. The erect penis, the unmistakable signal of his sex, shrunk on the next rotation. It became the clitoris. (pp. 59–60)

Yuck! But there's more:

> She lustily stands spreadeagle. Slowly she turns and bends to expose her dark buttocks and the wet genitals below, bidding him welcome. (p. 60)

These passages are chosen more or less at random. The pages drip with similar 'pornophilosophy', a genre which, I understand, originated with Sartre. Personally, if I must have porn, I prefer it shorn of pretension. One cannot help feeling that if Desmond Morris had written that paragraph (he wouldn't) he'd be noisily accused of flagrant, pot-boiling sensationalism. Should this book get away with it just because of who Lynn Margulis is?

Actually, although it has a joyless earnestness where Morris has a teasing sense of fun, *Mystery Dance* in one way approaches *The*

Naked Ape – the same charmingly cavalier lack of evidence for the same daringly speculative functional explanations. In some cases literally the same explanations, albeit unacknowledged. Margulis and Sagan attribute one of their theories of the female orgasm to an unpublished correspondent, a businessman inspired by a conversation with a sexually boastful US airman. In fact, as I told the same businessman when he wrote to me some years ago, an identical theory was clearly set out in *The Naked Ape* (distributed around the world in a mere 12 million copies).

Meanwhile, back at the dance, more and more veils are lasciviously removed until we expose the ancestral bacteria. Ah, you think, now Margulis must have something worthwhile to say. But no. With a deft gyration of the intellect, our scientific authoress rotates herself into her literary alter ego and, by the time he has given vent to another stupefying salvo of continental obscurantism, the moment has passed:

> But there is perhaps a deeper phase, the metaphysical plane of pure phenomena, continuous appearances. The evolutionary stripper is a curious creature: the G-string is not a thin cloth decorated with tassels but rather a word, a letter, a musical symbol for ultimate nakedness [honestly, I'm not making this up]. Paradoxically, when the G-string is removed – to the accompaniment of strange vibratory music consisting in part of a silent triangle and a gentle crash of cymbals – the nakedness is gone. S(he) stands before us as fully dressed as ever before. (p. 27)

What on earth, you may say, is all this about? Well, wait:

> We encounter our sexual ancestors along the slippery slope of signs and signifiers, via the medium of language. The use of signs of any kind necessarily obscures; words represent or replace signified things in absentia; they are little black masks. We postpone reality to discuss it; without this postponement, this instantaneous

replacement of our sexual ancestors, or things in general, by their signs there could be no possibility of language, of signification at all. (p. 28)

And then where would we be?

How a scientist of Margulis's calibre and rigour can be gulled by this pretentious drivel is as far beyond comprehension as the prose itself. Let us be charitable and hope that she had an argument with her co-author and lost. But if, rightly, you value Lynn Margulis and her reputation, do her a favour and ignore this book.

DETERMINISM AND DIALECTICS: A TALE OF SOUND AND FURY

If the previous review is my cruellest, this one – of *Not in Our Genes* by Steven Rose, Leon Kamin and Richard Lewontin, published in *New Scientist* in 1985 – is probably the most sarcastic I have published. I wouldn't call it cruel because the three authors of the book are all alpha males who can look after themselves, to put it mildly. One of them did just that when the review first came out: he threatened to sue me and *New Scientist*. The threat came to nothing. Nevertheless, not wishing to give needless offence, I have removed from this reprinting the particular paragraph to which he took exception.

The review appeared at the height of the so-called 'sociobiology controversy'. The history of that controversy has been perceptively chronicled by the sociologist Ullica Segerstråle. A modern consensus would agree that the battle was decisively lost by the side represented by the book under review. Its Marxist-inspired treatment of evolutionary biology now has a very dated feel and few followers among natural scientists. It's tempting to wonder, therefore, whether a reprint of this review is superfluous. At the time, however, the controversy was an exceptionally bitter one – so perhaps it has a salutary lesson to offer for the future?

Those of us with time to concentrate on our historic mission to exploit workers and oppress minorities have a great need to 'legitimate' our nefarious activities. The first legitimator we came up with was religion, which has worked pretty well through most of history; but 'the static world of social relations legitimated by God reflected, and was reflected by, the dominant view of the natural world as itself static'.

Latterly there has been an increasing need for a new legitimator. So we developed one: Science.

> The consequence was to change finally the form of the legitimating ideology of bourgeois society. No longer able to rely upon the myth of a deity ... the dominant class dethroned God and replaced him with science ... If anything, this new legitimator of the social order was more formidable than the one it replaced ... Science is the ultimate legitimator of bourgeois ideology.

Legitimation is also the primary purpose of universities:

> It is universities that have become the chief institutions for the creation of biological determinism ... Thus, universities serve as creators, propagators, and legitimators of the ideology of biological determinism. If biological determinism is a weapon in the struggle between classes, then the universities are weapons factories, and their teaching and research faculties are the engineers, designers, and production workers.

And to think that, through all these years working in universities, I had imagined that the purpose of science was to solve the riddles of the universe: to comprehend the nature of existence, of space and time and of eternity; of fundamental particles spread through 100 billion galaxies; of complexity and living organization and the slow dance through three billion years of geological time. No no, these trivial matters fade into insignificance beside the overriding need to legitimate bourgeois ideology.

How can I sum up this book? Imagine a sort of scientific Dave Spart trying to get into 'Pseud's Corner'.* Even the acknowledgements give us fair warning of what to expect. Where others might thank colleagues and friends, our authors acknowledge 'lovers' and 'comrades'. Actually, I suppose there is something rather sweet about this, in a passé, sixtiesish sort of way. And the 1960s have a mythic role to play in the authors' bizarre conspiracy theory of science. It was in response to that Arcadian decade (when 'students challenged the legitimacy of their universities . . . ') that 'the newest form of biological determinism, sociobiology, has been legitimated'.

Sociobiology, it seems, makes the two assertions 'that are *required* if it is to *serve as* a legitimization and perpetuation of the social order' (my emphasis). The 'panglossianism'† of sociobiology 'has played an important role in legitimation', but this is not its main feature:

> Sociobiology is a reductionist, biological determinist explanation of human existence. Its adherents claim, first, that the details of present and past social arrangements are the inevitable manifestations of the specific action of genes.

---

* To elucidate for readers outside Britain: the satirical fortnightly magazine *Private Eye* has a fictitious columnist called 'Dave Spart', an inarticulate left-wing student activist, whose cliché-ridden dronings usually begin with 'Er basically'. *Private Eye* also has a regular section called 'Pseud's Corner' whose title speaks for itself. My Oxford colleague the distinguished classical historian Robin Lane Fox once made a deliberate attempt, in his gardening column for the *Financial Times*, to get into Pseud's Corner. To his chagrin, that attempt at deliberate pretension failed, but the very next issue featured in Pseud's Corner a later gardening column written by Robin with no such aspiration.

† The well-known saying of Voltaire's satirical character Dr Pangloss was earlier introduced into evolutionary theory by J. B. S. Haldane as 'Pangloss's Theorem': 'Everything is for the best in this best of all possible worlds.'

Unfortunately, academic sociobiologists, unaccountably neglecting their responsibilities towards the class struggle, do not seem anywhere to have actually *said* that human social arrangements are the inevitable manifestations of genes. Rose, Kamin and Lewontin (henceforth RKL) have accordingly had to go further afield for their substantiating quotations, getting them from such respected sociobiologists as Mr Patrick Jenkin when he was minister for social services, and various dubious representatives of the National Front and the *Nouvelle Droite* whose works most of us would not ordinarily see (they are no doubt grateful for the publicity). The minister gives especially good value, by using a 'double legitimation of science and God . . . '.

Enough of this, let me speak plainly. RKL cannot substantiate their allegation about sociobiologists believing in inevitable genetic determination because the allegation is false. The myth of the 'inevitability' of genetic effects has nothing whatever to do with sociobiology, and everything to do with RKL's paranoiac and demonological theology of science. Sociobiologists, such as myself (much as I have always disliked the name, this book finally provokes me to stand up and be counted), are in the business of trying to work out the conditions under which Darwinian theory might be applicable to behaviour. If we tried to do our Darwinian theorizing without postulating genes affecting behaviour, we should get it wrong. *That* is why sociobiologists talk about genes so much, and that is all there is to it. The idea of 'inevitability' never enters their heads.

RKL have no clear idea of what they mean by biological determinism. 'Determinist', for them, is simply one half of a double-barrelled blunderbuss term, with much the same role and lack of content that 'Mendelist–Morganist' had in the vocabulary of an earlier generation of comrades. Today's other barrel, fired off with equal monotony and imprecision, is 'reductionist'.

[Reductionists] argue that the properties of a human society are . . . no more than the sums of the individual behaviours and tendencies of the individual humans of which that society is composed. Societies are 'aggressive' because the individuals who compose them are 'aggressive', for instance.

As I am described in the book as 'the most reductionist of sociobiologists', I can speak with authority here. I believe that Bach was a musical man. Therefore, of course, being a good reductionist, I must obviously believe that Bach's brain was made of musical atoms! Do Rose et al. sincerely think that *anybody* could be that silly? Presumably not, yet my Bach example is a precise analogy to 'Societies are "aggressive" because the individuals who compose them are "aggressive"'.

Why do RKL find it necessary to *reduce* a perfectly sensible belief (that complex wholes should be explained *in terms of* their parts) to an idiotic travesty (that the properties of a complex whole are simply the *sum* of those same properties in the parts)? 'In terms of' covers a multitude of highly sophisticated causal interactions, and mathematical relations of which summation is only the simplest. Reductionism, in the 'sum of the parts' sense, is obviously daft, and is nowhere to be found in the writings of real biologists. Reductionism, in the 'in terms of' sense, is, in the words of the Medawars,* 'the most successful research stratagem ever devised'.

RKL tell us that 'some of the most penetrating and scathing critiques of sociobiology have come from anthropologists'. The two most famous anthropologists cited are Marshall Sahlins and Sherwood Washburn, and their 'penetrating' critiques are, indeed, well worth looking up. Washburn thinks that, as all humans, regardless of kinship, share more than 99 per cent of their genes, 'genetics actually supports the beliefs of the social sciences, not the calculations of the sociobiologists'. Lewontin, the brilliant geneticist, could, if he wanted to, quickly clear up this pathetic little misunderstanding of

---

* In *Aristotle to Zoos*.

kin selection theory. Sahlins, in a book described as 'a withering attack' on sociobiology, thinks that the theory of kin selection cannot work because only a minority of human cultures have developed the concept of the fraction (necessary, you see, in order for people to calculate their coefficients of relatedness!). Lewontin the geneticist would not tolerate elementary blunders like this from a first-year undergraduate. But for Lewontin the 'radical scientist', apparently any criticism of sociobiology, no matter how bungling and ignorant, is penetrating, scathing and withering.*

RKL see their main role as a negative and purging one, even casting themselves as a gallant little fire brigade, 'constantly being called out in the middle of the night to put out the latest conflagration . . . All of these deterministic fires need to be doused with the cold water of reason before the entire intellectual neighborhood is in flames.' This dooms them to constant nay-saying, and they therefore now feel an obligation to produce 'some positive program for understanding human life'. What, then, is our authors' positive contribution to understanding life?

At this point, self-conscious throat-clearing becomes almost audible and the reader is led to anticipate some good embarrassing stuff. We are promised 'an alternative world view'. What will it be? 'Holistic biology'? 'Structuralistic biology'? Connoisseurs of the genre might have put their money on either of these, or perhaps on 'Deconstructionist biology'. But the alternative world-view turns out to be even better: 'Dialectical' biology! And what exactly is dialectical biology? Well – think, for example,

of the baking of a cake: the taste of the product is the result of a complex interaction of components – such as butter, sugar, and flour – exposed for various periods to elevated temperatures; it is

---

* Sahlins' and Washburns' fallacies are, respectively, Misunderstanding 3 and Misunderstanding 5 of my 'Twelve misunderstandings of kin selection', reprinted in *Science in the Soul*.

not dissociable into such-or-such a percent of flour, such-or-such of butter, etc., although each and every component . . . has its contribution to make to the final product.

When put like that, this dialectical biology seems to make a lot of sense. Perhaps even I can be a dialectical biologist. Come to think of it, isn't there something familiar about that cake? Yes, here it is, in a 1981 publication by the most reductionist of sociobiologists:

If we follow a particular recipe, word for word, in a cookery book, what finally emerges from the oven is a cake. We cannot now break the cake into its component crumbs and say: this crumb corresponds to the first word in the recipe, this crumb corresponds to the second word in the recipe etc. With minor exceptions such as the cherry on top, there is no one-to-one mapping from words of recipe to 'bits' of cake. The whole recipe maps onto the whole cake.

I am not, of course, interested in claiming priority for the cake (Pat Bateson had it first, in any case). But what I do hope is that this little coincidence may at least give RKL pause. Could it be that their targets are not quite the naively atomistic reductionists they would desperately like them to be?

So, life is complex and its causal factors interact. If that is 'dialectical', big deal. But no, it seems that 'interactionism', though good in its way, is not quite 'dialectical'. And what is the difference?

First [interactionism] supposes the alienation of organism and the environment . . . second, it accepts the ontological priority of the individual over the collectivity and therefore of the epistemological sufficiency of . . .

There is no need to go on. This sort of writing appears to be intended to communicate nothing. Is it intended to impress, while putting down smoke to conceal the fact that nothing is actually being said?

TUTORIAL-DRIVEN TEACHING

This essay was originally written for an Oxford ingroup readership: the 'parish magazine' of the university, the *Oxford University Gazette*. David Palfreyman, when he came to edit an anthology in praise of the Oxford tutorial in 2008, asked if he could reprint it and I was happy to agree. A modified version of it was also published in *Oxford Today*, the (now extinct, alas) magazine for alumni of the university. The sarcastic tone of the first paragraph will best be understood in the context of the original audience. British readers will know that there are no universities at Chipping Ongar, Herne Bay, Crichel Down etc. That was all part of the sarcasm. So was the refrain, 'What do the undergraduates think?' Student representation on faculty committees was just becoming all the rage at the time.

We in the biological faculty have been meeting to indulge in one of those satisfying orgies of communal breast-beating that afflict conscientious teachers from time to time. What are we doing wrong? How can we improve the way we teach and the way we examine? Certain other universities are said to do X; surely powerful evidence that X must be a good thing to do. The present examination system unfairly discriminates against students with a short attention span. Shouldn't we be doing continuous assessment like they do at Chipping Ongar? Could we make lecture attendance compulsory like they do at Herne Bay? What do the undergraduates think? (What do the individual undergraduate representatives on the committee think?) How about a clocking-in machine for practical classes? At Crichel Down they have 'tutorials' twenty strong, whereas we have a measly one student per tutorial which can't be good for their education. Why do the marks in the final examination cluster so monotonously about the middle of the second class instead of distributing themselves satisfyingly from extreme to extreme? What do the undergraduates think? The external examiners have ordered us in future to do Y (they do Y at their own universities and it works well enough) but, unfortunately, our Decrees are such that we'd need an Act of Parliament to be allowed to do Y. Here's a good idea: why don't we make our teaching very broad-based like they do at Budleigh Salterton? Excellent suggestion, and let's make it simultaneously very deep like they did at my old university. What do the undergraduates think?

This is all familiar enough – and laudable enough, for, despite my opening cynicism, there is certainly much that could be improved and it is undoubtedly our duty to seek it out. But one particular suggestion which has started to emerge recently has got to the roots of my hackles. Our teaching has been accused of being 'tutorial-driven'. The phrase originated with a pair of external examiners, who have been egged on to perceive their brief as widened from commenting on our examination to telling us how to run our university. It has been dutifully taken up by junior and senior members of Joint Consultative Committees, and is now echoing around departmental corridors. Our teaching is 'tutorial-driven' where it should be 'lecture-driven'. The meaning of tutorial-driven is best explained by reference to its remedy. The content of tutorial essays should be strictly limited to topics covered in official lectures. Tutors must be told the content of the lectures and must 'address' (to use the pretentious golfing jargon) these topics in their tutorials. Perhaps lecturers should distribute reading lists which all tutors must adopt and assign to their pupils.

I'll tell you what makes me so particularly sad about this philistinism. I was an undergraduate myself once; we were tutorial-driven (we didn't realize it at the time) and it has been the making of my whole life. Not just one particular tutor: it was the whole experience of the Oxford tutorial system. In my penultimate term Peter Brunet, my wise and humane college tutor,* managed to secure for me tutorials in animal behaviour with the great Niko Tinbergen, later to win the Nobel Prize for his part in founding the science of ethology. Tinbergen himself was solely responsible for all the lectures in animal behaviour, so he would have been well-placed to give 'lecture-driven' tutorials. I need hardly say that he did no such thing. Each week my tutorial assignment was to read one DPhil† thesis. My essay was to be a combination of DPhil examiner's report, proposal for follow-up research, review of the history of the subject in which the thesis fell, and theoretical and philosophical discussion of the issues that the thesis raised. Never for one moment did it occur to either of us to wonder whether this assignment would be directly useful to me in answering some exam question.

Another term my college tutor, recognizing that my bias in biology was more philosophical than his own, arranged for me to have tutorials with Arthur Cain, an effervescently brilliant young star of the department, who went on to become Professor of Zoology at Liverpool. Far from his tutorials being driven by any lectures then being offered for the Honour School of Zoology, Cain had me reading nothing but books on history and philosophy. It was up to me to work out the connections between zoology and the books that I was reading. I tried, and I loved the trying. I'm not saying that my

---

* A lovely man, he died very soon after retiring. His robust attitude to death was conveyed in the rhetorical question: 'Why would anyone want an old bugger like me to stick around?' I'd like to dedicate this essay to his memory.
† The Oxford name for PhD. Among the newer universities, Sussex and York have also adopted it, but most other universities prefer to follow Cambridge terminology, and even at Oxford the habit is growing of calling the degree, informally, 'PhD'.

juvenile essays in the philosophy of biology were any good – with hindsight I know that they were not – but I do know that I have never forgotten the exhilaration of writing them.

The same is true of my more mainline essays on standard zoological topics. I have not the slightest memory of whether we had a lecture on the water-vascular system of starfish. Probably we did, but I am happy to say that the fact had no bearing upon my tutor's decision to assign an essay on the topic. The starfish water-vascular system is one of many highly specialized topics in zoology that I now recall for the same reason – that I wrote an essay on them. Starfish don't have red blood, they have piped sea water instead. Sea water enters through a hole, and is constantly being circulated through an intricate system of tubes which form a ring around the centre of the star and lead off in branches down each of the five arms. The piped sea water embodies a unique hydraulic pressure system, operating the many hundreds of tiny tube feet arrayed along the five arms. Each tube foot ends in a little gripping sucker, and they grope and swing back and forth in collusion to pull the starfish along in a particular direction. The tube feet don't move in unison but are semi-autonomous and, if the circum-oral nerve ring that gives them their orders should chance to become severed, the tube feet in different arms can pull in opposite directions and tear the starfish in half.

I remember the bare facts about starfish hydraulics but it is not the facts that matter. What matters is the way in which we were encouraged to find them. We didn't just mug up a textbook: we went into the library and looked up books old and new; we followed trails of original research papers until we had made ourselves as near world authorities on the topic at hand as it is possible to become in one week. The encouragement provided by the weekly tutorial meant that one didn't just read about starfish hydraulics, or whatever the topic was. For that one week I remember that I slept, ate and dreamed starfish hydraulics. Tube feet marched behind my eyelids, hydraulic pedicellariae quested and sea water pulsed through my dozing brain. Writing my essay was the catharsis, and the tutorial was the justification for the entire week. And then the next week there would be a new topic and a new feast of images to be conjured in the library. We were being educated and our education was tutorial-driven.*

I don't mean to knock lectures. Lectures, too, can be inspiring, especially when the lecturer throws away utilitarian preoccupations with syllabuses and with 'imparting information'.† No zoologist of

---

* John Buxton, a senior colleague when I first arrived as a Fellow of New College, wrote this about his three tutors (in Classics, as it happens) in the college between the wars: 'They made us think that we were concerned, with them, in the discovery of the literature, the history and the thought of the ancient world. Nothing could have been more rewarding. We were not instructed by men who knew all the answers, we were partners in a quest and were being educated.' This memoir is not the only thing that makes me regret that I didn't take the trouble to get to know Mr Buxton better before his death. By the time I arrived in New College he had become a somewhat remote and forbidding figure, even unhappy: withdrawn, perhaps because of deafness. He was by then teaching English literature, despite having studied Classics as an undergraduate. He was also an accomplished ornithologist, author of the definitive book on the redstart. He had done part of the observational research for this while a prisoner of war in Germany, in which pursuit he had recruited the assistance of fellow prisoners. Clearly an interesting man and, from my point of view as a younger colleague, a missed opportunity.

† The purpose of a lecture should be not to inform but to inspire. Mark Twain is one of several alleged authors of the cynical witticism, 'College is a place where a professor's lecture notes go straight to the students' lecture notes, without passing through the brains of either.' As an undergraduate I ruined most of the benefit of lectures by taking notes. Obsessively. I never looked at my notes again. The one occasion when I forgot to bring a pen (and was too shy to borrow one from the hopelessly admired young woman sitting next to me) was the one occasion

my generation will forget Sir Alister Hardy's lightning works of art on the blackboard, his recitations of comic verse on larval forms and his mimed imitations of their antics, nor his evocation of the blooming plankton fields of the open sea. Others who took a more cerebral and less pyrotechnic approach were just as good in their way. But, however good the lectures, we didn't ask for our tutorials to be driven by them and we didn't expect our exam questions to be lecture-driven either. The whole field of zoology was fair game for the examiners and the only thing we could rely upon was a presumption that our question papers would not be too unfairly different from their recent predecessors. The examiners when setting the papers, and our tutors when handing out essay topics, neither knew nor cared which subjects had been covered in lectures.

Well, yes, of course I have just been indulging in a different kind of orgy, an orgy of nostalgia for my own university and its possibly unique method of education. I have no right to assume that the system under which I was educated is the best, any more than my incoming colleagues have a right to make the reciprocal assumption about their own excellent universities. We have to make individual arguments for the educational merits of whatever systems we wish to advocate. We must not assume that, because something is

---

when I actually remembered what the lecturer said, and I wrote a summary promptly on my return to my college room. In earlier centuries when books were hard to come by, the lecture lived up to its literal meaning. The lecturer had the book; the students didn't, and some of them might even have been unable to read. The lecturer stood in front of them with the book on the lectern, and read from it aloud. But our students can read. They can obtain information from the same source as the professor: by reading books and the primary research literature, in the internet age more readily available than ever before. The very best lecturers inspire students by thinking aloud, by grasping ideas out of the air in the presence of the students and encouraging them to do the same.

traditionally done at Oxford it is necessarily good (it is, in any case, amazing how *recent* most so-called ancient traditions often turn out to be).* But also we must beware of the opposite assumption that because something is traditionally done at Oxford it therefore is self-evidently bad. It is on its educational merits alone that I am prepared to argue the case that our Oxford education should continue to be 'tutorial-driven'. Or if it is not to be, and we decide to abolish the tutorial system, then let us at least know what it is that we are abolishing. If we replace the Oxford tutorial, let us do so in spite of its glories and because we think we have found something better, *not* because we never properly understood what a real tutorial was in the first place.†

---

* This was something I learned only after I had lived in Oxford for some time. A trivial example of an 'ancient tradition' which is actually very recent is the habit of throwing flour and champagne over friends emerging from their final examination. The Oxford tutorial itself, in its modern form, is a largely nineteenth-century invention, developed by Benjamin Jowett, Master of Balliol College from 1870 to 1893.

† In the second edition of David Palfreyman's collection I added a preamble. Among other things, I admitted that the one-to-one tutorial was expensive. I suggested that, rather than save money by increasing the student-to-teacher ratio so that a tutorial became a seminar, as at many universities, we should dispense with the idea that a tutor had to be an experienced authority on the subject. Instead, I advocated something like the American 'teaching assistant'. Tutors in my reformed system would be graduate students. While lacking the experience of a fully fledged tutorial fellow, they would make up for it in youthful enthusiasm. With hindsight I realized that many of my own tutors were actually graduate students or postdoctoral researchers, and little the worse for that. This practice was common in zoology, less so in traditional subjects like history which tended to be taught by older fellows of colleges.

LIFE AFTER LIGHT

I have long loved Daniel F. Galouye's novel *Dark Universe*, ever since it was recommended to me by a scientific mentor when I was a graduate student. I was therefore delighted when, in 2009, I was invited to introduce the audio edition. This is the text of my introduction, which I also spoke myself for the recording.

Bad science fiction is like a fairy tale, playing fast and loose with reality so that it degenerates lazily into magic spells. Good science fiction restrains itself, constrains itself within limits set by science, or licenses itself to change one part of science in a disciplined way, to explore the consequences. The very best science fiction succeeds in making us think about science in a new way; or philosophy in a new way; or, as in the case of this book, it makes us think about mythology and religion in a new way.

Imagine a world in which there is no light. Not a world in which there had never been light: a world in which there had been light but it disappeared. Daniel Galouye conceived an entirely plausible reason why this misfortune might have happened, to a colony of people, living deep underground, who were otherwise just like us, and whom we do not meet until many generations after the loss of light. We are not told the reason until the very end of the book, although readers may guess through tantalizing hints that are dropped in through the story. For example, the word for 'citizen' is 'survivor' (or 'survivoress'). Does that give a clue? And the religion of the people includes the twin demons 'Strontium' and 'Cobalt', together with the dreaded arch-devil 'Hydrogen Himself'. Or there's the common curse: 'Radiation!' 'Radiation take it!' 'Cobalt!'

When we join the story, light has been completely forgotten by everybody, except – fascinatingly – as an obscure race memory, which has become the basis for their worship. This is the conclusion that we gradually piece together as we are drawn into their world of darkness. Language is purged of all words involving sight, as the people shuffle about their subterranean world, using echolocation like bats, or 'zivving' (which we gradually work out must mean the use of infra-red radiation from warm bodies and hot springs). 'I hear what you mean', never 'I see what you mean'. They have no concept of a day or a year. The word 'light' is used frequently in everyday speech, but nobody knows what light is. References to light are religious only. It has become a myth, an atavistic relic of a time before a mythical Fall. 'Light Almighty' has become an oath. 'Light help us!' 'Oh, for Light's sake . . .'

The hero, Jared, embarks on a quest for 'darkness'. The word means nothing to those who experience nothing else, but Jared has an obscure theological intuition that it may be the key to understanding light. He attends a religious ceremony in which a 'sacred light bulb' is solemnly passed around for the whole congregation to feel, while they chant the catechism:

What is Light?
Light is a Spirit.
Where is Light?
If it weren't for the evil in man, Light would be everywhere.
Can we feel or hear Light?
No, but in the hereafter we shall see Him.

The resonances with Christian theology are haunting. Haunting, too, are eerily italicized interludes in which the hero visits, in dreams, three strange, presumably mutant, characters whose ghostly presence has visited him since childhood: 'the telepathic Kind Survivoress', through whom he is aware of 'Little Listener' (who has discovered how to use 'silent sound' from glow worms) and 'The Forever Man', a Methuselan character who cannot die, and who does nothing but tap a rock, into which his finger has, over the ages, bored a deep hole. The Forever Man is old enough to remember light, but his mind has gone. Only when Jared mentions his obsessive pilgrimage in search of 'darkness' does the full horror of the 'Fall from Light' come home to him. The awful darkness, the Forever Man remembers, is all around them – an observation which Jared, of course, does not understand.

The book ends with the repatriation of the 'survivors' into the outside world where light is real – our own world, which has become so familiar we forget how wonderful light is, and the gift of sight. Jared's first terrifying encounter with the sun – undoubtedly Hydrogen Himself – is movingly evoked.

Of all the novels I have read, this is perhaps the one that I find myself describing to others more often than any other. I am fascinated by the idea, and I find that others are too.* But as for the theological analogy, does it mean anything? Read it† and decide for yourself.

---

* It's not the only one of Daniel Galouye's novels of which I could
  say that. Another is *Counterfeit World* (1965), in which we discover
  that we live in a simulation programmed by a higher civilization.
  This is a postulate that it is hard, if not impossible, to disprove, as the
  philosopher Nick Bostrom has argued. It is also the plot of the
  rather silly film *The Matrix*.
† 'Listen to it', of course, in the case of the audiobook.

A SCIENTIFIC EDUCATION
AND THE DEEP PROBLEMS

I first read Fred Hoyle's *The Black Cloud* many years ago and
frequently recommended it to others, both privately and in print.
I was on good terms with Penguin Books and I persuaded them to
issue a new edition. They agreed, on condition that I contributed a
new afterword. This I was delighted to do, and the text (published
in 2010) is reproduced here. Later, when the audio edition came
out, I was invited to read my afterword for the recording.

In the previous piece I made a distinction between bad science
fiction, which plays fast and loose with reality, allowing what
amount to magic spells, and good science fiction, which admits
the constraints of real science. I mentioned that the very best
science fiction actually has the capacity to teach us science. *The
Black Cloud* is a prime example. I would now add an intermediate
category of science fiction into which I'd place, for example, H. G.
Wells's *The War of the Worlds* – and also many of the short stories
collected in anthologies of science fiction. I refer to stories that
may have merit as adventure yarns, thrillers, romances etc., but
whose connection with science or the scientific imagination
is incidental. An example might be an exciting shoot-'em-up
adventure tale that just happens to be set on Mars rather than in
the Wild West, and with ray-guns rather than revolvers. Michael
Crichton's *Timeline* is science fiction in the sense that the heroes
have the technology to travel in time, but when they get to the

middle ages the story becomes an ordinary thriller, albeit a good one, which might as well be set in the present.

Sir Fred Hoyle FRS (1915–2001) was a distinguished scientist, whose blunt, even abrasive, Yorkshire manner rubbed off on many of his science-fiction heroes, including Christopher Kingsley, the lead character of this, his first and best-known novel. As an astronomer, Hoyle was famous for being wrong about the Big Bang theory of the origin of the cosmos. He was against it – the very name is his own sarcastic coining – preferring his own elegant and pugnaciously defended 'Steady State' theory. He was spectacularly right in his theory of how the chemical elements are forged, ultimately from hydrogen, in the interiors of stars. Indeed, many scientists feel that a serious injustice was done to Hoyle when he was denied a share in the Nobel Prize that was eventually given to others for this foundational theory. About his incursions into theoretical biology and evolutionary theory, the less said the better.

As a novelist, I would say his output was mixed. *A for Andromeda*, co-authored with John Elliott, shares with *The Black Cloud* the enormous virtue of educating the reader in scientific principles at the same time as it entertains. In particular, the book expounds the important idea – later reprised by Carl Sagan in *Contact* – that, if an alien civilization wished to take over the Earth, they would most likely not visit us in person (galactic distances are too great) but would send coded information by radio, which would be deciphered as the instructions for building and programming a computer. The computer would then act as the aliens' proxy. To understand why this is so plausible is to understand some profound principles of science, and Hoyle brilliantly gets the point across.

Some of his other novels go to the other extreme, and are little more than pot-boilers. But *The Black Cloud* is, in my opinion, one of the greatest works of science fiction ever written, up there with the best of Isaac Asimov and Arthur C. Clarke. Right from the first page, it is what used to be called a 'rattling good yarn', one of those stories that grabs you on page one and doesn't let go until you finish it in the wee small hours. It helps that the book is set approximately in the present, and does not, like so much science fiction, bewilder us with strange, alien names and other-worldly customs which we don't begin to understand until we are well into the book, by which time our busy life might have shown us something better to do than go on reading it. Hoyle's characters love to think deep thoughts in their Cambridge rooms before a roaring log fire, and the recurring image is a delightfully comfortable one.

But the real virtue of *The Black Cloud* is this. Without ever preaching at us, Hoyle manages, as the story races along, to teach us some fascinating science along the way: not just scientific facts, but important scientific principles. We get to see how scientists work and how they think. We are even uplifted and inspired. Let me list just a few examples of the real science – and, indeed, philosophy – that the book spins off.

Scientific discoveries are often made, sometimes simultaneously, by a convergence of more than one method. Hoyle's black cloud is detected by direct observation through a Californian telescope, and simultaneously by indirect mathematical reasoning in Cambridge. The narrative in this early part of the book is ravishingly well handled, climaxing with a telegram sent from the Cambridge team to the California team. Neither side knows that the other has independently converged on the same alarming truth, and there is a goosepimpling moment when the words of the telegram 'seemed to swell to a gigantic size'.

The gradual elucidation of the true nature of the black cloud also gives fascinating insights into the way scientists think and argue among themselves. The hero, the Cambridge theoretical astronomer Christopher Kingsley, whom it is hard not to identify with Hoyle himself, and the Russian astronomer Alexandrov, who is the book's comic relief character, independently tumble to the startling truth – so startling that other characters stubbornly refuse to accept it. Kingsley and Alexandrov relentlessly insist that theories should be tested by *prediction*, and they gradually win over the sceptics. Once again, there is absorbing drama in the unfolding dialogue between cooperating and dissenting scientists.

Once the strange nature of the cloud is established, things move rapidly. I don't want to give anything away, but I can say that in this part of the story one of the scientific lessons we learn is about information theory. Information is a commodity, readily interchangeable from one medium to another. Beethoven moves us via our ears, but in principle there is no reason why an alien being, or an advanced computer, say, with no sense of hearing at all, shouldn't enjoy the music if supplied with the same temporal patternings (which might be hugely speeded up or slowed down), and the same mathematical relationships between frequencies – the ones that we interpret as melody and harmony. In information theory, the medium of transmission is arbitrary. This idea has been very influential on me, and I acknowledge that I first came to appreciate it through reading *The Black Cloud* as a young man.

A related point, of deep scientific and philosophical significance, is that the subjective individuality that each of us feels inside our skull depends upon the slowness and other imperfections of the channels of communication between us, for example language. If we could share our thoughts instantly by telepathy, fully and at the same rate as we can think them, we would cease to be separate individuals. Or, to put it another way, the very idea of separate individuality would lose its meaning. This, indeed, is arguably what *did* happen in the evolution of the nervous system. It is a thought that has intrigued me for much of my career as a biologist, and I was again led to it by reading *The Black Cloud*.

Arthur C. Clarke, a more consistent writer of good science fiction than Hoyle, although he only equalled Hoyle at his best, stated as his 'Third Law' that 'any sufficiently advanced technology is indistinguishable from magic'. *The Black Cloud* reinforces the message, in spades. Pizarro fired his cannon, and was taken for a god by the Incas. Imagine if he had arrived in a helicopter gunship instead of on a horse. Imagine the response of a medieval peasant, or even aristocrat, to a telephone, a television, a laptop computer, a jumbo jet. *The Black Cloud* vividly conveys to us what it would be like to be visited by an extraterrestrial being whose intelligence would seem god-like from our lowly point of view. Indeed, Hoyle's imagination far outperforms all religions known to me. Would such a super-intelligence then actually *be* a god?

An interesting question, perhaps the founding question of a new discipline of 'Scientific Theology'. The answer, it seems to me, turns not on what the super-intelligence is capable of doing, but on its provenance. Alien beings, no matter how advanced their intelligence and accomplishments, would presumably have evolved by something like the same gradual evolutionary process as gave rise to our kind of life. And this is where Hoyle makes this book's only scientific mistake, in my opinion. The eponymous super-intelligence of *The Black Cloud* is asked about the origin of the first member of its species, and it replies, 'I would not agree that there ever was a "first" member.' The response of the astronomers in the story is an in-joke by Hoyle: 'Kingsley and Marlowe exchanged a glance as if to say: "Oh-ho, there we go. That's one in the eye for the exploding-universe boys."' Never mind the astronomers, I must protest as a biologist. Even if Hoyle and his colleagues had been right that the universe has been in a steady state for ever, the same could not sensibly be claimed for the organized and apparently purposeful complexity that life epitomizes. Galaxies may spring spontaneously into existence, but complex life cannot. That is pretty much what complexity *means*!

There are other flaws in the novel. Despite the wonderfully true-to-life picture it paints of how scientists think, the dialogue occasionally becomes a little clunky, the jokes a little heavy. The character of the hero, Christopher Kingsley, always on the abrasive side, rises to heights – or descends to depths – of inhumane fanaticism in a horrifying scene near the end of the book, which one reviewer described as 'a fascinating glimpse into the scientific power dream' but which struck me as way over the top.

Ever since I first read this book, a phrase from it has haunted me: 'the Deep Problems'. These are the problems in science that we do not understand, perhaps can *never* understand, either because of the limitations of our evolved minds or because they are in principle insoluble. How did the universe begin, and how will it end? Can something come from nothing?* Whence the laws of physics? Why do the fundamental constants have the particular values that they do? What about other questions that are so far beyond us that we cannot even *ask*, let alone answer them? The idea of the Deep Problems, and the possibility that they might be understood by a superior intelligence but not by us, is humbling, but humbling in a way that is at the same time uplifting. Also challenging.

The tragic ending of the novel is moving, and deeply thought-provoking at the same time. It is followed by a gentle epilogue – again, the contemplation by the log fire – which pulls the threads together and leaves us on a high. The last words leave us exhilarated, even stunned, as we look back on this astonishing novel: 'Do we want to remain big people in a tiny world or to become a little people in a vaster world? This is the ultimate climax towards which I have directed my narrative.'

---

* Indeed it can, as Lawrence Krauss has explained. See my afterword to his *A Universe from Nothing* (pp. 353–7 below).

RATIONALIST, ICONOCLAST, RENAISSANCE MAN

The Ascent of Man was a thirteen-part BBC television series written and presented by Dr Jacob Bronowski, polymathic scientist, mathematician, poet and art connoisseur. Bertrand Russell, in his *History of Western Philosophy*, describes Omar Khayyam as 'the only man known to me who was both a poet and a mathematician'. He might have added Jacob Bronowski who, as a young man, was both published poet and poetry editor and critic, as well as being 'Senior Wrangler', i.e. the top mathematician of his year at Cambridge. *The Ascent of Man* series, first broadcast in 1973, was conceived and commissioned by David Attenborough when he was Controller of BBC2, and it is widely regarded as a candidate for the greatest television documentary series ever. Perhaps its only rivals are Kenneth Clark's *Civilisation*, also commissioned by David Attenborough, and of course Attenborough's own superlative natural history films. Like most such ambitious documentaries, Bronowski's was accompanied by a book. When a new edition was called for in 2011, his daughter, the historian and literary scholar Lisa Jardine, invited me to write the foreword. It is reproduced here.

'Last Renaissance man' has become a cliché, but we forgive a cliché on the rare occasion when it is true. Certainly it is hard to think of a better candidate for the accolade than Jacob Bronowski. You'll find other scientists who can parade a deep parallel knowledge of the arts, or – in one actual case – combine eminence in science with pre-eminence in Chinese history.* But who more than Bronowski weaves a deep knowledge of history, art, cultural anthropology, literature and philosophy into one seamless cloth with his science? And does it lightly, effortlessly, never sinking to pretension? Bronowski uses the English language – not his first language, which makes it all the more remarkable – as a painter uses his brush, with mastery all the way from broad canvas to exquisite miniature.

Inspired by the Mona Lisa, here is what he has to say about arguably the first and greatest Renaissance man, whose drawing of the baby in the womb introduced the television version of *The Ascent of Man*:

Man is unique not because he does science, and he is unique not because he does art, but because science and art equally are expressions of his marvellous plasticity of mind. And the Mona Lisa is a very good example, because after all what did Leonardo do for much of his life? He drew anatomical pictures, such as the baby in the womb in the Royal Collection at Windsor. And the brain and the baby is exactly where the plasticity of human behaviour begins.

How deftly Bronowski segues from Leonardo's drawing to the Taung baby: type specimen of our ancestral genus *Australopithecus*, victim – as we now know, though Bronowski didn't when he

---

* Joseph Needham, of course. He once unforgettably taught a lesson to the biology sixth form at my school. He brought along some ATP (adenosine triphosphate, the gasoline of life) and demonstrated its dramatic, dynamic effect on muscle fibres. We owed this privilege to his nephew, who was then a student teacher at the school. Nepotism isn't always a wholly bad thing.

performed his mathematical analysis on the tiny skull – of a giant eagle two million years ago.*

There's a quotable aphorism on every page of this book, something to treasure, something to stick on your door for all to see, an epitaph, perhaps, for the gravestone of a great scientist. 'Knowledge ... is an unending adventure at the edge of uncertainty.' Uplifting? Yes. Inspiring? Without doubt. But read it in context and it is shocking. The grave turns out to belong to an entire tradition of European scholarship, destroyed by Hitler and his allies almost overnight:

> Europe was no longer hospitable to the imagination – and not just the scientific imagination. A whole conception of culture was in retreat: the conception that human knowledge is personal and responsible, an unending adventure at the edge of uncertainty. Silence fell, as after the trial of Galileo. The great men went out into a threatened world. Max Born. Erwin Schrödinger. Albert Einstein. Sigmund Freud. Thomas Mann. Bertolt Brecht. Arturo Toscanini. Bruno Walter. Marc Chagall.

---

* Bronowski clearly felt an affinity with this diminutive skull. I too, for different reasons. The Taung Child's sad end moved me to one of my rare purple passages (with an oblique allusion to the Irish poet 'AE') in *The Greatest Show on Earth*: 'Poor little Taung Child, shrieking on the wind as you were borne aloft by the aquiline fury, you would have found no comfort in your destined fame, two and a half million years on, as the type specimen of *Australopithecus africanus*. Poor Taung mother, weeping in the Pliocene.' The Eagle and Child pub in Oxford is famous as the meeting place of the 'Inklings' literary circle which included J. R. R. Tolkien and C. S. Lewis. But its name also resonates with the tragic fate of the Taung Child, and I was happy to be invited to a ceremony there in 2006 at which the South African paleontologist Francis Thackeray unveiled a plaque commemorating the luckless young holotype.

Words so powerful don't need a raised voice or ostentatious tears. Bronowski's words gained impact from his calm, humane, understated tones, with the engagingly rolled Rs as he looked straight into the camera, spectacles flashing like beacons in the dark.

That was a rare dark passage in a book that is mostly filled with light, and genuinely uplifting. You can hear Bronowski's distinctive voice through this book, and you can see his expressive hand chopping down to cut through complexity and make a point. He stands before a great sculpture, Henry Moore's *The Knife Edge,* to tell us:

> The hand is the cutting edge of the mind. Civilisation is not a collection of finished artefacts, it is the elaboration of processes. In the end, the march of man is the refinement of the hand in action. The most powerful drive in the ascent of man is his pleasure in his own skill. He loves to do what he does well and, having done it well, he loves to do it better. You see it in his science. You see it in the magnificence with which he carves and builds, the loving care, the gaiety, the effrontery. The monuments are supposed to commemorate kings and religions, heroes, dogmas, but in the end the man they commemorate is the builder.

Bronowski was a rationalist and an iconoclast. He was not content to bask in the achievements of science but sought to provoke, to pique, to needle.

> That is the essence of science: ask an impertinent question, and you are on the way to a pertinent answer.

That applies not just to science but to all learning, epitomized for Bronowski by one of the world's oldest and greatest universities – in Germany, as it happens:

> The University is a Mecca to which students come with something less than perfect faith. It is important that students bring a certain

ragamuffin, barefoot irreverence to their studies; they are not here to worship what is known but to question it.*

Bronowski treated the magical speculations of primitive man with sympathy and understanding, but in the end

. . . magic is only a word, not an answer. In itself, magic is a word which explains nothing.

There is magic – the right kind of magic – in science. There is poetry too, and magical poetry on every page of this book. Science is the poetry of reality. If he didn't say that, it is the kind of thing he might have said, articulate polymath and gentle sage, whose wisdom and intelligence symbolizes all that is best in the ascent of man.

---

* How Bronowski would have hated today's fashion among students for blind faith in a temporary political orthodoxy, an orthodoxy which they not only hold themselves but force upon others. I refer to the habit of 'de-platforming' invited speakers. An especially egregious case was that of Germaine Greer, arguably the most distinguished feminist intellectual of our time. In 2015 she was invited to speak at Cardiff University. Thousands of students signed a petition to have her banned, because they disagreed with her opinion that a man should not be called a woman even if he's had his penis amputated. I express no opinion save a strong belief that Dr Greer should be allowed to express hers. The grounds offered by spokespeople of the Cardiff students was that 'trans women' would be 'offended' by her views. What else – I can almost hear Jacob Bronowski's measured, understated tones – is a university but a place where students go to be challenged by contrary views, even at the risk of 'offence'?

REVISITING
THE SELFISH GENE

This was my introduction to the thirtieth-anniversary edition of *The Selfish Gene*. Publishers love anniversaries. The fact that they go for multiples of ten is presumably an artefact of our possessing ten digits – most likely an arbitrary accident, although some evolutionists might argue the point. The same incidental fact leads us to treat particular birthdays with special significance – or foreboding. I think it was Fred Hoyle who speculated that if we had eight fingers (or some other power of two such as sixteen), computers might have been invented centuries earlier because binary arithmetic would have come more naturally to us. I'm not sure how plausible that is. What is plausible, however, is that we'd find it easier to deal with computers at the machine code level if we were accustomed to octal or hexadecimal reckoning. Anniversary editions every sixteen years might not satisfy publishers' zeal for commemoration, though, while octal milestones would seem over-indulgent.

It is sobering to realize that I have lived nearly half my life with *The Selfish Gene* – for better, for worse. Over the years, as each of my seven*

---

* More than seven now, of course. I won't specify a number because I hope, in any case, that it will increase as this anthology stays in print.

subsequent books has appeared, publishers have sent me on tour to promote it. Audiences respond to the new book, whichever one it is, with gratifying enthusiasm, applaud politely and ask intelligent questions. Then they line up to buy, and have me sign . . . *The Selfish Gene*. That is a bit of an exaggeration.* Some of them do buy the new book and, for the rest, my wife consoles me by arguing that people who newly discover an author will naturally tend to go back to his first book: having read *The Selfish Gene*, surely they'll work their way through to the latest and (to its fond parent) favourite baby?

I would mind more if I could claim that *The Selfish Gene* had become seriously outmoded and superseded. Unfortunately (from one point of view) I cannot. Details have changed and factual examples burgeoned mightily. But, with an exception that I shall discuss in a moment, there is little in the book that I would rush to take back now, or apologize for. Arthur Cain, late Professor of Zoology at Liverpool and one of my inspiring tutors at Oxford in the sixties, described *The Selfish Gene* in 1976 as a 'young man's book'. He was deliberately quoting a commentator on A. J. Ayer's *Language, Truth and Logic*. I was flattered by the comparison, although I knew that Ayer had recanted much of his first book and I could hardly miss Cain's pointed implication that I should, in the fullness of time, do the same.

Let me begin with some second thoughts about the title. In 1975, through the mediation of my friend Desmond Morris, I showed the partially completed book to Tom Maschler, doyen of London publishers, and we discussed it in his room at Jonathan Cape. He liked the book but not the title. 'Selfish', he said, was a 'down word'. Why not call it *The Immortal Gene*? Immortal was an 'up' word, the immortality of genetic information was a central theme of the book, and 'immortal gene' had almost the same intriguing ring as 'selfish

---

* Actually it's a substantial exaggeration. I wrote it somewhat tongue in cheek. Unfortunately many people take it seriously and apologize for asking me to sign the book. I now wish I hadn't made the joke.

gene' (neither of us, I think, noticed the resonance with Oscar Wilde's *The Selfish Giant*). I now think Maschler may have been right. Many critics, especially vociferous ones learned in philosophy as I have discovered, prefer to read a book by title only.* No doubt this works well enough for *The Tale of Benjamin Bunny* or *The Decline and Fall of the Roman Empire*, but I can readily see that *The Selfish Gene* on its own, without the large footnote of the book itself, might give an inadequate impression of its contents. Nowadays, an American publisher would in any case have insisted on a subtitle.

The best way to explain the title is by locating the emphasis. Emphasize 'selfish' and you will think the book is about selfishness, whereas, if anything, it devotes more attention to altruism. The correct word of the title to stress is 'gene': let me explain why. A central debate within Darwinism concerns the unit that is actually selected: what kind of entity is it that survives, or does not survive, as a consequence of natural selection. That unit will become, more or less by definition, 'selfish'. Altruism might well be favoured at other levels. Does natural selection choose between species? If so, we might expect individual organisms to behave altruistically 'for the good of the species'. They might limit their birth rates to avoid overpopulation, or restrain their hunting behaviour to conserve the species' future stocks of prey. It was such widely disseminated misunderstandings of Darwinism that originally provoked me to write the book.

Or does natural selection, as I urge instead here, choose between genes? In this case, we should not be surprised to find individual organisms behaving altruistically 'for the good of the genes', for example by feeding and protecting kin who are likely to share

---

* One philosopher went so far as to write: 'Genes cannot be selfish or unselfish, any more than atoms can be jealous, elephants abstract or biscuits teleological.' You'd think a philosopher, of all people, would have realized that the word 'selfish' was being used in a deliberately anthropomorphic sense, for a calculated reason.

copies of the same genes. Such kin altruism is only one way in which gene selfishness can translate itself into individual altruism. This book explains how it works, together with reciprocation, Darwinian theory's other main generator of altruism. If I were ever to rewrite the book, as a late convert to the Zahavi/Grafen 'handicap principle' I should also give some space to Amotz Zahavi's idea that altruistic donation might be a 'potlatch' style of dominance signal: see how superior to you I am, I can afford to make a donation to you!*

---

* It was obvious that flamboyant display organs like peacock fans and stag antlers were handicaps to individual survival. Most of us thought they increased a male's reproductive success *in spite of* being handicaps – because they impressed rivals or sexual partners enough to increase an individual's reproductive success even as they shortened his life. Zahavi controversially suggested that they were impressive precisely *because* they were handicaps. In his characteristic andromorphic language, males were, in effect, saying: 'Look at me, carrying this costly burden: mate with me (or be intimidated by me) because I have managed to survive in spite of this ostentatiously expensive handicap.' Almost all evolutionary biologists ridiculed Zahavi's idea and, in the first edition of *The Selfish Gene*, I was no exception. But in the second edition I had to climb down. Alan Grafen, my former pupil who has now evolved into my mentor, made an elegant mathematical model showing that the handicap principle is theoretically sound. And it applies not just to sexual selection but to advertising displays generally, including Zahavi's idea that dominant individuals donate gifts to subordinates to demonstrate their superiority. 'See how dominant I am, I can afford to give you a present.' It's important for the theory that the handicap is genuinely costly, not an ostentatious fake. My explanation of Grafen's model in non-mathematical language is given in a long endnote in the second edition of *The Selfish Gene*. You'd think Amotz Zahavi would have revelled in his eventual vindication at the hands of a brilliant mathematical biologist. Amusingly he seemed to prefer being an outsider. In a seminar, I said I agreed with him. 'No no, you don't

Let me repeat and expand the rationale for the word 'selfish' in the title. The critical question is: which level in the hierarchy of life will turn out to be the inevitably 'selfish' level, at which natural selection acts? The Selfish Species? The Selfish Group? The Selfish Organism? The Selfish Ecosystem? Most of these could be argued, and most have been uncritically assumed by one or another author, but all of them are wrong. Given that the Darwinian message is going to be pithily encapsulated as The Selfish *Something*, that something turns out to be the gene, for cogent reasons which the book argues. Whether or not you end up buying the argument itself, that is the explanation for the title.

I hope that takes care of the more serious misunderstandings. Nevertheless, I do with hindsight notice lapses of my own on the very same subject. These are to be found especially in chapter 1, epitomized by the sentence: 'Let us try to *teach* generosity and altruism, because we are born selfish.' There is nothing wrong with teaching generosity and altruism, but 'born selfish' is misleading. In partial explanation, it was not until 1978 that I began to think clearly about the distinction between 'vehicles' (usually organisms) and the 'replicators' that ride inside them (in practice, genes: the whole matter is explained in chapter 13, which was added in the second edition). Please mentally delete that rogue sentence and others like it, and substitute something along the lines of this paragraph.

Given the dangers of that style of error, I can readily see how the title could be misunderstood, and this is one reason why I should perhaps have gone for *The Immortal Gene. The Altruistic Vehicle* would have been another possibility. Perhaps it would have been too enigmatic; but, at all events, the apparent dispute between the gene and the organism as rival units of natural selection (a dispute that

---

understand' was his instant reply. Dear Amotz loved arguing, and he used to argue with John Maynard Smith in a style so relentless that the normally mild-mannered Mrs Maynard Smith in desperation banned him from the house. But he was right!

exercised the late Ernst Mayr to the end) is resolved. There are two kinds of unit of natural selection, and there is no dispute between them. The gene is the unit in the sense of replicator. The organism is the unit in the sense of vehicle. Both are important. Neither should be denigrated. They represent two completely distinct kinds of unit and we shall be hopelessly confused unless we recognize the distinction.

Another good alternative to *The Selfish Gene* would have been *The Cooperative Gene*. It sounds paradoxically opposite, but a central part of the book argues for a form of cooperation among self-interested genes. This emphatically does not mean that groups of genes prosper at the expense of their members, or at the expense of other groups. Rather, each gene is seen as pursuing its own self-interested agenda against the background of the other genes in the gene pool – the set of candidates for sexual shuffling within a species. Those other genes are part of the environment in which each gene survives, in the same way as the weather, predators and prey, supporting vegetation and soil bacteria are parts of the environment. From each gene's point of view, the 'background' genes are those with which it shares bodies in its journey down the generations. In the short term, that means the other members of the genome. In the long term, it means the other genes in the gene pool of the species. Natural selection therefore sees to it that gangs of mutually compatible – which is almost to say cooperating – genes are favoured in the presence of each other. At no time does this evolution of the 'cooperative gene' violate the fundamental principle of the selfish gene. Chapter 5 develops the idea, using the analogy of a rowing crew, and chapter 13 takes it further.

Now, given that natural selection for selfish genes tends to favour cooperation among genes, it has to be admitted that there are some genes that do no such thing and work against the interests of the rest of the genome. Some authors have called them outlaw genes, others ultra-selfish genes, yet others just 'selfish genes' – misunderstanding the subtle difference from genes that cooperate in self-interested cartels. Examples of ultra-selfish genes are meiotic drive genes and 'parasitic DNA' – the latter idea developed further by various authors under the catchphrase 'selfish DNA'. The uncovering of new and ever more bizarre examples of ultra-selfish genes has become a feature of the years since this book was first published.*

The Selfish Gene has been criticized for anthropomorphic personification and this too needs an explanation, if not an apology. I employ two levels of personification: of genes, and of organisms. Personification of genes really ought not to be a problem, because no sane person thinks DNA molecules have conscious personalities, and no sensible reader would impute such a delusion to an author. I once had the honour of hearing the great molecular biologist Jacques Monod talking about creativity in science. I have forgotten his exact words, but he said approximately that, when trying to think through a chemical problem, he would ask himself what he would do if he were an electron. Peter Atkins, in his wonderful book *Creation Revisited*, uses a similar personification when considering the refraction of a light beam, passing into a medium of higher refractive index which slows it down. The beam behaves as if trying to minimize the time taken to travel to an end point. Atkins imagines it as a lifeguard on a beach racing to rescue a drowning swimmer. Should he head straight for the swimmer? No, because he can run faster than he can swim and would be wise to increase the dry-land proportion of his travel time. Should he run to a point on the beach directly opposite his target, thereby minimizing his swimming time? Better, but still not the best. Calculation (if he had time to do it) would disclose to the lifeguard an optimum intermediate angle, yielding the ideal combination of fast running followed by inevitably slower swimming. Atkins concludes:

> That is exactly the behaviour of light passing into a denser medium. But how does light know, apparently in advance, which is the briefest path? And, anyway, why should it care?

---

* See Austin Burt and Robert Trivers' *Genes in Conflict*.

He develops these questions in a fascinating exposition, inspired by quantum theory.

Personification of this kind is not just a quaint didactic device. It can also help a professional scientist to get the right answer, in the face of tricky temptations to error. Such is the case with Darwinian calculations of altruism and selfishness, cooperation and spite. It is very easy to get the wrong answer. Personifying genes, if done with due care and caution, often turns out to be the shortest route to rescuing a Darwinian theorist drowning in muddle. While trying to exercise that caution, I was encouraged by the masterful precedent of W. D. Hamilton, one of the four named heroes of the book. In a paper of 1972 (the year in which I began to write *The Selfish Gene*) Hamilton wrote:

> A gene is being favoured in natural selection if the aggregate of its replicas forms an increasing fraction of the total gene pool. We are going to be concerned with genes supposed to affect the social behaviour of their bearers, so let us try to make the argument more vivid by attributing to the genes, temporarily, intelligence and a certain freedom of choice. Imagine that a gene is considering the problem of increasing the number of its replicas, and imagine that it can choose between . . .

That is exactly the right spirit in which to read much of *The Selfish Gene*.

Personifying an organism could be more problematic. This is because organisms, unlike genes, have brains,* and therefore really might have selfish or altruistic motives in something like the subjective sense we would recognize. A book called *The Selfish Lion* might actually confuse, in a way that *The Selfish Gene* should not. Just as one can put oneself in the position of an imaginary light beam,

---

* Well, not all of them have brains, but it's the ones that do that could be problematic in the sense of the previous sentence.

intelligently choosing the optimal route through a cascade of lenses and prisms, or an imaginary gene choosing an optimal route through the generations, so one can postulate an individual lioness, calculating an optimal behavioural strategy for the long-term future survival of her genes.* Hamilton's first gift to biology was the precise mathematics that a truly Darwinian individual such as a lion would, in effect, have to employ, when taking decisions calculated to maximize the long-term survival of its genes. In *The Selfish Gene* I used informal verbal equivalents of such calculations – on the two levels.

In the following passage we switch rapidly from one level to the other:

> We have considered the conditions under which it would actually pay a mother to let a runt die. We might suppose intuitively that the runt himself should go on struggling to the last, but the theory does not necessarily predict this. As soon as a runt becomes so small and weak that his expectation of life is reduced to the point where benefit to him due to parental investment is less than half the benefit that the same investment could potentially confer on the other babies, the runt should die gracefully and willingly. He can benefit his genes most by doing so.

That is all individual-level introspection. The assumption is not that the runt chooses what gives him pleasure, or what feels good. Rather, individuals in a Darwinian world are assumed to be making an *as-if* calculation of what would be best for their genes. This

---

* The point is that the imagined lioness would not calculate 'How can I best survive?', nor even 'How can I best ensure the survival of my children and grandchildren?' Instead, natural selection favours lionesses who do the equivalent of calculating 'How can I best ensure the survival of my genes?' The quantity that such an optimizing lioness would need to calculate, if she could, is what W. D. Hamilton called 'inclusive fitness'.

particular paragraph goes on to make it explicit by a quick change to gene-level personification:

> That is to say, a gene that gives the instruction 'Body, if you are very much smaller than your litter-mates, give up the struggle and die' could be successful in the gene pool, because it has a 50 per cent chance of being in the body of each brother and sister saved, and its chances of surviving in the body of the runt are very small anyway.

And then the paragraph immediately switches back to the introspective runt:

> There should be a point of no return in the career of a runt. Before he reaches this point he should go on struggling. As soon as he reaches it he should give up and preferably let himself be eaten by his litter-mates or his parents.

I really believe that these two levels of personification are not confusing if read in context and in full. The two levels of 'as-if calculation' come to exactly the same conclusion if done correctly: that, indeed, is the criterion for judging their correctness. So, I don't think personification is something I would undo if I were to write the book again today.

Unwriting a book is one thing. Unreading it is something else. What are we to make of the following verdict, from a reader in Australia?

> Fascinating, but at times I wish I could unread it . . . On one level, I can share in the sense of wonder Dawkins so evidently sees in the workings-out of such complex processes . . . But at the same time, I largely blame *The Selfish Gene* for a series of bouts of depression I suffered from for more than a decade . . . Never sure of my spiritual outlook on life, but trying to find something deeper – trying to believe, but not quite being able

to – I found that this book just about blew away any vague ideas I had along these lines, and prevented them from coalescing any further. This created quite a strong personal crisis for me some years ago.

I have previously described a pair of similar responses from readers:

A foreign publisher of my first book confessed that he could not sleep for three nights after reading it, so troubled was he by what he saw as its cold, bleak message. Others have asked me how I can bear to get up in the mornings. A teacher from a distant country wrote to me reproachfully that a pupil had come to him in tears after reading the same book, because it had persuaded her that life was empty and purposeless. He advised her not to show the book to any of her friends, for fear of contaminating them with the same nihilistic pessimism (*Unweaving the Rainbow*).

If something is true, no amount of wishful thinking can undo it. That is the first thing to say, but the second is almost as important. As I went on to write:

Presumably there is indeed no purpose in the ultimate fate of the cosmos, but do any of us really tie our life's hopes to the ultimate fate of the cosmos anyway? Of course we don't; not if we are sane. Our lives are ruled by all sorts of closer, warmer, human ambitions and perceptions. To accuse science of robbing life of the warmth that makes it worth living is so preposterously mistaken, so diametrically opposite to my own feelings and those of most working scientists, I am almost driven to the despair of which I am wrongly suspected.

A similar tendency to shoot the messenger is displayed by other critics who have objected to what they see as the disagreeable social, political or economic implications of *The Selfish Gene*. Soon after

Mrs Thatcher won her first election victory in 1979, my friend Steven Rose wrote the following in *New Scientist*:

> I am not implying that Saatchi and Saatchi engaged a team of sociobiologists to write the Thatcher scripts, nor even that certain Oxford and Sussex dons are beginning to rejoice at this practical expression of the simple truths of selfish genery they have been struggling to convey to us. The coincidence of fashionable theory with political events is messier than that. I do believe though, that when the history of the move to the right of the late 1970s comes to be written, from law and order to monetarism and to the (more contradictory) attack on statism, then the switch in scientific fashion, if only from group to kin selection models in evolutionary theory, will come to be seen as part of the tide which has rolled the Thatcherites and their concept of a fixed, 19th century competitive and xenophobic human nature into power.

The 'Sussex don' was the late John Maynard Smith, admired by Steven Rose and me alike,* and he replied characteristically in a letter to *New Scientist*: 'What should we have done, fiddled the equations?' One of the dominant messages of *The Selfish Gene* (reinforced by the title essay of *A Devil's Chaplain*) is that we should not derive our values from Darwinism, unless it is with a negative sign. Our brains have evolved to the point where we are capable of rebelling against our selfish genes. The fact that we can do so is made obvious by our use of contraceptives. The same principle can and should work on a wider scale.

Unlike the second edition of 1989, this anniversary edition adds no new material except this introduction, and some extracts from reviews chosen by my three-times editor and champion, Latha

---

* I treasure the moment – and I don't think Steven would mind my recalling it – when John, at a Royal Society conference, said: 'Steven, you have just said something which you *know* to be stupid.'

Menon. Nobody but Latha could have filled the shoes of Michael Rodgers, K-selected Editor Extraordinary, whose indomitable belief in this book was the booster rocket of its first edition's trajectory.

This edition does, however – and it is a source of particular joy to me – restore the original foreword by Robert Trivers. I have mentioned Bill Hamilton as one of the four intellectual heroes of the book. Bob Trivers is another. His ideas dominate large parts of chapters 9, 10 and 12, and the whole of chapter 8. Not only is his foreword a beautifully crafted introduction to the book; unusually, he chose the medium to announce to the world a brilliant new idea, his theory of the evolution of self-deception. I am most grateful to him for giving permission for the original foreword to grace this anniversary edition.

II

WORLDS BEYOND WORDS: CELEBRATING NATURE

IN CONVERSATION WITH ADAM HART-DAVIS

EVOLUTION AND PLAIN WRITING IN SCIENCE

Adam Hart-Davis is a publisher, broadcaster and television presenter, famous in Britain for his quirky dressing style and his excellence as an explainer of science, often with a historical slant. Of all the hundreds of interviews that I have done, all around the world, this one perhaps came closest to a pithy summary of all my science. It is part of a series of interviews with scientists that Adam conducted as part of the Magrack *Maximum Science* series, broadcast on AMC Television in the United States in 2002. The text reproduced here is an abbreviated transcript of our conversation, published in Britain in *Talking Science* (2004).

Can science answer the 'why?' questions? What's so crucial about the gene? Can we appreciate science in the way we can appreciate music?

AH-D: Richard, you say that we live for only a tiny piece of the enormous life of the universe, and that we should spend our lives trying to understand why we're here and indeed why the universe is here. But surely science can't answer these 'why' questions, can it?

RD: There are many meanings of the word 'why'. To me, as a scientist, it can mean two things. One is, what is the sequence of events that leads to us being here and being the way we are? And science answers that. The second kind of why is, what is it for? Science can't answer that; indeed, I think it's a meaningless question, it's not a question that should be asked, except for those cases where we're dealing with something that's been designed by people. You can say what a corkscrew is for, what a fountain pen is for, but you can't say what life is for, or what a mountain is for, or what the universe is for. Living things are a special case: you can ask what a bird's wing is for, what a dog's tooth is for. That has a special meaning within the context of Darwinian natural selection; it means, specifically, what has that thing done to help this creature's ancestors survive and reproduce? That's a special kind of why.

But in the colloquial sense of why, I think it's perfectly reasonable to say, why are we here? Meaning, what is the sequence of events, what is the set of antecedent conditions that leads to us being here? I can't imagine a better way of spending my brief time in the sun than trying to understand how to answer such questions.

AH-D: The way we perhaps all know you best is through your book *The Selfish Gene*, which is what, twenty-five years old now? Are you surprised at how successful that book was and is?

RD: When I was writing it, I did jokingly refer to it as 'my bestseller', not really thinking that it would be. And it was never a mega-bestseller in the first six months, the way one thinks of a sort of blockbuster bestseller, but I've been pleased by the way it's been selling steadily in all the years since. That, in the long run, is a better sort of bestseller than the sort that sells madly in the first six months and then is never heard of again.

AH-D: When you were writing it, were you just telling a simple version of Darwin's ideas of evolution?

RD: I thought I was doing that at the time, and in many ways I still think I was, but I believe there's a difference between popularizing, which means taking something that's already familiar to the scientific community and making it comprehensible – and I've certainly done a fair bit of that – and what I like to think that I've been doing, which is a bit more 'changing the way people think' – and that includes not just lay people but my fellow scientists, my colleagues. I've been told (often enough that I believe it) that even scientists in the field have changed the way they think as a consequence not of anything I've discovered but of my way of putting things, which was sufficiently unfamiliar that it really did turn upside down the way people thought about familiar things.

AH-D: Why did you choose the gene as your unit?

RD: I don't think it's right to say that I chose the gene. Choosing the gene was done for me by nature. What I was doing was taking the existing neo-Darwinian theory and saying, this is a gene's-eye view; that's what some people hadn't quite realized. They had been focused on the level of the individual organism. The individual organism was the agent of life: the rabbit, or the elephant, or whatever it was. It's true that the rabbit works to survive and reproduce, but if you ask why the rabbit works to survive and reproduce, what's actually being reproduced are its genes.

What happens is that over many generations the genes that are good at making rabbits are the ones that are still around. Therefore, when you look at rabbits, you see that they are made by genes that are good at making rabbits, genes that have come down through many generations. The gene is the only part of the rabbit, or the elephant, or the human, or any other creature you care to mention, which goes on from generation to generation, in principle indefinitely. In principle, the information in a gene is immortal, and what that means is that good ones are immortal, bad ones are not, and so the world becomes filled with the good ones. Those are the ones that survive; that is the Darwinian process. But individuals, individual bodies, individual rabbits, elephants or humans, they die anyway; they don't survive, not in the long run. Information does survive, and DNA is the information that living things have.

AH-D: You're taking a sort of science-fiction view.

RD: It's funny you should say that; in a way, I'm just re-expressing what's in neo-Darwinism, but it is a sort of science-fiction view. To describe human beings I used the phrase 'lumbering robots', which caused a certain amount of controversy. It just means that the genes are the information; they're the bit that passes down through the generations. What they use to survive is the body, and the body can be thought of in this sort of science-fiction way, as a robot that the genes build for themselves and then ride around inside. And it's because they ride around inside their robot that the survival of the robot is intimately bound up with the survival of the genes. It's a machine that carries its own blueprint around with it, and therefore if it survives and does its job of reproducing, the blueprint will survive.

AH-D: Your thesis essentially says that it's gene survival that matters. But genes have to cooperate at some levels, don't they, even in one individual, let alone in a species?

RD: Cooperation is immensely important. I'm not talking now about cooperation between individuals, which is also important, at least in certain kinds of animal. But cooperation in genes is immensely important, because it's meaningless to talk about single genes in the business of building a body. The business of building a body, which is embryology, is a hugely cooperative enterprise, but it's wrong to think of the cooperative bunch of genes as a kind of unit that goes around in time together. They don't – they are constantly split up, separated, and recombined in sexual reproduction. But they do cooperate within each body in making that body. The way to think about it is that those genes survive best which are good at cooperating with the other genes that they're most likely to meet, and that means the other genes that are in the gene pool of the species. The gene pool of a species means all the genes in a species; you can call them a pool, because they're constantly being mixed up, stirred up together in sexual reproduction. The genes in the gene pool build the bodies characteristic of that species, rabbit bodies, elephant bodies, or whatever it is, and the ones that survive in the gene pool are the ones that are good at collaborating with the other ones in the gene pool, to make rabbits or elephants.

AH-D: So you mean as well as being selfish, they also have instructions to cooperate with other chaps?* You make it sound simple, but it really is difficult to get hold of this idea, isn't it? People generally find it hard to understand Darwinism . . .

You have said we should spend a chunk of our lives trying to understand why we're here, and you also say we should have a sense of wonder: we should look at science in the same sort of way as we listen to music. Surely you can't do that? You need years of training before you can understand the language of science.

RD: The point about the analogy with music is that we need people to *play* music, and they need years of experience, they need to spend hours a day practising their instrument or they won't play it well enough. In the same way, practising scientists do need years of training. But you can enjoy music, appreciate music even at quite a sophisticated level, without being able to play a note. Similarly, I think you can appreciate and enjoy science at quite a sophisticated level without being able to do science. I want to encourage people to treat science in the same kinds of way they would treat music or art or literature: as something to be enjoyed, not at a superficial level, but at quite a deep level, without necessarily

---

* I believe that by 'other chaps' he meant 'other genes'. If I had understood him to mean 'other individuals' I would surely have pulled him up at the time.

being able to tell one Bunsen burner from another or integrate a function.

AH-D: But isn't language an impossible barrier? I can go and switch on the radio and listen to music and enjoy it, and I don't need to know what it is, who's playing, who the composer was, any of that – I can just hear the noise.

RD: You might be surprised; you need to acculturate yourself to the music you're listening to. We are brought up in Western music, and so we do find it easier to understand because of that.

AH-D: But if I listened to, say, Japanese music, I might not instantly understand, but it would be interesting to listen to, whereas if I went into the coffee room in the Zoology Department at Oxford – and I'm interested in science, I understand quite a lot – I probably wouldn't understand what two people are talking about.

RD: Two people in the coffee room in a scientific department, talking about research, are indeed using a sort of shorthand language, which they use for the sake of brevity. But it isn't very difficult to switch from that language to one that can be understood much more widely. I feel I have a mission to persuade my scientific colleagues to write their science as if they had a lay person looking over their shoulder, not to write in a language which is completely opaque to other people. I believe they'll do better science if they do that, I think they'll communicate with other scientists better if they do that. I even think they'll understand better the science that they themselves are doing.

The other thing is that different sciences are not easily intelligible to other scientists. I don't understand physics very well, and I think I'm right in saying a lot of physicists don't understand biology very well, though it's probably easier to understand biology. There are aspects of especially modern physics, quantum theory in particular, that are exceedingly hard to understand, totally counter-intuitive. Many physicists have said that they don't understand it either; they do the maths, and they use their mathematics to predict results, and they find that the predictions are fulfilled with astonishing accuracy. The predictions of quantum theory are said to be fulfilled with an accuracy equivalent to predicting the distance between New York and Los Angeles to within the thickness of one human hair. That's what quantum theory can do, and yet many of the people who do it would agree that they don't really understand it at a gut level, because it is so counter-intuitive. If *they* can't understand it, it's not surprising that you and I can't either.

Nevertheless, there are books that really make a strong effort to explain quantum physics, relativity and, in biology, evolution. Evolution has its own difficulties of being understood. It's not as difficult as quantum theory, but there are difficulties. You have to get your mind around enormous quantities of time, otherwise you just can't believe that you could go from a bacterium to a human.

AH-D: William Paley said, if I walk along and I pick up a rock, it's not surprising, but if I pick up a watch, I cannot believe that this was not designed by a watchmaker. What's wrong with Paley's argument of a designer?

RD: I've satirized it as the Argument from Personal Incredulity. But you've only got to think about it a little bit and you realize that there's an infinite regression. It's actually not an explanation at all to say that the watch had a designer, because the designer himself needs an explanation.

That's what's elegant, that's what's beautiful about the Darwinian explanation for living watches: for eyes and elbows and hearts. Because we start from very simple beginnings – simple things are by definition easier to understand than complex things – and we work up by slow, gradual degrees, over many, many generations; and because at every step of the way it's comprehensible and the explanation really works. We end up with a complete and satisfying explanation for where the complexity, for where the living watches come from. By living watches I mean eyes, heart, ears and so on.

We understand where they come from, and we don't need to postulate anything mystical or mysterious. The problem with Paley's explanation, with the designer, is that it explains precisely nothing, because the designer himself is presumably even more complicated than the watch, even more complicated than the heart or the eye. So it's a non-explanation.

AH-D: Now, many great thinkers have one thing in their life that they're most proud of. James Watt, who designed lots of steam engines, was most proud of the parallel motion, a lovely bit of applied mathematics. What are you most proud of having found, discovered, invented?

RD: I now think that the book I am most proud of, as a book for lay people, is *Climbing Mount Improbable*. But as a contribution to knowledge I am most proud of *The Extended Phenotype*. It was my second book and it was, as you might guess from its title, not primarily aimed at lay people, though a lot of lay people have read it. It's aimed at my professional colleagues, and I'm most proud of it because the idea goes beyond what others had already done. A phenotype is the manifestation of a genotype. Genes are DNA, a phenotype is something like blue eyes or red hair – or you could think of it even as the whole person. It's that which is manifested. When talking about a particular gene, you would talk about the phenotypic expression of the gene. You would say that this particular gene on the chromosome produces a big nose, the phenotype. So, the genes in me produce my phenotype and the genes in you produce your phenotype . . .

AH-D: . . . which goes as far as my fingertips, but no further.

RD: Yes, that's right for conventional phenotypes. But *The Extended Phenotype* says that genes can have a phenotypic effect outside the body in which they sit. I argue this in a kind of step-by-step softening-up way, starting with animal artefacts, with things that animals make, like birds' nests or caddis larvae houses. Caddis larvae are little insects that live in streams and build houses for themselves out of stones or sticks or snails' shells or leaves, depending on the species. So the outer shell of this insect is not part of its body; it's made by the animal. The stone ones are the nicest, because they really do cement the stones with cement that they make. You can watch the insect building; cementing little bricks into a stone wall with great skill.

That is part of the extended phenotype of genes in the caddis larva, and the justification for it is a Darwinian one. It's clearly a Darwinian adaptation. It's well designed in just the same way as the shell of a snail or the beak of a parrot is well designed. It's a product of natural selection.

Natural selection works only by the differential survival of genes, so there have to be genes for caddis larvae houses of various shapes. There have to be good houses and bad houses, and natural selection favours the good ones. It is genes that determine the improvements in the houses – you could say improvements in the building behaviour – but we're accustomed to the idea that when you talk about a gene for something, for some kind of phenotypic effect, it's already at the end of a long chain of causation. The only thing that a gene is really *for* is a protein. That protein then interacts cooperatively with other proteins made by other genes to produce a complicated sequence of embryological events, to produce the building behaviour – like a recipe when you're cooking something.

Well, if you're going to say the genes are genes for building behaviour, you might as well go one step further and say they are genes for the house that is built. So the house is an extended phenotype. If you accept that, you could say, what about a bird's nest – or maybe a bower would be a better example. A bower bird is a bird that lives in Australia or New Guinea; the male attracts the female not by having a gorgeous tail, as in a peacock, but by building a kind of external tail, which is a bower made of grass, decorated with coloured berries or flowers or Foster's beer cans. This bower is what lures the female. It's not a nest, it's not a house that you live in, it's an external tail attracting females. It's an extended phenotype. They're clearly shaped in this case by sexual selection, and the shaping has to have come about by the selection of genes. That means there have to have been genes for changes in the shape of the bower – the extended phenotype again.

Now imagine a parasite, say a tiny worm, living inside a host, say an ant. The worm manipulates the ant's behaviour for the worm's benefit. Like many parasites, the worm needs to get out of this host, the ant, into the next host of its life-cycle, which is a sheep. And in order to do this, the sheep has to accidentally eat the ant.

The worm makes the ant more likely to be eaten by changing its behaviour. It makes the ant climb up to the top of grass stems, whereas the ant would normally go down to the bottom of grass stems. The ant is manipulated like a puppet, like a lumbering robot. The worm sits in the brain of the ant, and makes a lesion, or changes the ant's brain in some way so that the ant no longer goes down to the bottom of grass stems, but goes to the top of grass stems. So the worm gets itself eaten in the lumbering robot that is the ant.

That too is a Darwinian adaptation. I'm entitled to say that because by the normal Darwinian logic we always say that Darwinian selection favours a phenotype, and as a consequence of that, the genes for making that phenotype survive. In this case, the proximal phenotype is in the worm, but the phenotype that really matters is the change in behaviour of the ant.

So now I've softened you up. By going from the caddis, where the extended phenotype is an inanimate house made of stones, we've now got the 'house' being a living ant, but the logic is the same. So the extended phenotype is now allowed to be a living creature, but it's not the living creature whose cells actually contain the genes, it's another living creature.

The final step, having softened you up so far, is to take a parasite that doesn't live inside its host, and the best example of this is a cuckoo. The cuckoo nestling, the baby cuckoo, manipulates the behaviour of the foster parents so that they feed it. Birds have even been seen flying to their own nest and then diverting to another nest where they see baby cuckoos and they feed them, because the colour of the gape is so irresistible. So now, instead of a worm sitting inside an ant and manipulating its behaviour, we have a baby cuckoo not sitting inside its foster parent, but removed from it by some distance, and manipulating it by light – by the foster parent's sense organs. That too, the changed behaviour of the host, the foster parent, is the extended phenotype of genes in the cuckoo. So it's a long, developing, incremental argument, and you have to read the book.

AH-D: All right, I'll read the book. You obviously love this stuff; you get totally wrapped up in it. You also write about beauty: tell me about the rainbow.

RD: The title of my book *Unweaving the Rainbow* comes from Keats, who complained in a poem that Newton, by explaining the rainbow, removed the magic and removed the joy and reduced it to something dull. I think the opposite. I think, and I think any scientist would, that to explain something is not to destroy the beauty. In many ways, it enhances the beauty. I like lying on my back in the tropics, looking up at the night sky and seeing the Milky Way; that's a beautiful, transporting experience. And it's not in any way reduced by such knowledge as I have – which is by no means the knowledge of a proper astronomer – about what the Milky Way is. The fact that I know in my limited way that when I look at the Milky Way I'm looking (backwards in time, which is even more wonderful) at our own galaxy, and that there are other galaxies, indeed billions of other galaxies, with the same general properties as ours, only increases the beauty.

CLOSE ENCOUNTERS
WITH THE TRUTH

This review of *The Demon-Haunted World* was published in *The Times* in February 1996. I wish I had known Carl Sagan. I met him once only, over coffee with Stephen Jay Gould, in London where we were all invited speakers at the 1994 meeting convened by John Maddox to celebrate *Nature*'s 125th anniversary. His eloquence came over in private conversation no less than on television or in print. Why didn't Dr Sagan receive the Nobel Prize in Literature? He seems such an obvious candidate. Was it for the same snobbish reason he wasn't elected to the US National Academy of Sciences? No, more probably it just never occurred to the Swedish Academy that scientists write literature. Of all his beautiful books, I think *The Demon-Haunted World* is my personal favourite. But it's a hard choice to make.

As I close this eloquent and fascinating book, I recall the final chapter title from one of Carl Sagan's earlier works, *Cosmos*. 'Who speaks for Earth?' is a rhetorical question that expects no particular answer, but I presume to give it one. My candidate for planetary ambassador, my own nominee to present our credentials in galactic chancelleries, can be none other than Carl Sagan himself. He is wise, humane, polymathic, gentle, witty, well-read, and incapable of composing a dull sentence. I confess to the habit, when reading books, of underlining occasional sentences that I particularly like. *The Demon-Haunted World* forced me to desist, simply to save on ink. But how can I not quote Sagan's answer to the question why he bothers to work at explaining science? 'Not explaining science seems to me perverse. When you're in love, you want to tell the world. This book is a personal statement, reflecting my lifelong love affair with science.'

Buoyant and uplifting though much of the book is, its subtitle is 'Science as a candle in the dark',* and it ends in foreboding. Science – not the facts of science but the scientific method of critical thought – 'may be all that stands between us and the enveloping darkness'. The dark is the dark of medieval and modern witch-hunts, of the pathological dread of non-existent demons and UFOs, of humanity's wanton gullibility in the face of fat-cat mystics and the obscurantist gurus of postmodern metatwaddle. One of Sagan's most chilling quotes is a call to arms against science, from a book published in 1995, which concludes: 'Science itself is irrational or mystical. It's just another faith or belief system or myth, with no more justification than any other. It doesn't matter whether beliefs are true or not, as long as they're meaningful to you.'

Truth has its enemies, as Sagan documents. But, perhaps because he doesn't live in Britain, he overlooks a separate problem faced by science in our culture: a philistine double standard. When the *Daily Telegraph* reported a survey finding that a high percentage of adults think the sun goes round the Earth, the then editor inserted, 'Doesn't it? Ed.' One immediately thinks of Bernard Levin's preening delight in his own ignorance, or of the patronizing snigger with which television announcers render science stories as the concluding 'joke' item at the end of the news. If a survey found that 50 per cent of adults believe Shakespeare wrote the *Iliad*, what editor would find it funny to insert a parenthetical 'Didn't he? Ed.'? That's

---

* I was later to borrow his phrase, amalgamating it with one from *Macbeth*, for the second volume of my own autobiography, *Brief Candle in the Dark*.

the double standard. Again, when the aggressive habits of rott-weilers were being excitedly promoted by the news media a while ago, the responsible government minister went on the radio to reveal the disturbing extent of the problem. Dogs, she explained patiently, don't have DNA. Ignorance on such a scale would not be countenanced in a minister of the Crown, were the subject any-thing other than science.

Among the gifts science has to offer is, in Sagan's words, a balo-ney detection kit. His book is in part a manual for using the kit. Here is how to test the credentials of the superhuman extraterrestri-als who annually swarm to Earth in UFOs and abduct humans for sexual experiments (to the victims' considerable profit when they sell their stories to the inexhaustibly gullible – or cynical – press).

> Occasionally, I get a letter from someone who is in 'contact' with extraterrestrials. I am invited to 'ask them anything'. And so over the years I've prepared a little list of questions. The extraterrestri-als are very advanced, remember. So I ask things like, 'Please provide a short proof of Fermat's Last Theorem'. Or the Goldbach Conjecture . . . I never get an answer. On the other hand, if I ask something like 'Should we be good?' I almost always get an answer. Anything vague, especially involving conventional moral judge-ments, these aliens are extremely happy to respond to. But on anything specific, where there is a chance to find out if they actu-ally know anything beyond what most humans know, there is only silence.

Scientists are sometimes suspected of arrogance. Sagan com-mends to us by contrast the humility of the Roman Catholic Church which, as early as 1992, was ready to pardon Galileo and admit pub-licly that the Earth does revolve around the sun. We must hope that this outspoken magnanimity will not cause offence or 'hurt' to 'the supreme religious authority of Saudi Arabia, Sheik Abdel-Aziz Ibn Baaz', who, in 1993, 'issued an edict, or fatwa, declaring that the world is flat. Anyone of the round persuasion does not believe in God and should be punished.' Arrogance? Scientists are amateurs in arrogance. Moreover, they have a modicum to be arrogant about. Scientists

can routinely predict a solar eclipse, to the minute, a millennium in advance. You can go to the witch doctor to lift the spell that causes your pernicious anaemia, or you can take Vitamin B12. If you want to save your child from polio, you can pray or you can inoculate. If you're interested in the sex of your unborn child, you can consult plumb-bob danglers all you want . . . but they'll be right, on average, only one time in two. If you want real accuracy . . . try amniocentesis and sonograms. Try science.

I wish I had written *The Demon-Haunted World*. Having failed to do so, the least I can do is press it upon my friends. Please read this book.

CONSERVING COMMUNITIES

In an age of extinction, still photographs, alongside films like the documentaries of David Attenborough, may turn out to be a priceless archive of what we shall have lost. One of our great photographers is Art Wolfe. Too splendid to be called a coffee-table book, *The Living Wild* is a 2000 collection of his wildlife photographs. The publishers invited five scientists, including the primatologists George Schaller and Jane Goodall, to write essays extolling in words the beauties of the living world to accompany Art Wolfe's magnificent pictures. I was one of the five, and this is my contribution.

The third planet is unique. Luxuriating over our sphere's surface, thinning up into the air, and etching its way down into the rocks is a layer in which something rich and new is added to the physics that unremarkably pervades the rest of the solar system. That special layer is, of course, the layer of life. It is not that the laws of physics are disobeyed at the planet's rim: vanish the thought. But living matter deploys physics in unusual ways. So unusual – 'emergent' – that the error of believing the laws of physics to be defied is forgivable. Which is just as well, because everyone has been tempted by that error, most people through history have succumbed to it, and many still do.

Darwin may not have been quite the first to resist the temptation, but the comprehensiveness with which he repudiated it entitles him to most of the honour. In spite of its title, his great book is less on the origin of species than on the origin of adaptation. That is to say, it is on the origin of the design illusion, that powerful simulacrum that led people to suspect, wrongly, that material causes are not enough to explain biology.

The illusion of design is at its strongest in the tissues and organs, the cells and molecules, of individual creatures. The individuals of every species, without exception, show it powerfully, and it springs forth from every picture in this book. But there is another illusion of design that we notice at a higher level – also so splendidly displayed in these pages: the level of species diversity. Design seems to reappear in the disposition of species themselves, in their arrangement into communities and ecosystems, in the dovetailing of species with species in the habitats that they share. There is a pattern in the intricate jigsaw of rainforest, say, or coral reef, which leads rhetoricians to preach disaster if but one component should be untimely ripp'd from the whole. In extreme cases, such rhetoric takes on mystical tones. The womb is of an earth goddess, all life her body, the species her parts. Yet, without giving in to such extravagance, there is a strong illusion of design at the community level, less compelling than within the individual organism but worth attention.

The animals and plants that live together in an area seem to fit one another with something like the glovelike intimacy with which the parts of an animal mesh with other parts of the same organism. A Florida panther has the teeth of a carnivore; the claws of a carnivore; the eyes, ears, nose and brain of a carnivore; leg muscles that are suitable for chasing meat, and guts that are primed to digest it. Its parts are choreographed in a dance of carnivorous unity. Every sinew and cell of the big cat has meat eater inscribed through its very texture, and we can be sure that this extends deep into the details of biochemistry. The corresponding parts of, say, a bighorn sheep are equally unified with each other, but to different ends. Guts designed to digest plant roughage would be ill served by claws and instincts designed to catch prey. And vice versa. A hybrid between a panther and a sheep would fall flat on its evolutionary face. Tricks of the trade cannot be cut from one and pasted into the other.* Their compatibility is with other tricks of the same trade.

Something similar can be said of communities of species. The language of the ecologist reflects this. Plants are primary producers. They trap energy from the sun, and make it available to the rest of the community, via a chain of primary, secondary and even tertiary consumers, culminating in scavengers. Scavengers play a recycling 'role' in the community, and I use quotation marks advisedly. Every species, in this view of life, has a role to play. In some cases, if the performers of some role, such as scavengers, were removed, the whole community would collapse. Or its 'balance' would be upset and it might fluctuate wildly, out of 'control' until a new balance is set up, perhaps with different species playing the same roles. Desert communities are different from rainforest communities, and their component parts are mutually ill suited, just as – or so it seems – herbivorous colons are ill suited to carnivorous habits. Coral-reef communities are different from sea-bottom communities, and their parts cannot be exchanged. Species become adapted to their community, not just to a particular physical region and climate. They become adapted to each other. The other species of the community are an important – perhaps the most important – feature of the environment to which each species becomes adapted.

The harmonious role-playing of species in a community, then, resembles the harmony of the parts of a single individual organism. The resemblance is deceptive and must be treated with caution. Yet it is not completely without foundation. There is an ecology within the individual organism, a community of genes in the gene pool of a species. The forces that produce harmony among the parts of an organism's body are not wholly unlike the forces that produce the

---

* Actually, such is the amazing power of genomic science, there might come a time when cutting and pasting of this kind could be done. But the important point is that the results wouldn't work well functionally even if the transplanting were technically doable.

illusion of harmony among the species of a community. There is balance in a rainforest, structure in a reef community, an elegant meshing of parts that recalls coadaptation within an animal body. In neither case is the balanced unit favoured as a unit by Darwinian selection. In both cases the balance comes about through selection at a lower level. Selection doesn't favour a harmonious whole. Instead, harmonious parts flourish in the presence of each other, and the illusion of a harmonious whole emerges.

At the individual level, to rehearse an earlier example in genetic language, genes that make carnivorous teeth flourish in a gene pool containing genes that make carnivorous guts and carnivorous brains, but not in a gene pool containing genes for herbivorous guts and brains. At the community level, an area that lacks carnivorous species might experience something similar to a human economy's 'gap in the market'. Carnivorous species that enter the area find themselves flourishing. If the area is a remote island that no carnivorous species has reached, or if a recent mass extinction has devastated the land and created a similar gap in the market, natural selection will favour individuals within non-carnivorous species that change their habits and become carnivores. After a long enough period of evolution, specialist carnivore species will descend from omnivorous or herbivorous ancestors.

Carnivores flourish in the presence of herbivores, and herbivores flourish in the presence of plants. But what about the other way around? Do plants flourish in the presence of herbivores? Do herbivores flourish in the presence of carnivores? Do animals and plants need enemies to eat them in order to flourish? Not in the straightforward way that is suggested by the rhetoric of some ecological activists. No creature normally benefits from being eaten. But grasses that can withstand being cropped better than rival plants can really flourish in the presence of grazers – on the principle of 'my enemy's enemy'. And something like the same story might be told of some animal victims of parasites – and predators, although here the story is more complicated. It is still misleading to say that a community 'needs' its parasites and predators like a polar bear needs its liver or its teeth. But the enemy's enemy principle does lead to something like the same result. It can be right to see a community of species as a kind of balanced entity, which is potentially threatened by removal of any of its parts.

This idea of community, as made up of lower-level units that flourish in the presence of each other, pervades life. Even within the single cell, the principle applies. Most animal cells are communities of hundreds or thousands of bacteria, which have become so comprehensively integrated into the smooth working of the cell that their bacterial origins have only recently become understood. Mitochondria, once free-living bacteria, are as essential to the workings of our cells as our cells are to them. Their genes have flourished in the presence of ours as ours have flourished in the presence of theirs. Plant cells by themselves are incapable of photosynthesis. That chemical wizardry is performed by guest workers within the cells, originally bacteria and now relabelled chloroplasts. Plant eaters, such as ruminants and termites, are themselves largely incapable of digesting cellulose. But they are good at finding and chewing plants. The gap in the market offered by their plant-filled guts is exploited by symbiotic micro-organisms that possess the biochemical expertise necessary to digest plant material efficiently. Creatures with complementary skills flourish in each other's presence.

And the process is mirrored at the level of every species' 'own' genes. The entire genome of a polar bear or a penguin, of a caiman or a guanaco, is a set of genes that flourish in each other's presence. The immediate arena of this flourishing is the interior of an individual's cells. But the long-term arena is the gene pool of the species. Given sexual reproduction, the gene pool is the habitat of every gene as it is recopied and recombined down the generations.

This gives the species its singular status in the taxonomic hierarchy. Nobody knows how many separate species there are in the world, but we at least know what it would mean to count them. Arguments about whether there are thirty million separate species, as some have estimated, or only five million, are real arguments.

The answer matters. Arguments about how many genera there are, or how many orders, families, classes, or phyla have no more status than arguments about how many tall men there are. It's up to you how you define tall and how you define a genus or a family. But – as long as reproduction is sexual – the species has a definition that goes beyond individual taste and does so in a way that is really important. Fellow members of a species participate in the same shared gene pool. The species is defined as the community whose genes share that most intimate of cohabiting arenas, the cell nucleus – a succession of cell nuclei through generations.

When a species splits off a daughter species, usually after a period of accidental geographical isolation, the new gene pool constitutes a new arena for intergene cooperation to evolve. All the diversity on Earth has come about through such splittings. Every species is a unique entity, a unique set of coadapted genes, cooperating with each other in the enterprise of building individual organisms. The gene pool of a species is an edifice of harmonious cooperators, built up through a unique history. Any gene pool, as I have argued elsewhere, is a unique written record of ancestral history. Slightly fanciful perhaps, but it follows indirectly from Darwinian natural selection. A well-adapted animal reflects, in minute detail even down to the biochemical, the environments in which its ancestors survived. A gene pool is carved and whittled through generations of ancestral natural selection to fit that environment. In theory a knowledgeable zoologist, presented with the complete transcript of a genome, should be able to reconstruct the environmental circumstances that did the carving.* In this sense the DNA is a coded description of ancestral environments, a 'genetic book of the dead'.

The extinction of a species therefore diminishes us in a sense that the death of an individual perhaps does not. To be sure, every individual is unique, and to that extent irreplaceable. But the set of genes

---

* On this point, see also 'Worlds in microcosm', pp. 199–220 below.

in a species' gene pool represents a unique solution to the problem of survival. An individual organism, by contrast, is only a permutation of the units of that solution: unique, but not unique in an interesting way. If an individual dies, there are lots more where it came from. It is just another deal from the same pack of cards. When the last individual of a species dies, the whole pack has been destroyed. No doubt other species will arise to take its place, but they will take time to build up an equivalently intricate collection of mutually compatible genes, and their new solution to the problem of DNA preservation will always be different from the old. When the last (probably) Tasmanian wolf* died in Hobart Zoo in 1936, we lost tens of millions of years' worth of carnivorous research and development.

It is possible to take a robust view of extinction, even mass extinction. We can tough-mindedly point out that extinction is the norm for species throughout geological history. Even our own swath of chainsaw and concrete devastation is only the latest in a long series of cleanouts from which life has always bounced back. What are we and our domination of the world but another natural process, no worse than many before? The catastrophe that ended the dinosaurs had a consequence that might lead us to take a positively cheerful attitude towards it: us. From a more dispassionate point of view, every mass extinction opens up yawning gaps in the market, and the headlong rush to fill them is what, time after time, has enriched the diversity of our planet.

Even the most devastating of mass extinctions can be defended as the necessary purging that makes rebirth possible. No doubt it is fascinating to wonder whether rats or starlings might provide the ancestral stock for a new radiation of giant predators, in the event that the whole order Carnivora was wiped out. But none of us would ever know, for we

---

* I reject the irritatingly popular misnomer 'Tasmanian Tiger'. Stripes alone are too superficial to outweigh the thylacine's massive resemblance to a dog or wolf, one of the most spectacular examples I know of convergent evolution.

do not live on the evolutionary timescale. It is an aesthetic argument, an argument of feeling, not reason, and I confess that my own feelings recoil. I find my aesthetics incapable of quite such a long view.

The dinosaurs are gone. I mourn them and I mourn the giant ammonites, and before them the mammal-like reptiles and the club moss and tree fern forests of the coal measures, and before them the trilobites and eurypterids: but they are beyond recall. What we have now is a new set of communities, our own contemporary buildup of mutually compatible mammals and birds, flowering plants and pollinating insects. They are not better than the communities that preceded them. But they are here, we have the privilege of studying them, they took agonizing ages to build up, and if we destroy them we shall not see them replaced. Not in our lifetime, not in five million years. If we destroy the ecosystems of which we are a part, we condemn not just our own generation, but all the generations of descendants that we could realistically hope to succeed us, to a world of devastation and impoverishment.

The case for conserving wild nature is sometimes made in terms of the crudest self-interest. We need the diversity of the rainforests because who knows where our next set of medicines and crop plants will come from? Well, if that is what it takes to mobilize support, so be it, though it rings hollow to me, hollow and even ignoble. The justification for conservation to which I return is an aesthetic one, and what is wrong with that? Who, having looked through the pages of this book, could contemplate with anything but sorrow the extinction of any one of the species here pictured?

But the best is the enemy of the good. We live in an economic world (interestingly, the Darwinian world of wild nature is an economic one too) where everything has its price. It seems all too easy to take the aesthetic high ground and look down on selfish, utilitarian motives for saving rainforests and rare species. But what proportion of our own wealth, or our own time, are we prepared to sacrifice to such an end? Not much. Even before we get selfish, there are other calls on our charitable generosity. What about the victims of the latest earthquake, famine, tornado or other human catastrophe? Many of the habitats of endangered species are also the homelands of human poor, who can be forgiven for seeing wild animals as competitors rather than as enhancers of life's richness. 'Life's richness' can ring hollow, as hollow perhaps as your own child's stomach. The aesthetic view of wild nature is, from such a low vantage point, a luxury that the hungry cannot afford.

Let us not be too ready, then, to condemn those attempts by southern African states to make game parks 'pay their way', perhaps by turning the need to 'cull' into an excuse to sell big-game shooting licences. Of course it seems obscene to gratify human blood-lust, especially in those cases in which the animals themselves are tame and trusting, and the 'hunters' safely ensconced in vehicles, with experienced rangers on hand to protect them if they miss. But this too is an aesthetic judgement, and the practice can be defended on grounds of economic practicality, a defence that is not available to, for example, the mincing, strutting, primping bullfighter. Disagreeable as it sounds, I sometimes find it hard to maintain my confidence that the southern African solution to the problem of saving the elephant and the black rhinoceros is not the most practical one. Nevertheless, on balance I still prefer the solution of a total ban on all trading in ivory and rhinoceros horn.

Such inconclusive meditation is a sure sign that I have no easy solution to offer. I return to aesthetics; but this book persuades me that here are more than ordinary aesthetics. It is not just the streamlined beauty of a swimming whale, the muscular tautness of a stalking big cat, the iridescent extravaganza of peacock or scarlet macaw. It is not just the pleasure of gazing at a spectacle, and of reflecting on the privilege of being able to do what future generations may be denied. Evolutionary thinking can give our aesthetic a new depth. We are not just looking at an animal as if it were an ordinary work of art. If it is a work of art, it is one that has been perhaps ten million years in the crafting. This seems to me to make a difference.

DARWIN ON THE SLAB

Inside Nature's Giants was a popular TV programme, broadcast by Channel 4 in four series from 2009 to 2012, in which the internal anatomy of a series of large animals was displayed by dissection. The lead dissector was the American comparative anatomist Joy Reidenberg. I was invited to comment, both for television and for the live audience of veterinary students, on evolutionary aspects. In the case of the giraffe I was allowed to don overalls and rubber boots to assist Joy and the rest of the dissection team. When a book to accompany the series was published in 2011, I was invited to contribute a foreword. This is what I wrote.

When I was first approached to take part in *Inside Nature's Giants*, I immediately suggested that, if they ever had the opportunity to dissect a giraffe, they should try to hunt for an extraordinary bit of anatomy called the left recurrent laryngeal nerve. If you were designing a nerve connecting the brain to the voicebox, would you send it on a detour down into the chest to loop around one of the large arteries there, and then back up to its target organ at the top of the neck? Of course not. Yet that is exactly what the recurrent laryngeal nerve does, not because it has other business with branches in the chest – it doesn't. The detour makes perfect sense as a relic of evolutionary history and our fishy ancestry, but from a design point of view in a mammal it is a botched job, pure and simple: no self-respecting engineer would perpetrate such a thing. It's bad enough in humans, but in a giraffe . . . that I had to see!

Almost a year later I found myself in the Royal Veterinary College, blinking in amazement at a state-of-the-art dissecting theatre, one entire wall made of sheer plate glass, behind which, in near-darkness, tiers of mesmerized students stared out of the gloom at an electrifying scene: a work of art that would leave Damien Hirst stranded drably in formaldehyde. Bright arc lights bore down on a young giraffe – which had unfortunately died in a zoo – on a huge dissecting table, one leg winched towards the ceiling in an attitude of stark hyper-realism. Its yellow and brown patchwork coat seemed to glow to match the bright orange overalls and white rubber boots of the dissecting team, a surreal uniform which I was also required to don when I joined them under the lights. Almost euphoric with the coincidence, I realized that it was Darwin Day 2009 – his 200th birthday – and I was privileged to spend it with the expert team of comparative anatomists and veterinary pathologists as they carefully traced that paragon of Darwinian paradox, the laryngeal nerve of the giraffe.

For me it was the start of a fascinating association with the *Inside Nature's Giants* team, as they unveiled the intricate internal complexity – a characteristically contrarian mixture of clumsy and elegant – inside some of the most amazing animals ever to evolve: elephant, lion, whale, cassowary, crocodile, python, polar bear, shark . . . and more. The overwhelming impression I get from surveying internal anatomy is that it is a beautifully honed mess! Every organ and structure has a function, but this has evolved gradually and sometimes imperfectly, with vestigial weaknesses reflecting the unimaginably long journey of the animal's DNA through geological deep time. From sea to land, from deserts to jungles, from shredding leaves to slaughtering wildebeest, the anatomies of animals tell us not only what they do now, but what their ancestors did in the past.

Inside Nature's Giants opens a bright window on each animal's life and also its evolutionary story. Each chapter of this book gives a unique anatomical insight into a different animal. The orange-suited explorers never cease to be surprised by what's under the skins of nature's giants. Engagingly, they not only demonstrate what we already know, but they also share with us the exhilarating experience of learning on the job. It has been my privilege to join them, and it is my pleasure to introduce this book.

LIFE WITHIN LIFE

The Extended Phenotype has long been a candidate for the book of which I am most proud, and I always hoped that experimental biologists would take it seriously. Among those who have done so is the Irish scientist David Hughes, now working in America. He went so far as to organize a whole conference near Copenhagen on the subject of *The Extended Phenotype*. A diverse group of biologists attended, including the distinguished population geneticist Marc Feldman and our leading authority on animal artefacts, Michael Hansell. Hughes' own field is parasitology. I was invited to write the foreword to the excellent book on *Host Manipulation by Parasites* (2012) which he edited together with Jacques Brodeur and Frédéric Thomas.

If I were asked to nominate my personal epitome of Darwinian adaptation, the *ne plus ultra* of natural selection in all its merciless glory, I might hesitate between the spectacle of a cheetah outsprinting a jinking Tommie in a flurry of African dust, or the effortless streamlining of a dolphin, or the sculptured invisibility of a stick caterpillar, or a pitcher plant silently and insensibly drowning flies. But I think I'd finally come down on the side of a parasite manipulating the behaviour of its host – subverting it to the benefit of the parasite in ways that arouse admiration for the subtlety, and horror at the ruthlessness, in equal measure. No need to single out any particular example, they abound on every page of this splendid book and they are thrilling in their dark ingenuity.

Hughes, Brodeur and Thomas have assembled a team of experts to fill eleven chapters with different aspects of this engrossing topic, skilfully edited to achieve a rare lightness of style and with the added bonus – which I have not seen before – of an afterword by yet another expert, reflecting on each chapter and giving a fresh perspective. My own perspective is scarcely fresh – it is exactly thirty years since *The Extended Phenotype* was published – but forgive me for hoping it is worth briefly restating. I shall refrain from detailed citations of that book in what follows. Parasitic manipulations of host behaviour are all extended phenotypes of parasite genes.

We take the *organism* for granted as the level in the hierarchy of life at which we attribute *agency*. It is the organism – not the gene or the cell or the species – that chases, pounces or flees; the organism that patiently rasps its way through the shell of another organism in order to devour it; the organism that inserts its proboscis (or its penis) inside the body of another organism in order to suck blood (or inject sperm). But my thesis is that the organism is a phenomenon whose very existence needs explaining, no less than the phenomenon of sex or the phenomenon of multicellularity.

One could imagine a form of life – on another planet, say – where living matter is not bunched up into discrete bodies – organisms. Just as life didn't have to become multicellular, and just as reproduction didn't have to be sexual, so living material didn't have to become packaged into discrete, individual organisms, bounded by a skin, distinct and separate from other individual organisms, behaving as unitary, purposeful agents. The only thing that is really fundamental to Darwinian life is self-replicating, coded information – genes, in the terminology of life on this planet. For reasons that should be positively explained, not just accepted, genes cooperated to develop the organism as their primary vehicle of propagation. Thinking about parasites, and especially parasitic manipulation of hosts, leads us directly to an understanding of the nature and significance of the organism – indeed, to what I shall offer as the very definition of an organism.

I wrote in 1990 that 'an organism is an entity, all of whose genes share the same stochastic expectations of the distant future'. What does this mean? A typical organism, such as a pheasant or a wombat, is a collection of trillions of cells, each containing near-identical copies of the same genome, which is itself a collection of tens of thousands of genes. Why do all the genes 'work together' for the good of the organism rather than going their separate ways as a pure 'selfish gene' view of life might expect? It is because they are all constrained to share the same exit route from the present organism into the future: the sperms or eggs or propagules of the present organism. To be more accurate, all have arrived in the present organism by travelling through a large succession of ancestral organisms, leaving each one via shared propagules. All have therefore been selected to 'agree' on the same 'interests': the organism must do whatever is necessary, given the species' way of life, to survive long enough to reproduce; it must successfully woo a member of the opposite sex; it must be good at caring for its young, if that is what the species happens to do. That is why the genes within an organism work harmoniously as a cooperative, rather than as an anarchistic rabble with every gene out for its own selfish survival.

If there is a parasite within the organism, say a fluke, the genes of the parasite potentially can affect the host phenotype – the extended phenotype. As abundantly demonstrated in this book, they normally bend the host phenotype in a direction hostile to that favoured by the host's 'own' genes. We take this for granted, but it is true only because they usually do not share a common route into the genetic future and therefore do not stand to gain from the same outcomes. But, to the extent that parasite genes and host genes do share future expectations, to that extent will the parasite become benign. In extreme cases, the distinction between parasite and host will be blurred, and eventually the two merge to form a single organism – as with lichens, for example, or mitochondria in eucaryotic cells.

Consider two kinds of parasite,* extremes at the ends of a continuum. To clarify the distinction, we can make them both bacteria and both feed inside the cells of the host. They differ crucially by virtue of their route into their next host. *Verticobacter* can only infect the progeny of its host, and it does so by travelling inside the eggs or sperms of the host. *Horizontobacter* can infect any members of the host species, with no special preference for the individual host's own children.

The point of the difference is this. The genes of *Horizontobacter* have no particular interest in the survival of the individual host, and especially no interest in the host's successful reproduction. The genes of *Horizontobacter* might happily evolve towards causing its bacterial phenotypes to eat the host to death from within and finally cause the host to explode, releasing a cloud of bacterial spores to the four winds, to be breathed in by new hosts. The genes of *Verticobacter*, on the other hand, have identical interests to the genes of the host. Both 'want' the host to survive long enough to reproduce, both 'want' the host to be attractive to the opposite sex, both 'want' the host to be a good parent, and so on. There is no difference between the interests of host genes and parasite genes, because both stand to gain from exactly the same outcome: the successful passing on of gametes in which both sets of genes travel together into the future. The extended phenotypic effects of the genes of *Verticobacter* will tend to improve the survival, sexual attractiveness and parental effectiveness of the host, in exactly the same way as the host's 'own' genes, and they will work in cooperation with the host's 'own' genes in exactly the same way as the host's 'own' genes cooperate with each other. My repetition of quotation marks may seem like overkill, but it is very deliberate. Remember,

* I hope it is clear, although I forgot to state it explicitly, that my two kinds of parasite are hypothetical.

my thesis is that the only reason an organism's own genes cooperate with each other is that they share their exit route from the present organism into the future.

So closely bound are the genes of host and *Verticobacter*, it is likely that the bacteria will cease to be clearly distinguishable. They will eventually merge into the corporate identity of the organism. That, indeed, is exactly what happened long ago with the bacterial ancestors of mitochondria and chloroplasts, which is why these organelles so long escaped recognition as bacteria at all. The same process of merging identities has been re-enacted more recently a number of times in evolution, for example in some of the large protozoans that inhabit the guts of termites.*

What does it mean to say that genes 'want' something? It means that, as a consequence of natural selection, they can be expected to influence phenotypes in the direction of that something, whether or not the phenotype in question can be regarded as belonging to their 'own' individual organism. Indeed, it is part of the logic of the extended phenotype argument that the very idea of an 'own' individual organism has no necessary validity. The genes of a 'parasite' at the *Verticobacter* end of the continuum have just as much right to claim the phenotype of the 'host' as the host's 'own' genes.

These were the thirty-year-old thoughts that were revived in me as I read this book, and this was the thread with which I bound its provocative wonders together in my mind. But other readers may find other threads, and this foreword has already been self-indulgent enough. The book itself is fascinating. It should be read not just by parasitologists but by ethologists, ecologists, evolutionary biologists, poets and – importantly – doctors, vets and all involved in the health professions. As Nesse and Williams explain in their pathbreaking book,† many of the symptoms of disease are manifestations

---

* See 'The Mixotrich's Tale' in *The Ancestor's Tale*.
† *Why We Get Sick: the new science of Darwinian medicine.*

of parasitic manipulation of patients, and many are countermanipulations by patients. This book should be on the syllabus of medical schools the world over.

Poets? Yes, poets. And philosophers and theologians too. Samuel Taylor Coleridge attended Humphry Davy's science lectures at the Royal Institution in London 'to improve my stock of metaphors'. The distinguished philosopher Daniel Dennett begins his magisterial account of religion as a natural phenomenon* with a description of the behaviour of an ant, possessed by the brainworm *Dicrocoelium dendriticum*. It is such a powerful metaphor for a human brain, possessed by a powerful, parasitic idea and forgoing all the normal benefits and comforts of life in its service. What might the author of 'Yea, slimy things did crawl with legs upon the slimy sea' not have done with the rich materials furnished by this book?

---

* *Breaking the Spell: religion as a natural phenomenon.*

PURE DELIGHT IN A
GODLESS UNIVERSE

This essay, previously unpublished, was commissioned – and titled – by the publishers as part of the promotion campaign for *The Greatest Show on Earth*, my 2009 exposition of the evidence for Darwinian evolution. Some years earlier, an anonymous donor had given me a favourite T-shirt bearing the slogan 'Evolution, the Greatest Show on Earth, the Only Game in Town'. I wanted to use the whole thing as the title of my book but the publishers thought it too long. Maybe one day I'll use *The Only Game in Town* as my title for another book. It would be appropriate for a book on 'Universal Darwinism'.

'Pure delight' is a good phrase. It well expresses a state of mind that is one of the joys of modernity. Everything that happens happens for a reason. Frogs don't turn into princes, water doesn't turn into wine, and men don't walk on water. The universe is governed by simple regularities, which we call the fundamental laws of physics and which are obeyed everywhere. Although the laws themselves are simple and inviolate, their consequences can be extremely complex, notably in those extraordinary entities, which may be unique to our one planet, and which we call living things. There is delight in the fact that life exists, and something close to ecstasy in our ability to understand why.

When did the pure delight begin? I can't pinpoint any precise moment in history and, alas, there are still many people who have yet to taste it. Perhaps I should begin by looking at what it replaces, going back to prehistory.

Our wild ancestors struggled to survive in a hostile world. Danger crouched behind every rock or tree, lurked just below the dark surface of every lake or river. Mysterious pestilences swept through the tribe without warning, like biblical plagues, to smite a loved child or carry off a young warrior in his prime. Droughts and famines, floods and forest fires, blizzards and earthquakes: you never knew when the next disaster would strike, nor from which direction.

There were good times too, of course: good years when the rains came as they should, and the tribal lands bloomed and swarmed with game; good days when a lucky escape from a lion, or an unexpected recovery from a snakebite or a fever, provoked a spontaneous outburst of thankfulness.

There is a reason for everything, and understanding it is a part of the pure delight that science gifts to us. Some fraction of the delight remains, even when what we understand is a terrible misfortune. Smallpox is a loathsome disease, but there is consolation in understanding that it has an intelligible cause (a virus) and that it can be prevented (by tricking the body into 'learning' how to deal with the virus, through vaccination with a weakened strain). Epilepsy is not possession by a devil that needs to be cast out, but a comprehensible malfunctioning of the brain, which science can control with drugs such as sodium valproate. Earthquakes and tsunamis are not punishment for sin but the repercussions of shifts in tectonic plates.

In pre-scientific times, apparently random events, which have complex and multifariously interacting causes, defied explanation and invited wild hypothesizing. 'Once, during a terrible drought, we sacrificed a goat, and the very next day rain clouds gathered and the blessed rains came. Therefore, whenever we want rain we had better sacrifice a goat. Maybe it doesn't always work, but who will dare stop the practice to test the hypothesis?' 'I have had a run of bad luck, so this must mean that somebody is practising witchcraft against me.' 'A black cat just crossed my path, so I must take extra special care today lest evil befall.' Tribal peoples today, and presumably the ancestors of all of us, lived their whole lives tormented by such fearful imaginings.

Our tribal ancestors had no way of knowing the true causes of seemingly random events, so it was natural to ascribe *agency* to them. An infant son developed a raging fever and died. The grieving parents had no way of knowing that two weeks earlier, in the silence of the night, a female *Anopheles gambiae* mosquito had bitten him and injected into his bloodstream a deadly culture of *Plasmodium falciparum*. What could they think, but that the tragedy must surely have been the work of a malevolent agent – perhaps an enemy practising witchcraft against the family, in revenge for the father's bribing a shaman to curse his land and cause his crops to fail?

The human mind eagerly sees agency wherever it looks, even where there is none. A stalking leopard really is an agent. It will follow you, chase you, pounce on you, throttle you and eat you. It takes deliberate steps to destroy you, actively countering your attempts to evade it. You must beware of leopards, and adopt the sound working hypothesis that a leopard is out to get you. Lightning is not like a leopard, not an agent. Lightning does not seek out prey and devour it. But how is the primitive mind to know the difference?

It is one of the insights of the unfairly maligned field of evolutionary psychology that there is some benefit in assuming agency, even where the assumption is in fact mistaken. Agents, such as leopards, are potentially more dangerous than non-agents, such as stones. That dappled pattern over there is probably only pebbles. But it could be a cunningly camouflaged leopard, waiting to spring. Best to assume the worst and take evasive action on the assumption that it is a leopard. There are costs in getting it wrong, and benefits in getting it right, either way, but there is an asymmetry. A 'false positive' error (assuming it is a leopard, when really it is only pebbles)

wastes time and energy in needless evasive action. But a 'false negative' error (assuming pebbles when it really is a leopard) is likely to
be fatal. There are admittedly costs in being too risk-averse – an
individual who is too fearful, too flighty, never gets any of the business of life done. But there are greater costs in being too recklessly
risk-prone, and the balance, according to this hypothesis, led to an
unwarranted (as it might seem to us in our modern complacency)
predilection for assuming agency. A tree branch is not an agent of
malevolent action against humans, but a python is. That 'branch'
over there just might be a python. Better give it a wide berth, even
though that wastes a little time. Natural selection favoured a pre-
cautionary principle, whereby the mind's realistic estimate of the
likelihood of true agency is inflated into an unrealistic, but nevertheless prudent, overestimate.

It was this tendency to see agents everywhere that led to our
ancestors inventing spirits of the forest, river sprites, thunder gods,
sun gods, rain gods, fire gods, evil spirits, devils and malevolent
hobgoblins. Such superstitious descrying of agents where there
were none led directly to the animistic religions that dominate the
minds of tribal peoples to this day. Those pagan religions were the
ancestors of the epic polytheisms of the Greeks, Romans, Norse-
men and Hindus. And in turn, through various cultural innovations,
for example in ancient Egypt and the deserts of the Middle East,
they were subsumed into the monotheisms that now dominate –
and arguably threaten – the world.

The tendency to infer agency, wherever we look, became especially
beguiling in the face of our natural curiosity about *origins*. How did
the world begin? Where did our ancestors originally come from?
How did the plants and animals on which we depend come into
existence? What about the sun, the moon and the stars? What started
it all off? Creation myths, of dazzling variety and often poetic beauty,
pervade the world. And every single one of them is dominated by the
same prudently evolved presumption of agency: agency, agency,
agency; never chance, never luck, never a real explanation that

actually explains anything, but always an unexplained agency that begs more questions than it answers.

Just as primitive minds saw lightning, and the wind, fire and the sun, as deliberate agents, so virtually all minds saw agency behind 'creation' – and most minds, it has to be said, still do. Especially when they contemplate the stunning beauty, complexity, and apparent design of the living kingdoms. Who could look at the wing of a swallow, the trunk of an elephant, the eye of a hawk, the athletic grace of a jinking gazelle or a sprinting cheetah in pursuit, and not see overwhelming evidence of a purposeful agent?

The answer is – nobody before Charles Darwin, give or take a few partial and equivocal predecessors. Everywhere you look, the illusion of design seems to shout 'agent'. Just as a leopard's purposeful stalking compels the conclusion that it is an agent, out to get you, so the world's apparently purposeful plenitude of crawling, squirming, running, jumping, breathing, hunting, buzzing life appears to cry out, 'This was all the work of a creative, designing agent.'

Yet Darwin showed that it is not so. Life, by its very existence, could have qualified for the title of a great show. What raises it to the accolade of *The Greatest Show on Earth* is the pure delight of understanding, of explanation: the thrill of running down the quarry of final comprehension and final banishing of superstition. We still have a long way to go. But humanity is on the verge of waking up and revelling in the pure delight of a comprehensible universe.

TRAVELLING WITH DARWIN

This is an abbreviated version of my introduction to the 2003
Everyman edition of Darwin's *Origin of Species* and *The Voyage of
the Beagle*.

What is the present status of Darwinism? Should we today read Darwin's works for their historic interest only, or are they still regarded as scientifically true? We can now see that, to a remarkable extent, Darwin got it right. Above all, we still hold to his solution of the really big question, the question of how the laws of blind physics could generate ever-increasing prodigies of statistically improbable, organized complexity (a feat that, as we shall see, has even – erroneously – been thought to fly in the face of one of the most important of those laws, the second law of thermodynamics). Quite apart from the correctness of his fundamental point, Darwin showed astonishing prescience when it came to detail. His passages on ecology, on the tangled webs of interaction between animals and plants, sound uncannily modern. It is his ideas on the detailed nature of heredity that we can now see to have been most seriously wrong. Indeed, reforms in ideas on heredity have led, in two successive episodes, to announcements of the births of 'neo'-Darwinisms.

The first neo-Darwinism dates from the 1890s and was mainly associated with the name of the great German biologist August Weismann. Weismann's most important contribution was the idea of the genetic material as an autonomous river flowing through time, with bodies – you and me – as temporary conduits for it. Weismannism is diametrically opposed to Lamarckism. Weismann's genetic river flows on inviolate, and acquired characteristics are never inherited. The second neo-Darwinism, the one to which the name is applied today, is the population genetic version of Darwinism initiated by R. A. Fisher, J. B. S. Haldane and Sewall Wright. This is conventionally dated from about 1930, when Fisher published *The Genetical Theory of Natural Selection*, although his writings on the subject go back at least a decade earlier.

Darwin's view of heredity was that of his time. There seemed no particular reason to doubt that environmentally induced changes in the body impressed themselves on the hereditary substance and so were passed on to future generations. Darwin did not doubt it, and when it suited him he freely made use of the idea. Indeed, in *Variation under Domestication*, he developed his own theory of acquired heredity – 'Pangenesis'. Tiny particles called gemmules circulated in the blood, becoming imprinted with information from all parts of the body and conveying it to the sex cells and hence into the next generation. This is one of the very few of Darwin's ideas that we can say with confidence is utterly wrong, not just in detail but in spirit.

Darwin also accepted the assumption of blending inheritance prevalent at that period. Children clearly resembled both their parents, but some seemed to take more after the father, others more after the mother. Presumably some substance from each parent was blended with that of the other, though not necessarily in equal proportions. The incompatibility of all forms of blending inheritance with natural selection was pointed out by Fleeming Jenkin, later Professor of Engineering at Edinburgh, in 1867 in the *North British Review*. Jenkin, who was hostile to Darwin's ideas, produced an ingenious and apparently knockdown argument against the theory of natural selection. He used a human example to express his point, and couched it in the racialist terms that were almost universal then, so I'll put it in a less offensive way. If children are intermediate between their parents – a blend like a mixture of two pots of paint – new variations will inevitably tend to disappear as the generations go by. No matter how strongly natural selection favours a characteristic, that characteristic is bound to become successively diluted as time passes. Nothing can save the beneficial characteristic from being swamped out of existence.

It is tempting to say – and frequently said – that it was Jenkin's challenge that goaded Darwin into importing a large measure of Lamarckian heredity into the fifth and sixth editions of the *Origin*. After all, it is known that Darwin was worried by Jenkin's argument, and the assumption of blending inheritance is indeed well-nigh fatal to natural selection. Although this feels right, it has been pointed out that in fact most of the fifth edition revisions had already been made before Darwin could have read Jenkin's article. Whatever the stimulus for these revisions, they are among the most important reasons for preferring the first edition for this Everyman reprinting. The first edition was written without benefit of criticism from Darwin's lesser contemporaries, and it was the better for it!

Incidentally, the *Origin* is not the only one of Darwin's books of which this might be said. In the second edition of *The Descent of Man*, Darwin raised the question of the 1:1 sex ratio, only to drop it again:

> I formerly thought that when a tendency to produce the two sexes in equal numbers was advantageous to the species, it would follow from natural selection, but I now see that the whole problem is so intricate that it is safer to leave its solution to the future.

It was the splendidly clear-thinking Ronald Fisher who came to the rescue in 1930. He introduced his discussion by quoting the above passage from Darwin, and then went on to his own elegant solution:

> In organisms of all kinds the young are launched upon their careers endowed with a certain amount of biological capital

derived from their parents ... Let us consider the reproductive value of these offspring at the moment when this parental expenditure on their behalf has just ceased. If we consider the aggregate of an entire generation of such offspring it is clear that the total reproductive value of the males in this group is exactly equal to the total value of all the females, because each sex must supply half the ancestry of all future generations of the species. From this it follows that the sex ratio will so adjust itself, under the influence of Natural Selection, that the total parental expenditure incurred in respect of children of each sex, shall be equal; for if this were not so and the total expenditure incurred in producing males, for instance, were less than the total expenditure incurred in producing females, then since the total reproductive value of the males is equal to that of the females, it would follow that those parents, the innate tendencies of which caused them to produce males in excess, would, for the same expenditure, produce a greater amount of reproductive value; and in consequence would be the progenitors of a larger fraction of future generations than would parents having a congenital bias towards the production of females. Selection would thus raise the sex-ratio until the expenditure upon males became equal to that upon females.

But what Fisher and many others overlooked is that Darwin himself came much closer to the correct, Fisherian solution in the *first* edition of *Descent* than in the second edition, which is the one Fisher quoted.

Let us now take the case of a species producing, from the unknown causes just alluded to, an excess of one sex – we will say of males – these being superfluous and useless, or nearly useless. Could the sexes be equalized through natural selection? We may feel sure, from all characters being variable, that certain pairs would produce a somewhat less excess of males over females than other pairs. The former, supposing the actual number of the offspring to remain constant, would necessarily produce more females, and would therefore be more productive. On the doctrine of chances a greater number of the offspring of the more productive pairs would survive; and these would inherit a tendency to procreate fewer males and more females. Thus a tendency toward equalization of the sexes would be brought about.

This really is a remarkably close approximation to Fisher's own formulation. If only Darwin had had the courage of his first edition convictions! The whole question of Fisher's predecessors in sex ratio theory is admirably discussed by one of Fisher's most distinguished pupils, A. W. F. Edwards.*

Returning to the *Origin*, and Fleeming Jenkin's riddle of the apparently disappearing variance, the solution eventually turned out to lie in the fact that heredity is not blending at all, as Jenkin and Darwin thought, but 'Mendelian'; a matter not of blending substances but of reshuffling indivisible particles. One of the famous ironies of the history of science is that Mendel's experiments had already been published in 1865, two years before Jenkin's troubling assault. Properly interpreted, Mendel would eventually turn out to be Darwin's saviour from Jenkin's erroneous criticism. But, far from being properly interpreted, Mendel was not even read, by Darwin or apparently anyone else, until 1900. And even after the rise of Mendelism, its leading exponents, the 'Mendelians' William Bateson, Hugo de Vries and others, so misunderstood its significance as to think it positively opposed to Darwinism. This was largely because the most visible mutations were too large and dramatic to do the job that Darwinism would demand of them.

The matter was not definitively resolved until it was finally taken in hand, again by R. A. Fisher. Incidentally, there is a pleasing continuity in the fact that Fisher was backed and supported in his

---

* See his 'Natural selection and the sex ratio'.

young days by one of Charles Darwin's sons, the eugenicist Major Leonard Darwin. Fisher's daughter, Joan Fisher Box, was later to write, in her biography of her father:

> Major Darwin, by his faith and his personal generosity, stood godfather to Fisher's scientific career, and in the 30 years of deepening friendship until Darwin's death in 1942, Fisher honoured him as a father.

W. Weinberg in Germany and the eccentric English mathematician G. H. Hardy had shown by elementary algebra that there would be no inherent tendency for genes to change in frequency. In the same way as Newton argued that bodies in motion would stay in motion unless positively resisted, Hardy and Weinberg showed, with almost embarrassingly simple algebra, that gene frequencies in populations would remain the same unless positively perturbed by some force such as selection. Starting from the so-called Hardy/Weinberg Law, Fisher, Haldane and Wright built up a new way of thinking about Darwinism which is pretty much the way we think about it today. Evolution became seen as changes in frequencies of rival ('allelic') particles in a pool of particles, the gene pool. As some genes become more frequent in the population and others less frequent, so the shapes and sizes of organisms change. As I would put it today, geological time is the stage for a continually renewed dance of immortal genes, changing partners and casting off an unending string of mortal bodies. So the spectre of Fleeming Jenkin was finally laid.

What might have been pointed out in Jenkin's own time was that his argument was only incidentally anti-Darwinian. It was also an argument against manifest fact. If Jenkin had been right, variation should not be conserved and should decrease with dramatic rapidity. Any cohort of grandchildren should be visibly less variable than their grandparents were. Mix black paint and white, and the result is grey paint. Mix grey with grey and you'll get more grey: all the king's horses and all the king's men will never reconstitute the original black and white. Yet with living things, as the generations go by, it is perfectly obvious that this is not happening. The descendants of variable ancestors do not merge into grey uniformity. They remain, on the whole, just as variable as earlier generations ever were. Jenkin thought his argument anti-Darwinian. It was actually anti-known-fact and therefore clearly wrong.

Fisher pointed out that Darwin, as early as 1857, had a dim perception of some sort of particulate inheritance. In a letter to Huxley, Darwin had written:

> I have lately been inclined to speculate, very crudely and indistinctly, that propagation by true fertilization will turn out to be a sort of mixture, and not true fusion, of two distinct individuals, as each parent has its parents and ancestors. I can understand on no other view the way in which crossed forms go back to so large an extent to ancestral forms. But all this, of course, is infinitely crude.

Being wise after the event, one can say that the principle of particulate or Mendelian inheritance, and its significance in refuting Jenkin, is staring us in the face every time we contemplate sex itself. We all have a mother and a father but we are not intermediate hermaphrodites. If maleness and femaleness do not blend, Jenkin-style, why should we assume that anything else does? Some things are so obvious that it requires a Fisher to see beyond the obviousness.

But, yet again, what even Fisher did not realize is how tantalizingly close Darwin himself came. In an 1866 letter to Wallace,* which Fisher would surely have quoted had he known of it, Darwin said:

---

* Letter dated 'Tuesday, February, 1866', published in James Marchant's edition of Alfred Russel Wallace's *Letters and Reminiscences*, vol. 1. I am extremely grateful to Dr Seymour J. Garte of New York University, who found this letter by chance in a volume of correspondence between Darwin and Wallace in the British Library in London,

My dear Wallace

... I do not think you understand what I mean by the non-blending of certain varieties. It does not refer to fertility; an instance will explain. I crossed the Painted Lady and Purple sweetpeas, which are very differently coloured varieties, and got, even out of the same pod, both varieties perfect but none intermediate. Something of this kind I should think must occur at least with your butterflies & the three forms of Lythrum; tho' these cases are in appearance so wonderful, I do not know that they are really more so than every female in the world producing distinct male and female offspring ...

Believe me, yours very sincerely

Ch. Darwin

Here Darwin comes closer to anticipating Mendel than in the passage quoted by Fisher, and he even mentions his own Mendel-like experiments on sweet peas.

Fleeming Jenkin was not the only physical scientist to give Darwin a hard time. Sir William Thomson, Lord Kelvin – incidentally, a former collaborator of Fleeming Jenkin in laying the transatlantic cable – was seen by many as the leading physicist of his generation. He pronounced, with the lofty certainty of a physicist talking down to a mere biologist, that the Earth was probably only a few tens of millions of years old, and certainly far too young to allow evolution to achieve what Darwin supposed. Darwin quailed a little before such massively authoritative pontification, but he remained true to his own, more deeply reasoned convictions.

In our century, of course, he was triumphantly vindicated.

---

immediately recognized its significance and sent a copy to me. I have previously published this letter in *A Devil's Chaplain*.

Radioactive dating methods show the Earth's age to be measured in thousands of millions of years, ample – indeed, almost too ample – for Darwin's purposes. Even by the end of the nineteenth century Kelvin's conclusions were already being questioned by physical scientists, among them Darwin's son George. It is easy to forgive Kelvin his error; harder to forgive his assumption that, if the best physicist of the day says one thing and the best biologist the opposite, it is axiomatic that the physicist must be right. In our own time the eminent astrophysicist Sir Fred Hoyle ran an idiosyncratic anti-Darwinian campaign which led him, together with another astronomer, C. Wickramasinghe, even to question the authenticity of one of our best-known fossils, the primitive bird *Archaeopteryx*, and to claim that 'any *physicist* looking at it would have worries'.* Hoyle and Wickramasinghe have also written:

> At one stage we received a suggestion by telephone that a panel of four sedimentary geologists be appointed, two by the Museum, two by us, to pronounce on the authenticity of the fossil. Our reply was typical of all that is objectionable in our way of expressing things.

The remarks about sedimentary geologists that follow do indeed bear out this judgement, though why they are so proud of being objectionable is less clear. They conclude (emphasis mine):

> Besides which, controversies are not settled in the *quantitative sciences* in such a ludicrous way.

The same assumption about the hierarchy of sciences, with physics at the top, can be detected in the following letter to *The Times* of

---

* Quoted in *New Scientist*, 14 April 1985, p. 3 (my emphasis).

19 December 1981 by a professional physicist (and a Fellow of the Royal Society):

> As a physicist, I cannot accept [natural selection] . . . It seems to me to be impossible that chance variation should have produced the remarkable machine that is the human body. Take only one example – the eye . . .

The objection to Darwinism that it is a theory of 'chance' is one of the most popular and most foolish of all. Mutation is indeed random, although only in the sense that it is not directed towards improvement. Natural selection is quintessentially the opposite of random. Any fool can see that random chance cannot put together living complexity. That is precisely why Darwinism is necessary. It is a preposterous irony that the statistical improbability of living organization is regularly advanced as though it counted *against* Darwinism, rather than in favour of it.

Other alleged objections were mostly disposed of by Darwin himself in his 'Difficulties on Theory' chapter of the *Origin*, and the three subsequent chapters on 'Instinct', 'Hybridism' and 'The Imperfection of the Geological Record'. Some more modern objections Darwin would have rumbled with ease. I'll mention two, 'thermodynamics' and 'tautology', which are based upon half-understood smatterings of science and philosophy.

The second law of thermodynamics states that any closed system tends to become more disordered. This is fundamentally because there are more ways of being disordered than ways of being in a state that we would recognize as ordered. Evolution represents an increase in the ordering of matter. Therefore, the argument triumphantly concludes, evolution violates the second law and evolution cannot occur! The physicist's simple reply to this argument is that we are not dealing with a closed system: external energy is fed in. But a more cutting retort than this is called for. It is the very existence of life – whether it comes about by evolution, divine creation or by any other means – that constitutes, if anything does, an increase in the ordering of matter. If the thermodynamic argument could be used to prove the non-existence of evolution, it could equally be used to prove the non-existence of life. The idea that evolution violates the second law is not merely mistaken – it turns out to be actively foolish.

Half-understood philosophy is no less pernicious than half-understood physics. Most of us nowadays have picked up the philosopher's nostrum that a good scientific theory should be 'falsifiable'. Any theory that turns out to be a necessary truth, a tautology, is of no value according to this philosophy. Darwinian natural selection is sometimes summed up in Herbert Spencer's phrase 'survival of the fittest'. The 'fittest' are defined as those that survive. So the phrase, and by triumphant implication the theory, is circular, tautological and empty. Once again, we can give a scholarly reply and also a more scathing one. The scholarly reply is that it is only recently that 'fittest' has come to be defined in such a way as to make 'survival of the fittest' circular. For Darwin, the fittest were the strongest, the fastest, the sharpest-eyed. As part of the neo-Darwinian revolution, theorists needed a name for the quantity that natural selection tends to maximize, a mathematical quantity sometimes symbolized as *W*. They lit upon 'fitness'. If fitness is *defined* in this way, it is obvious that the phrase 'survival of the fittest' has got to become a tautology.

To come to the more scathing reply, 'survival of the fittest' is a tautology in the same sense as it is tautological to say: 'If a horse runs at twice the speed of a man, the horse will cover a given distance in half the time as a man.' This follows from the definition of 'speed'. Only someone 'educated far beyond his capacity to undertake analytical thought'* would conclude that because the statement is tautological neither horse nor man will complete the course! Darwin himself would perhaps have appealed, as he so often did, to

---

* The quotation is from Peter Medawar (see p. 14 above).

the facts of domestication. Nobody denies that selective breeding has been extremely effective in making racehorses run as fast as they do, or in improving the milk yield of cows. If we define the 'fittest' racehorse as the type most likely to be selected for breeding, then what is going on in a stud farm is a process of (artificial) 'survival of the fittest' which is every bit as tautological as survival of the fittest in wild nature. But nobody ever tried to persuade a horse breeder or a cattle breeder: 'You're wasting your time trying to improve the breed; your whole enterprise is a circular tautology and therefore doomed to failure.'

On the Origin of Species (the 'On' was dropped from the sixth edition) is not the only important book that Darwin wrote. I have already mentioned *The Descent of Man*. But the one that Everyman have chosen to reprint together with the *Origin* is Darwin's first and most charming book, commonly known as *The Voyage of the Beagle* or *A Naturalist's Voyage*, which was based upon his diaries kept during the voyage of HMS *Beagle*. I say 'commonly known as', because the Victorians had a more fluid view of book titles than we do; it is even possible to be unsure which title applies to which book. The full title of the first edition, published in 1839 by Henry Colburn, London, was *Voyages of the Adventure and Beagle. Narrative of the Surveying Voyages of His Majesty's Ships Adventure and Beagle between the years 1826 and 1836 describing their Examination of the Southern Shores of South America and the Beagle's circumnavigation of the globe. In Three Volumes.* Volumes I and II were by Captains King and Fitzroy. Darwin's was Volume III, entitled *Journal and Remarks, 1832–1836*. Soon afterwards, Colburn reissued Darwin's volume separately, under a new title. For his second edition, in 1845, Darwin went alone to a new publisher, John Murray, with whom he was to stay for the rest of his writing life. This second edition's title, slightly modified from Colburn's second title, was *Journal of Researches into the Natural History and Geology of the Countries Visited during the Voyage of HMS 'Beagle' round the*

World. It is that second edition, originally published in 1845, which is reprinted here.

Several further editions have appeared, mostly versions of the second, 1845 edition. Many of these have an abbreviated title on the spine, and sometimes on the title page as well: *A Naturalist's Voyage* or *A Naturalist's Voyage Around the World*. Others, including my own copy published in 1891 by Ward Lock, have a different abbreviated title on the cover: *Darwin's Journal During the Voyage of HMS Beagle round the World*. To add one more layer to the complication, in 1933 Darwin's granddaughter Nora Barlow edited *Charles Darwin's Diary of the Voyage of the Beagle* for Cambridge University Press. This is a reprinting of the original diaries on which the earlier book was based.

By whatever name, it is a delightful and beautifully written travel book which always makes me think of my earlier hero, Hugh Lofting's 'Doctor Dolittle' (now, alas, banned by righteous librarians for 'racism' – regrettable, no doubt, but surely a peccadillo in comparison to the rampant speciesism of our society which Dr Dolittle could do so much to remove from childish minds).

Darwin's travel notes offer tantalizing glimpses of how his mind was working towards the great truth that he was later to unveil for us. On every page, the book gives us hints of the great sage that the young Darwin was to become. We are left with a picture of a clever, sensitive and humane observer, a restlessly curious intellect, and a tireless collector whose youthful energy makes a poignant contrast with his invalided maturity and with the brooding melancholy that stares across to us from the portraits of his old age. There is adventure here. And humour, as in the story of how Darwin tried to use the bolas, the South American hunting weapon consisting of a weight on the end of a rope, but succeeded only in capturing the legs of his own horse – to the huge amusement of the gauchos.

We also can appreciate the liberal decency of the young Darwin, his passionate opposition to slavery (which caused a falling-out with Fitzroy), and his bewildered disgust at, for example, the indiscriminate hunting of South American Indians by the Spanish:

> This is a dark picture; but how much more shocking is the unquestionable fact, that all the women who appear above twenty years old are massacred in cold blood! When I exclaimed that this appeared rather inhuman, he answered, 'Why, what can be done? They breed so!'
>
> Everyone is convinced that this is the most just war, because it is against barbarians. Who would believe in this age that such atrocities could be committed in a Christian civilised country?

But most of his attention is focused upon his beloved natural history, and geology. *The Voyage of the Beagle* is a travel book by one of the most intelligently observant and thoughtful travellers of all time. What a privilege to see the world through such eyes. What a bonus to glimpse the mind of a developing genius in the flower of his robust and energetic youth.

Charles Darwin should be read because, give or take a few minor details, the evolutionary world-view that he has given us, that worldview which is so utterly different from anything that went before,* is, according to all that we have learnt since his time, literally true and unlikely ever to be superseded. Evolution as a fact has the same status as the fact that the Earth is round and not flat (though people can be found to call either 'mere theories'). It is Darwin who conclusively showed to the world that evolution is true.

He also gave us by far the most plausible theory for *how* evolution has taken place, the theory of natural selection. I would go further. I have argued elsewhere[†] that, at least as far as

---

* To be sure, Darwin was not the first evolutionist. But he was the first to present a full, rounded, evolutionary world-view.

† For example in my essay on 'Universal Darwinism', reprinted in *Science in the Soul*.

adaptation – the strong illusion of design that pervades all living things – is concerned, not only is natural selection the correct explanation for adaptive evolution on this planet; it is the only explanation ever suggested that *in principle* is capable of explaining adaptation. We therefore have grounds for believing that Darwinism has a universality that otherwise we expect to find only in the great theories of physics. If there is life – if there is any form of organized complexity – it will have evolved by some form of Darwinian natural selection. If that sounds like a rash or an arrogant thing to say – who, after all, would have thought that Newton would be superseded by Einstein and Planck? – reflect that there are truths that everybody would accept are simply true and will never be superseded.

It is sometimes suggested that Darwin is one of a trio of nineteenth-century giants, along with Marx and Freud. But are Marx and Freud really in his league? If we are ever contacted by alien beings from anywhere else in the universe sufficiently advanced technologically to reach us, will we have anything to say to them, when they step from their celestial *Beagle*? We shall surely share with them at least parts of mathematics and physics. They will have computed the same value of π; they will have the geometrical theorem that we attribute to Pythagoras; they will revere their equivalents of Einstein and of Planck. But there is no reason to suppose that they will have a Marx or a Freud. Why should these men's discoveries have any applicability outside the narrow confines of one species of animal on one planet in one galaxy? It is only the anthropologists on the alien expedition who will have any time for Marx and Freud (let alone Derrida and Foucault!). But if I am right about the universal significance of Darwinian natural selection, our aliens *will* revere the immortal memory of their Charles Darwin.

PICTURES OF PARADISE

For an evolutionary biologist, I came rather late in my career to Galápagos. So late that, when the science-loving philanthropist Victoria Getty heard, at a dinner party, that I had never been there, she resolved, then and there, to remedy the situation. She chartered a boat called (as it happened) *The Beagle*, and we spent a delightful couple of weeks sailing from island to island, going ashore in each case in an inflated rubber dinghy. I have now made up for lost time by visiting the biologists' paradise that is Galápagos no fewer than five times. Each time I have learned more, partly from the animals themselves and partly from the wonderfully knowledgeable Ecuadorian naturalist guides who accompany all such expeditions. From Darwin's *Voyage of the Beagle* on, the archipelago has inspired lots of books. Paul Stewart's *Galápagos: the islands that changed the world* (2006) is an especially lyrical one, both in its writing style and in its photographs, and I was pleased to contribute this foreword.

The good science fairy flew right round the world, looking for a favoured spot to touch with her magic wand and turn into a scientific paradise, a geological and biological Eden, the evolutionary scientists' Arcadia. You may question her motives or her existence, but of the place she lit upon there is no doubt. It lies beneath the eastern Pacific, approximately 91 degrees west and 1 degree south, some 1,170 kilometres west of the coast of Ecuador – Darwin's 'Republic of the Equator'. She blessed this spot with her wand and turned it into a volcanic hot spot (or hell-mouth as Paul Stewart describes it). Meanwhile, her collaborator, the science fairy godmother, arranged for the Nazca tectonic plate to move in an orderly fashion towards the mainland, at a stately pace of 4 centimetres per year. The result of these two science-friendly circumstances – plate moving over hot spot – is that the Galápagos archipelago was spun out on what Paul Stewart calls a geological conveyor belt. It is consequently an almost perfect natural laboratory of evolution – scene of an experiment planned in scientific heaven. The experimental plots (or islands as we mortals call them) are laid out, as they should be, in order of age, from the ringing black clinker of young Fernandina in the west to Española in the east – the latter soon (by geological standards) to go the way of its vanished predecessors and disappear beneath the waves.

I was stimulated to think along these lines, not by visiting the islands themselves – although I did go twice within one year – but by reading Paul Stewart's marvellous book. I expected it to contain some stunning photographs. How could it not, when produced by an award-winning cameraman and members of the BBC's peerless Natural History Unit – what we might call the David Attenborough stable of film-makers? What I didn't know is that Paul Stewart has a writing style to match. The publishers sent me a typescript before I saw any of the pictures. I read it in one day (which I only do with well-written books) and closed it feeling, 'Who needs photographs when the author can paint word pictures like that?' Later the photographs arrived, and I took it all back. Maybe we don't *need* the photographs but it is wonderful to have them.

To return to my fancy of the good science fairy and her wand, the hot spot needed to be carefully placed, just the right distance from the South American mainland. Too close, and the archipelago would have been overrun by South American immigrants. Faunistically, it would just be a suburb of the mainland. Too far, and it would be too depauperate to tell us anything. But with the volcanic hell-mouth placed – as it is – at just the right distance from the mainland, and with the islands themselves neatly spaced out, all evolutionary hell breaks loose. But it breaks loose with that controlled moderation that marks the well-designed experiment: just enough richness to be interesting and revealing, not enough to confuse and smother the take-home message.

Darwin himself, as this book makes clear, didn't take home quite as much as he might have. Although he described Galápagos as the origin of all his views, those views were not at the time matured enough to motivate him to label his specimens. He muddled his finches (and the *Beagle*'s crew ate the adult tortoises) and Darwin later had to rely on the finches that Fitzroy and Covington took home, in order to reconstruct the Darwinian take-home message. The story of Darwin and Galápagos is the topic of one of the chapters. Of equal human interest is the history of the archipelago's discovery and some of the macabre tales of subsequent immigrants. So are the chapters on the geology, the life on land and in the surrounding waters, and the pressing need to conserve this priceless, natural museum of evolution – and redeem our past failings for, as Stewart eloquently puts it, 'greed and natural history repeatedly proved the twinned incubus and succubus of misery'.

To anyone contemplating a visit to Las Encantadas – the Enchanted Isles – the gazetteer alone is worth the price of this book. Paul Stewart's *Galápagos* will be my treasured companion on my next visit, and I shall take along an extra copy to present to the boat's library. If you are not able to go in person, reading this book and savouring its pictures is, if not a substitute, a delight to be going on with.

III

INSIDE THE SURVIVAL MACHINE: EXPLORING HUMANITY

IN CONVERSATION WITH STEVEN PINKER

LANGUAGE, LEARNING AND DEBUGGING THE BRAIN

This is an edited transcript of a conversation recorded for the Channel 4 Television programme *The Genius of Charles Darwin*, produced and directed by Russell Barnes and first broadcast in 2008, which won the British Broadcasting Award for 'Best Documentary Series' of that year. A fuller version appears in the audiobook. Steven Pinker is a distinguished psychologist, now at Harvard. He is the author of beautifully written books on linguistics, cognitive and evolutionary psychology, history and much else. Though he is younger than me, I regard him as an intellectual mentor and hero.

Why do we enjoy music? Why do we dance? Is the mind formed as much by nature as nurture? Why do we get anxious and depressed? Why are there some things we just find so very hard to understand?

RD: Steve, astonishingly modern though Darwin is to us when we read him today, we do have to remember that part of what he was doing was having arguments that maybe we don't have any more. At least so I thought. If you read *The Descent of Man* and *The*

Expression of the Emotions, for example, it becomes really clear that what he's doing a lot of the time is trying to make clear the continuity between human and non-human animals. Trying to minimize the distinctions between them. And perhaps especially in psychological respects in *The Expression of the Emotions*. Is that a book that you feel is seminal to what you're doing?

SP: It's a fascinating book, filled with insight, which I still write about and use in class. But it's interesting to see that he was fighting on two fronts. On the one hand he had to satisfy his fellow biologists that he could account for biological complexity: where the eye came from and why the body is so well engineered. On the other hand, that played into the hands of the creationists of his day. So, on another front, he had to point out how clunky and maladapted many of our traits were because of their origin in animal precursors. And in *The Expression of the Emotions*, it's interesting, there is no discussion of any adaptive function of facial expressions or body postures. Now it just seems as plain as day that we wear our emotions on our face in order to broadcast signals to other people. That was not part of his explanation at all. There was an upwelling of psychic energy. A kind of hydraulic model of the nervous system that he used to explain a number of our expressions. And he even had a Lamarckian theory in there that, for example, the angry expression where we bare our canine teeth, as if unsheathing them, even though they're no longer long enough that we would have to do that, he explained as a kind of acquired habit passed on to offspring. And the reason is that he was battling against the creationists of his day who offered the expressions as proof of God's existence because God gave them to us in order that we could express our moral sentiments. This drove Darwin crazy. It's clear it really got his goat. And so he hammered as hard as he could on the idea that our emotional expressions are just welling up with energy, they're pushes and pulls of some mechanical model of the nervous system he had, they're holdovers of our animal past. Anything but actual signals of internal states.

RD: That is very interesting. Do you think that some of the rather extreme hostility to evolutionary psychology is because people still are reluctant to accept the continuity between humans and non-humans? Perhaps you should begin by saying a bit about what evolutionary psychology is.

SP: Yes. Evolutionary psychology is the attempt to offer an ultimate explanation for the way we are.* The organization . . .

RD: You mean the Darwinian functional explanation – what's it good for?

SP: Exactly. What's it good for, in the sense of how did it increase the rate of reproduction of our ancestors in the kind of environment in which we evolved. So it's one of many components of a psychological explanation to explain why we perceive and remember and feel as we do. You have to look at the nervous system, how the brain computes it. You've got to look at how it develops in the child. But I think you also have to look at why it's organized the way it is as opposed to all the other ways it could've been organized, and that requires asking the question: 'What function does it serve, or did it serve, in our ancestors?'

RD: And when we ask such functional questions we have to be a bit sophisticated about it. If we take a particular one which on first sight is difficult, like, say, 'Why do we enjoy music?' – how would you go about giving a functional explanation for that?

SP: Well, I don't think everything has a functional explanation. In fact, it's pretty important that not everything has a functional explanation or it would be too easy. I think a proper functional explanation has to start with some kind of engineering analysis, completely independent

---

* See his *How the Mind Works*. This book is relevant to much that follows in this conversation.

of what we know about human beings or whatever the organism is. So let's take, for example, stereo vision, the ability to combine the images from the two eyes, to get an impression of depth. One can do a bit of geometry on a piece of paper and show why it would make sense to combine the images from two cameras at slightly different vantage points. And, indeed, if one were to build a robot that had to navigate an environment it would make sense to build in stereo vision. We can look at the design specs for a stereo vision system, then bring our human being into the lab, look at how human stereo vision works, and see how well the design specs from the engineering analysis match up with what we empirically observe in humans. The closer the fit, the more confidence we have that the function of stereo vision is to see in depth. And if you don't have the engineering analysis when you just observe that something is present in humans, it's impossible, I think, to make a proper functional or adaptationist explanation. And music might be an example. As important as music is in our lives, as much pleasure as it gives us, it's completely unclear to me why it would have to be music that would do any of those things. Why would notes and rhythmic and harmonic relations be a solution to any engineering problem? And it's possible, if it's an empirical issue, that music is actually a by-product of other adaptations. Perhaps our sensitivity to speech, a harmonically rich sound that the brain has to analyse into its frequency components in order to understand. Perhaps a by-product of emotional calls that go way back in primate evolutionary history: sighs, moans, laughs, cries and so on. Possibly a by-product of motor control: keeping your bodily actions at a constant, optimal rhythm. And maybe what music does is it combines bits and pieces of all these other parts of the brain, packs them into a supernormal stimulus, something that actually presses our buttons harder than anything in the natural environment would, and we enjoy it. At least, that's another hypothesis that shows not everything has to be an adaptation.

RD: If I've got that right, you're saying that in order to analyse speech, the brain has to have certain mechanisms for noting frequencies, analysing harmonics and things. Otherwise we couldn't tell the difference between, say, vowel sounds. The difference between 'ah' and 'ooh' and 'uh' – that's not in itself musical, but the same software mechanisms, the same brain mechanisms that you need in order to tell whether you've got an 'ah' or an 'uh' or an 'ooh' can't help being supernormally stimulated by, say, pure tones, or notes that stay the same for a long time, or harmonies, chords, harmonious chords. Is that something like what you're saying?

SP: Yes, exactly. At least, it's certainly a viable hypothesis and I think it's more plausible than any hypothesis that tries to find some function in music.

Now, I've got to add that people in music hate this theory.

RD: Oh really?

SP: They are ultra-adaptationists. Because I think there's a confusion between senses of 'adaptation'. In the technical sense of biology, 'adaptive' means a trait that, compared to alternatives, enhanced the rate of reproduction of our ancestors. In everyday conversation, if something's adaptive, it means it's healthy, it's valuable, it allows you to improve, it makes life worth living . . . And of course music is all those things, but it's just a different sense of that term 'adaptation'.

RD: Life-enhancing, yes.

SP: Life-enhancing, exactly. And I like to say that it's a mistake to equate adaptation in the biologist's sense, which is what we were just talking about, with what is worthy in life. In fact, they are often at cross-purposes. So reading, for example, which is almost certainly not an adaptation, because it just appeared too recently in human evolutionary history for it to have left its mark on the genome, is one of the things that makes life worth living. And it's not an adaptation. And one could argue: perhaps genocide is an adaptation. That one tribe, whenever it's convenient and has the means, will wipe out another tribe: that could've been selected for. And of course that is, well, . . .

RD: Not life-enhancing. To put it mildly.

These theories you've been putting forward are what you might call 'by-product' theories, and I think it's very important not to forget about by-products whenever we're giving adaptive explanations. It's so easy to pick on something that the animal we're looking at does and say 'What's the good of that?', and so often it turns out that, actually, it's not good for anything, but the underlying mechanism or whatever is producing that thing is also producing something else. And so, in this case, you're saying that music, or the enjoyment of music, is almost an inevitable by-product of speech-analysing mechanisms. And so we've always got to think 'by-product' when we're dealing with adaptation explanations. And maybe that's often an answer to the many rather virulent criticisms that we get in evolutionary psychology: that the critics are not thinking 'by-product' and they don't think that we're thinking 'by-product'. Would that be fair, do you think?

SP: Oh, absolutely. And in fact, I don't think one could make an argument that anything is an adaptation unless one had an alternative explanation. Such as, that it's a by-product or arose by chance. That's what gives the field its empirical content. Namely, you can test these alternative hypotheses and you might be right or you might be wrong. And they can be falsified.

RD: When we teach about evolution, we naturally tend to focus on anatomy but you can equally well say that psychology, that our minds, are evolved organs or organ systems, couldn't you?

SP: Well, yes. And we have every reason to believe that the mind is a product of the activity of the brain. The brain is an organ, it's got an evolutionary history. All of the parts in the human brain you can find in the brain of a chimpanzee and other mammals. And we also know that the brain is not just a random neural network; it has quite an intricate structure. And we have reason to believe that a lot of the products of the brain – our perceptions, our emotions, our language, our ways of thinking – are strategies for negotiating the world,

surviving, bringing up children, finding mates, negotiating relationships, deterring enemies, forming alliances, and so on. Why do we
believe that? Well, by many criteria, the design specs for an optimal
system uncannily match what we observe in human emotions. For
example, in reciprocal altruism, in a computer tournament of the
strategies that do well in trading favours without being exploited,
they are almost exactly what you find for human emotions like sympathy, trust, anger, gratitude and so on, so that if one were to lay out
what the computer algorithm ought to do in a competition to best
take advantage of exchange of favours, and then look at, empirically,
social psychology studies – What circumstances lead to guilt? What
leads to gratitude? What leads to righteous anger? – they match up
very, very closely. And one can make predictions about features of
human emotion that hadn't been observed before. Bring people into
the lab, test them, and see: do the people have those traits?

RD: Very nice. I suppose we can all easily understand why sexual
lust has Darwinian survival value. Are you now saying that the
mechanisms for guilt, trust etc., those sorts of things which are
mediated into reciprocal altruism, that those are a bit like lust? I
mean there's 'lust to trust' or . . .

SP: Indeed, I think they are. People have no problem accepting
Darwinian explanations for emotions that are triggered by the
physical world – like fear of heights and snakes and spiders and the
dark and deep water . . . or disgust at bodily secretions that might
be carrying parasites, or rotting meat and so on . . . but often feel a
little more surprised and even resistant to the idea that some of our
moral emotions might have an evolutionary basis. Like trust, like
sympathy, like gratitude. But I think that, just as it is clear that fear
has an evolutionary basis, so our moral emotions can be analysed
in the same way.

RD: There is, of course, notoriously, a lot of controversy about how
much is predetermined by the genes and how much the mind is a

blank slate, and obviously you're a protagonist in that.* Do you think there are actual evolutionary Darwinian reasons for some genetic predispositions in the mind?

SP: Well, yes. For one thing, blank slates don't do anything, and unless there's some kind of set of motives, one can't have an intelligent system that does anything – that processes some inputs and ignores others, that treats certain inputs as reinforcers.

RD: It's got to have some criterion for deciding when it's learning. You can't do everything by learning, there's got to be some criterion for deciding what should be taken as positive, reward, what should be taken as punishment. Something of that sort. If nothing else, that's got to be built in.

SP: Exactly, and also the way in which one analyses the sensory input. Imagine a mimicking bird that, when exposed to human speech, would be able to mindlessly reproduce the sound sequence. That's clearly not what our children do – although of course they have to take an input from the world, they have to listen to whether their community is speaking English or Japanese or Swahili. What they do in listening to those sense impressions is they chop them into words, they look for the grammatical regularities, then group them into phrases. They don't just reproduce the sound like parrots. So there are multiple ways of learning which depend on what species you belong to.

RD: You could imagine a continuum between having almost everything learned, with just a very bare minimum criterion for what to learn, and building in just about everything. And I suppose people have always been a bit biased towards thinking that humans rely on learning more than other species. Is it true to say that evolutionary psychologists are seeing more and more that there is rather more detail built in from the start than we had supposed?

---

* See his *The Blank Slate*.

SP: Yes, I would argue that that's the case – which doesn't mean that humans are insensitive to input or don't engage in learning. The question is: how many different kinds of learning are there and to what extent is each form of learning tailored to a particular problem? Do we learn, say, an eye for beauty in a different way from how we learn the grammatical rules of our language? Which, in turn, is different from how we learn about the physical world, like what makes objects fall and bounce and roll the way they do. So I don't see it so much as a conflict between nature and nurture, but rather fleshing out how nurture is rooted in nature.* That is, what do we pay attention to, what conclusions do we draw, how do we analyse the input, what motivates us? And that, I think, throws you back into nature because you can't learn everything. A learning system has to have the machinery that allows it to learn. That machinery, I think, ultimately has to be explained in terms of evolution.

RD: Human children, uniquely, learn language, and I think it's right to say, isn't it, that they have evolved brain mechanisms which enable them to learn language remarkably swiftly. Other species don't have that. And this leads linguists like you to say that there's some deep structure which is common across all languages. I think lay people often find that hard to believe because they hear languages sounding so very different. What does it mean to say that there's some deep commonness between all languages?†

SP: Well, in fact, I think, laypeople's intuitions can go in both directions because on the one hand, you do notice diversity across human languages, especially when you learn a second one. On the other hand, when you look at your kids and you see that by the age of two and a half or three they're stringing words together into fluent grammatical sentences without having been explicitly taught, a lot of parents say, 'This is

---

* See also Matt Ridley, *Nature via Nurture*.
† See Steven Pinker, *The Language Instinct*.

just a miracle. Now I know what you mean by the idea that children are wired for language.' So clearly they can't be wired for words, they can't be wired for grammatical constructions, at least not the particular ones in any language ... What they can be wired for is an ability to find words. An ability to ferret out the rules that combine them into meaningful phrases and sentences. And that can't be taken for granted. Just listening to speech, which is what a tape recorder can do, isn't enough to get you to produce and understand new sentences, ones that you haven't heard before. Sentences like the journalist's cliché, 'When a man bites a dog, that's a worthy story'. Fantasies like 'The cow jumped over the moon and the dish ran away with the spoon'. Brand new combinations, words that we didn't simply memorize from our parents, but that any child can understand on first hearing and produce new ones of their own. And we see this when children start to make errors like 'We holded the baby rabbits' or 'Why did the waiter disadappear?' that is, make it disappear. When children are clearly analysing quite deeply the speech that they hear, and are already showing off the talent of recombining words into grammatical combinations. They adapt an ability that matches that of their parents, and on the way their errors show us that they're engaging in this process of grammatical analysis.

RD: What about some more examples of behaviours or psychological dispositions in humans that are probably innate – I mean, are hard-wired?

SP: Well, I think one has a strong suspicion that's the case when people do things that aren't particularly good for them – in fact are really inexplicable in terms of just maximizing their well-being – but that have a very clear rationale in terms of what would have enhanced the survival of our ancestors. So ... an eye for beauty might be one of them. We all know in romantic attraction that we are sometimes oddly attracted to people who have a particular set of facial features, they're really not best for us, but the magnetism is very difficult to turn off. And the signs of beauty, it's been found, correspond to typical signs of fertility and health in ancestral conditions. Youth, and youth more so in men looking at women than women looking at men, and of course youth corresponds with fertility. Signs of not being infected with diseases and parasites. Signs of having a normal set of hormones and developmental history. Signs of tracking the ideal type in one's population, of having a set of facial features that are close to the average. All of these were predictions based on what ought to be attractive in people seeking one another and have been shown through the computer-generated faces to really be the things that make a face, an artificial face, look beautiful.

RD: What about evolutionary psychiatry, I suppose you could call it? Is there a Darwinian explanation for mental disorders like, say, depression?

SP: A lot of unpleasant emotions may not be maladaptive in the sense of, ultimately, what was good for our ancestors in evolutionary time or even what's good for us today. An analogy would be pain: pain is something that obviously we would like to have as little as possible, but people who are congenitally unable to feel pain die very young because they chew through their lips, they scald themselves with boiling hot coffee that they don't mind drinking; they fail to change positions while sitting, and so they suffer from chronic inflammation of their joints, whereas when you feel pain you're prompted to constantly shift to allow the blood to circulate. So, at first glance you might imagine how wonderful it would be never having to feel pain, but in fact it would be quite a curse. And I suspect that might be true for some of our unpleasant emotions as well. Anxiety, which can be debilitating in high doses, in lower doses is what prompts us to finish work, to run away from an impending storm, to finish our papers on time, to prepare for a lecture or a date. Sadness, which in extreme form can lead to a debilitating depression, in a lesser form is what allows us to value our children and our mates and to avoid the kind of loss for which sadness is a kind of internal punishment. And a number of psychiatric disorders might be, as with chronic pain, too much of an unpleasant reaction that, when it's better modulated by environmental events, actually is necessary to human flourishing.

RD: Do you think that evolutionary psychology can even explain why we find some things hard to understand? A lot of people find Darwinism itself hard to understand, and almost everybody finds quantum physics hard to understand. Can you give a kind of evolutionary spin on why we find things hard to understand?

SP: Well, there's no guarantee that our minds will, at least intuitively, understand phenomena at scales of time and space that are very different from our own experience: hundreds of millions of years, billions of light years, angstroms, where our best science tells us that the laws of reality work very differently from those at the scales that we're used to thinking about. And so disciplines like quantum mechanics, and evolutionary biology, and cosmology, and relativity, and even neuroscience might be deeply baffling. In fact, a lot of science education consists not so much of filling children's minds with new information as debugging the information that they bring into the classroom by virtue of being human beings. It's very hard to think of.

RD: Even Newtonian physics.

SP: Even Newtonian physics, the idea that, for example, an object left to its own devices . . .

RD: Can go on moving for ever. Yes, that is very counter-intuitive. And it does suggest, if this is true, that you're right about evolutionary psychology, that we come into the world with an intuitive physics, which is pre-Newtonian physics. I mean, things come to rest, rather than go on for ever. We understand things that move at relatively slow speeds as opposed to near the speed of light, that's why we can't understand Einstein unless we're very specially trained.

SP: We have workarounds for all of these. Largely, that's what education consists of, namely, figuring out the analogies, which when selectively applied, allow you to understand something very unfamiliar in terms of something that is much more familiar. So you understand evolution like pigeon breeding, only iterated hundreds of thousands of times. Or you change your intuitions about physics by trying to imagine what it would be like to ride along a beam of light. It may even be that ultimately, there are domains that are permanently baffling, because of the way our minds work, such as why a brain firing in patterns feels like something to the person who is that brain.

It's quite possible that in areas where philosophers seem to have gone around in circles for thousands of years, in which every possible solution seems deeply unsatisfying yet all the phenomena remain accounted for, there we might be running up against the limitations of what our own minds find intuitive.

RD: If our ancestors had been in the habit of travelling near the speed of light, Einstein would be obvious. If our ancestors had been the size of neutrinos, quantum physics would be obvious. But we live in a middle world between those extremes, and so we're limited to understanding things at that level.

SP: And we have ways of analysing reality that are very easily misapplied, such as thinking that evolution must be something that has a purpose in the same way that we have a purpose.

RD: Because purpose is so endemic in our human life and in our social life.

SP: Indeed, we use it to make sense of each other's behaviour. 'Why did John just get on a bus?' Well, it's not because of some big magnet that pulled him up. It's because he wanted to get somewhere and he knew the bus would take him there, which is why he won't get on a bus that has a slightly different number. That kind of explanation is indispensable to our common-sense social life. When we misapply it to questions like 'Why are there humans?', 'Why is there a planet Earth?', we're apt to be misled and have to have that tendency debugged.

RD: Steve, thank you very much. That was a fascinating conversation. Many thanks.

OLD BRAIN, NEW BRAIN

Jeff Hawkins is a highly successful innovator in the world of computer technology who turned his attention to the brain and how it works, and founded an institute to study it. His talk at the 2019 CSICon meeting in Las Vegas sponsored by the Center for Inquiry (CFI) was a sensation – not so much food for thought as a banquet for thought on steroids. I was honoured to be invited to write the foreword to his book *A Thousand Brains*.

Don't read this book at bedtime. Not that it's frightening. It won't give you nightmares. But it is so exhilarating, so stimulating, it'll turn your mind into a whirling maelstrom of excitingly provocative ideas – you'll want to rush out and tell someone rather than go to sleep. It is a victim of this maelstrom who writes the foreword, and I expect it'll show.

Charles Darwin was unusual among scientists in having the means to work outside universities and without government research grants. Jeff Hawkins might not relish being called the Silicon Valley equivalent of a gentleman scientist but – well, you get the parallel.

Darwin's powerful idea was too revolutionary to catch on when expressed in a brief article, and the Darwin–Wallace joint papers of 1858 were all but ignored. As Darwin himself said, the idea needed to be expressed at book length; and, sure enough, it was his great book that shook Victorian foundations a year later. Book-length treatment, too, is needed for Jeff Hawkins's 'Thousand Brains Theory'. And for his notion of 'frames of reference' – 'the very act of thinking is a form of movement' – bullseye! These two ideas are each profound enough to fill a book. But that's not all Jeff Hawkins gives us.

T. H. Huxley famously said, on closing *On the Origin of Species*, 'How extremely stupid of me not to have thought of that.' I'm not suggesting that brain scientists will necessarily say the same when they close this book. It is a book of many exciting ideas, rather than one huge idea like Darwin's. Still, I suspect that not just T. H. Huxley but his three brilliant grandsons would have loved it: Andrew because he discovered how the nerve impulse works (Hodgkin and Huxley are the Watson and Crick of the nervous system); Aldous because of his visionary and poetic voyages to the mind's furthest reaches; and Julian because he wrote this poem, extolling the brain's capacity to construct a model of reality, a microcosm of the universe:

The world of things entered your infant mind
    To populate that crystal cabinet,
    Within its walls the strangest partners met,
And things turned thoughts did propagate their kind.
For, once within, corporeal fact could find
    A spirit. Fact and you in mutual debt
    Built there your little microcosm—which yet
Had hugest tasks to its small self assigned.
Dead men can live there, and converse with stars:
    Equator speaks with pole, and night with day:
Spirit dissolves the world's material bars—
    A million isolations burn away.
The Universe can live and work and plan,
At last made God within the mind of man.

The brain sits in darkness, apprehending the outside world only through a hailstorm of Andrew Huxley's nerve impulses. A nerve impulse from the eye is no different from one from the ear or the big toe. It's where they end up in the brain that sorts them out. Jeff Hawkins is not the first scientist or philosopher to suggest that the reality we perceive is a constructed reality, a model, updated and informed by bulletins streaming in from the senses. But Hawkins is, I think, the first to articulate, eloquently and at length, the idea that there is not one such model – there are thousands, one in each of the many neatly stacked columns that constitute the brain's cortex. There are about 150,000 of these columns and they are the stars of the first section of the book, along with what he calls 'frames of reference'. Hawkins's thesis about both of these is provocative, and it'll be interesting to see how it is received by other brain scientists: well, I suspect. Not the least fascinating of his ideas here is that the cortical columns, in their world-modelling activities, work semi-autonomously. What 'we' perceive is a kind of democratic consensus from among them.

Democracy in the brain? Consensus, and even dispute? What an amazing idea. It is a major theme of the book. We human mammals are the victims of a recurrent dispute: a tussle between the old reptilian brain, which unconsciously runs the survival machine, and the mammalian neo-cortex sitting in a kind of driver's seat atop it. This new mammalian brain, the cerebral cortex, thinks. It is the seat of consciousness. It is aware of past, present and future, and it sends instructions to the old brain which executes them.

The old brain, schooled by natural selection over millions of years when sugar was scarce and valuable for survival, says: 'Cake. Want cake. Mmmm, cake. Gimme.' The new brain, schooled by books and doctors over mere tens of years in which sugar has been over-plentiful, says: 'No, no. Not cake. Mustn't. Please don't eat that cake.' Old brain says: 'Pain, pain, horrible pain, stop the pain *immediately*.' New brain says: 'No, no, bear the torture, don't betray your country by surrendering to it. Loyalty to country and comrades comes before even your own life.'

The conflict between the old reptilian and the new mammalian brain furnishes the answer to such riddles as 'Why does pain have to be so damn painful?' What, after all, is pain for? Pain is a proxy for death. It is a warning to the brain: 'Don't do that again: don't tease a snake, pick up a hot ember, jump from a great height. This time it only hurt, next time it might kill you.' But now a designing engineer might say, what we need here is the equivalent of a painless flag in the brain. When the flag shoots up, don't repeat whatever you just did. But instead of the engineer's easy and painless flag, what we actually get is pain, often excruciating, unbearable pain. Why? What's wrong with the sensible flag?

The answer probably lies in the disputatious nature of the brain's decision-making processes: the tussle between old brain and new brain. It being too easy for the new brain to over-rule the vote of the old brain, the painless flag system wouldn't work. Nor would torture.

The new brain would feel free to ignore my hypothetical flag and endure any number of bee stings or sprained ankles or torturers' thumbscrews if, for some reason, it 'wanted to'. The old brain, which really 'cares' about surviving to pass on the genes, might 'protest' in vain. Maybe natural selection, in the interests of survival, has ensured 'victory' for the old brain by making pain so damn painful that the new brain cannot overrule it. As another example, if the old brain were 'aware' of the betrayal of sex's Darwinian purpose, the act of donning a condom would be unbearably painful.

Hawkins is on the side of the majority of informed scientists and philosophers who will have no truck with dualism: there is no ghost in the machine, no spooky soul so detached from hardware that it survives the hardware's death, no Cartesian Theatre (Dan Dennett's term) where a colour screen displays a movie of the world to a watching self. Instead, Hawkins proposes multiple models of the world, constructed microcosms, informed and adjusted by the rain of nerve impulses pouring in from the senses. By the way, Hawkins doesn't totally rule out the long-term future possibility of escaping death by uploading your brain to a computer, but he doesn't think it would be much fun.

Among the more important of the brain's models are models of the body itself, coping, as they must, with how the body's own movement changes our perspective on the world outside the prison wall of the skull. And this is relevant to the major preoccupation of the middle section of the book, the intelligence of machines. Jeff Hawkins has great respect, as do I, for those smart people, friends of his and mine, who fear the approach of super-intelligent machines to supersede us, subjugate us or even dispose of us altogether. But Hawkins doesn't fear them, partly because the faculties that make for mastery of Chess or Go are not those that can cope with the complexity of the real world. Children who can't play chess 'know how liquids spill, balls roll, and dogs bark. They know how to use pencils, markers, paper, and glue. They know how to open books and that paper can rip.' And they have a self-image, a body image that emplaces them in the world of physical reality and allows them effortlessly to navigate through it.

It is not that Hawkins underestimates the power of artificial intelligence and the robots of the future. On the contrary. But he thinks most present-day research is going about its investigations the wrong way. The right way, in his view, is to understand how the brain works and borrow its ways but hugely speed them up.

And there is no reason to (indeed, please let's not) borrow the ways of the old brain, its lusts and hungers, cravings and angers, feelings and fears, which can drive us along paths seen as harmful by the new brain – harmful at least from the perspective that Hawkins (and I, and almost certainly you) hold. For he is very clear that our enlightened values must, and do, diverge sharply from the primary and primitive value of our selfish genes – the raw imperative to reproduce at all costs. Without an old brain, in his view (which I suspect may be controversial) there is no reason to expect an AI to harbour malevolent feelings towards us. By the same token, and also perhaps controversially, he doesn't think switching off a conscious AI would be murder: without an old brain, why would it feel fear or sadness? Why would it want to survive?

In the chapter 'Genes versus knowledge' we are left in no doubt about the disparity between the goals of the old brain (serving self- ish genes) and the new brain (knowledge). It is the glory of the human cerebral cortex that it – unique among all animals and unprecedented in all geological time – has the power to defy the dictates of the selfish genes. We can enjoy sex without procreation, we can devote our lives to philosophy, mathematics, poetry, astro- physics, music, geology, the warmth of human love, in defiance of the old brain's genetic urging that these are a waste of time – time that 'should' be spent fighting rivals and pursuing multiple sexual partners.

As I see it, we have a profound choice to make. It is a choice between favoring the old brain or favoring the new brain. More specifically, do we want our future to be driven by the processes that got us here, namely, natural selection, competition, and the drive of selfish genes? Or, do we want our future to be driven by intelligence and its desire to understand the world?

I began by quoting T. H. Huxley's endearingly humble remark on closing Darwin's *Origin*. I'll end with just one of Jeff Hawkins's many fascinating ideas – he wraps it up in a mere couple of pages – which had me echoing Huxley. Feeling the need for a cosmic tombstone, something to let the galaxy know that we were once here and capable of announcing the fact, Hawkins notes that all civiliza- tions are ephemeral. On the scale of universal time, the interval between a civilization's invention of electromagnetic communica- tion and its extinction is like the flash of a firefly. The chance of any one flash coinciding with another is unhappily small. What we need, then – it's why I called it a tombstone – is a message that says, not 'We are here' but 'We were once here'. And the tombstone must have cosmic-scale duration: not only must it be visible from parsecs away, it must last for millions if not billions of years so that it is still proclaiming its message when other flashes of intellect intercept it long after our extinction. Broadcasting prime numbers or the digits of π won't cut it. Not as a radio signal or a pulsed laser beam, anyway. They certainly proclaim biological intelligence, which is why they are the stock-in-trade of SETI (the Search for Extraterrestrial Intelligence) and science fiction, but they are too brief, too in-the-present. So, what signal would last long enough and be detectable from a very great distance in any direction? This is where Hawkins provoked my inner Huxley.

It's beyond us today, but in the future, before our firefly flash is spent, we could put into orbit around the Sun a series of satellites 'that block a bit of the Sun's light in a pattern that would not occur naturally. These orbiting Sun blockers would continue to orbit the Sun for millions of years, long after we are gone, and they could be detected from far away.' Even if the spacing of these umbral satellites is not literally a series of prime numbers, the message could be made unmistakable: 'Intelligent Life Woz 'Ere.'

What I find rather pleasing – and I offer the vignette to Jeff Hawkins to thank him for the pleasure his brilliant book has given me – is that a cosmic message coded in the form of a pattern of intervals between spikes (or in this case anti-spikes, as his satellites dim the sun) would be using the same kind of code as a neuron.

This is a book about how the brain works. It works the brain in a way that is nothing short of exhilarating.

BREAKING THE
SPECIES BARRIER

John Brockman is much more than a literary agent – though he is that and a highly successful one. He is a godfatherly impresario for science, going in to bat for scientific writers among editors and literary intellectuals. He hosts a kind of online salon (*The Edge*) where scientists, and other scholars interested in science, meet to exchange and debate ideas. An aspect of this is his annual question, sent out around Christmas time to his Edge circle. I've never fully understood why busy people are so willing to drop everything to answer these questions. It's not as though John exerts undue pressure. I think it may be that we all respect the other members of his address list so highly that we can't bear not to join in. Whatever the reason, I have seldom missed a year since the series started. This was my contribution to the 2009 symposium, *This Will Change Everything: ideas that will shape the future.*

Our ethics and our politics assume, largely without question or serious discussion, that the division between human and 'animal' is absolute. 'Pro-life', to take just one example, is a potent political badge, associated with a gamut of ethical issues such as opposition to abortion and euthanasia. What it really means is pro-*human*-life. Abortion clinic bombers are not known for their veganism, nor do Roman Catholics show any particular reluctance to have their suffering pets 'put to sleep'. In the minds of many confused people, a single-celled human zygote, which has no nerves and cannot suffer, is infinitely sacred, simply because it is 'human'. No other cells enjoy this exalted status.

But such 'essentialism' is deeply unevolutionary. If there were a heaven in which all the animals who ever lived could frolic, we would find an interbreeding continuum between every species and every other. For example, I could interbreed with a female who could interbreed with a male who could . . . fill in a few gaps, probably not very many in this case . . . who could interbreed with a chimpanzee. We could construct longer but still unbroken chains of interbreeding individuals to connect a human with a warthog, a kangaroo, a catfish. This is not a matter of speculative conjecture; it necessarily follows from the fact of evolution.

Theoretically, we understand this. But what would change everything is a practical demonstration, such as one of the following:

1   The discovery of relict populations of extinct hominins, such as *Homo erectus* or *Australopithecus*. Yeti enthusiasts notwithstanding, I don't think this is going to happen. The world is now too well explored for us to have overlooked a large, savannah-dwelling primate. Even *Homo floresiensis* has been extinct for seventeen thousand years.* But if it did happen, it would change everything.

2   A successful hybridization between a human and a chimpanzee. Even if the hybrid were infertile, like a mule, the shock waves that would be sent through society would be salutary. This is why a distinguished biologist described this possibility as the most immoral scientific experiment he could imagine:† It would change everything! It cannot be ruled out as impossible, but it would be surprising.

---

* And, according to some plausible experts, never existed at all as a separate species anyway.
† He can't have let his imagination roam very far. More immoral than

3   An experimental chimera in an embryology lab, consisting of approximately equal numbers of human and chimpanzee cells. Chimeras of human and mouse cells are now constructed in the laboratory as a matter of course, but they don't survive to term. Another example of our speciesist ethics is the fuss now made about mouse embryos containing some proportion of human cells: 'How human must a chimera be before more stringent research rules should kick in?' So far, the question is merely theological, since the chimeras don't come anywhere near being born, and there is nothing resembling a human brain. But, to venture down the slippery slope so beloved of ethicists, what if we were to fashion a chimera of 50 per cent human and 50 per cent chimpanzee cells and grow it to adulthood? That would change everything. Maybe it will.

4   The human genome and the chimpanzee genome are now known in full. Intermediate genomes of varying proportions can be interpolated on paper. Moving from paper to flesh and blood would require embryological technologies that will probably come onstream during the lifetime of some of my readers. I think it will be done, and an approximate reconstruction of the common ancestor of ourselves and chimpanzees will be brought to life. The intermediate genome between this reconstituted 'ancestor' and modern humans would, if implanted in an embryo, grow into something like a reborn *Australopithecus*: Lucy the Second. And that would (dare I say 'will'?) change everything.

Dr Mengele's experiments? Why actually, when you bother to think about it, is it immoral at all? The only reason I can think of is that the unfortunate creature would feel lonely and out of place in the world, and would be the victim of prurient curiosity, like 'Mr Savage' in Aldous Huxley's *Brave New World* but more so.

I have laid out four possibilities that would, if realized, change everything. I have not said that I hope any of them will be realized. That would require further thought. But I will admit to a *frisson* of enjoyment whenever we are forced to question the hitherto unquestioned.

BRANCHING OUT

This review of *Extinct Humans* by Ian Tattersall and Jeffrey H. Schwartz appeared in the *New York Times* in 2000.

This undeniably informative and beautifully illustrated book has pretensions to be more. It aspires to be iconoclastic and revolutionary – what the publisher calls a 'radical reinterpretation'. The straw man that the authors enthusiastically demolish, and repeatedly attribute to the ornithologist Ernst Mayr and the geneticist Theodosius Dobzhansky, is the idea that fossil hominids are arranged in an evolutionary order, each one evolving into the next until the penultimate one finally turns into us. It would follow from this view – if anybody held it in the extreme form energetically savaged by Ian Tattersall and Jeffrey H. Schwartz in *Extinct Humans* – that you need only know how old a fossil is to classify it. If it's a million years old it must be *Homo erectus*. Their own view is that there have been more than fifteen species of humans, many of which coexisted with one another, and all of which are now extinct except *Homo sapiens*.

The longing to be revolutionary is something of an occupational disease among paleontologists. Perhaps it stems from an unwarranted fear that their subject might be thought pedestrian and atheoretical. In Tattersall and Schwartz's own words, 'the Synthesis effectively left paleontology without a theoretical framework because paleontologists were relegated to the essentially clerical task of sorting out the mundane historical details of what had evolved into what'.

The reference is to the neo-Darwinian Modern Synthesis, that triumphant amalgam of Darwinism with Mendelian genetics, achieved in the 1930s and 1940s by, among others, Mayr and Dobzhansky, and this may account for the book's tiresomely rebarbative nose-thumbing. 'But what else would you expect given the stranglehold that Mayr and Dobzhansky's pronouncements of a single-evolving human lineage had on paleoanthropology, wherein the only players are *Australopithecus, Homo habilis, Homo erectus* and *Homo sapiens*?' At times sneering spills over almost into paranoia:

> But how could anyone, much less a bunch of paleoanthropologists who were not equipped with the supposedly more biologically informed backgrounds of Mayr and Dobzhansky, disagree with them? No one could. Or, at least, no one would who didn't want to be accused of being anti-evolutionary – or, much worse, anti-Darwinian.

But that's enough about the tone of the argumentative parts of this book. What of their substance? What is actually at stake when claims are made about numbers of extinct species of hominid? Where fossils are concerned, rather little.

Whether they belonged to fifteen species or one, individual males mated with individual females and had individual children. Some local populations were far from other populations, perhaps separated by rivers or mountains, or even just by cultural barriers, and so did not interbreed. There was a continuum of variation among local populations scattered around Africa, and later around the world. Names like 'race', 'subspecies', 'species' and 'genus' signify, in increasing degrees, the genetic separation that tends to arise when populations do not interbreed for a while.

With one exception, such names are arbitrary, in the same way as 'tall' or 'fat'. Little therefore rides on statements like 'Fifteen species of hominid have gone extinct.' We find it handy to say 'John Cleese is a tall man; Mickey Rooney is a short man.' But it is not clever to get embroiled in passionate arguments over how many categories of tallness or shortness deserve a name (giant, dwarf, average, etc.). Similarly, it follows from evolution that if all the hominids who ever lived were available to us in a gigantic fossil museum, all attempts to segregate them into non-overlapping species or genera would be futile. No matter how richly branched the evolutionary tree, every fossil would be connected to every other by unbroken chains of potential intermarriage. The only reason we can indulge our penchant for discontinuous names at all is that we are mercifully spared sight of the extinct intermediates. In the case of living animals, we see only the tips of the evolutionary twigs. For paleontologists, the mercy* is that so few individuals fossilize. We may believe that the genus *Homo* is descended from the genus *Australopithecus*. But it is ludicrous to suggest that there must once have been a *Homo* child at the breast of an *Australopithecus* mother. It necessarily follows from the fact of evolution that discontinuous naming must ultimately break down.

The species is the one exception to the rule that all taxonomic names are subjective. Following Ernst Mayr himself, most biologists recognize an objective criterion for deciding whether two animals belong in the same species. Do they interbreed? Not 'Can they interbreed in a zoo or with the help of artificial insemination?' but 'Do they interbreed under natural conditions?' Even in the case of surviving animals, it is not an easy criterion to apply. It is frequently infeasible for taxonomists to test interfertility, and they must judge whether the anatomical and physiological differences between two animals are such as we would normally expect to

---

* This paradox applies, of course, only to paleontologists who feel forced to slap a Linnaean binomial name on every fossil in the museum drawer. Paleontologists with a true interest in biology and evolution cannot regard it as a mercy. For them, the more fossils the merrier, including unbroken series of intermediates.

accompany inability to interbreed. Where we are dealing with fossils, such subjective judgements are inevitable.

It follows that arguments over whether there have been fifteen species of extinct hominid or some other number are not worth the fuss. They amount to no more than the traditional taxonomic disputes between 'lumpers' and 'splitters'. Lumpers say that what they call *Homo erectus* lived in both Africa and Asia. Splitters say the African specimens belong to a different species, *Homo ergaster*, if not (as the authors of this book suspect) more than one. Implicitly, splitters are committing themselves to the belief that Asian specimens would not have interbred with African specimens if given the opportunity, the lumpers to the opposite belief. It is a matter of judgement. Maybe the splitters' judgement is sounder than the lumpers', but judgement is all it is.

'Some paleoanthropologists were – bizarrely – heard muttering that there was not enough "morphological space" between gracile australopiths and *H. erectus* to admit a third species.' Why is that bizarre? What else could one mutter (if we must use such a loaded word) in reaching a judgement of this kind? What else are Tattersall, a curator at the American Museum of Natural History, and Schwartz, an anthropologist at the University of Pittsburgh, doing than making the opposite judgement, that there is more than enough 'morphological space' separating the specimens?

Is this to say that taxonomic disagreements always lack real substance? Certainly not. A group of molecular taxonomists has recently claimed that hippopotamuses are closer cousins to whales than they are to pigs. This is an astounding assertion. A world in which it was true would be meaningfully different from the world in which zoologists had hitherto thought they were living. The same cannot be said of a world in which there were fifteen species of extinct hominid as opposed to five. Here, the difference would lie not in the real world but in the predilections of taxonomists. But if the molecular claim about hippos and whales is upheld, it will be as though we had suddenly discovered that humans are closer cousins to bushbabies than they are to chimpanzees. Now that really would be revolutionary. It would mean a difference in the real world.

The same is true of a recent claim about mitochondria, by the great John Maynard Smith, no less. Mitochondria are tiny organelles that swarm inside our cells, originally descended from free-living bacteria but now an indispensable component of our metabolic machinery. Like bacteria, mitochondria have their own DNA, which passes down the generations as a separate stream, parallel to the main river of our nuclear DNA. The orthodox view, espoused by Tattersall and Schwartz (and by me in everything I have hitherto written on the subject), is that mitochondrial DNA is inherited purely in the maternal line, and partakes of no sexual recombination. This makes it exceptionally useful for taxonomists, and it underlies the famous 'African Eve' theory, well described in this book. But now Maynard Smith and his colleagues have made the radical claim that, contrary to all previous belief, mitochondrial DNA, like nuclear DNA, does recombine sexually. This is a substantial claim about the real world. If true, it will have far-reaching implications. For instance, it will markedly change estimates of the likely age of 'African Eve'.

I learned a lot about fossils from this book, and shall continually go back to it with profit. But a radical reinterpretation it is not. Paleontologists should shake the chip off their collective shoulder and recognize that their subject is too intrinsically fascinating to need any boost from gratuitous iconoclasm.

DARWINISM AND
HUMAN PURPOSE

This is a slightly abbreviated version of my contribution to John R. Durant's 1989 collection of essays on *Human Origins*.

There is an ambiguity in the way we use the language of purpose. When we say 'the purpose of an aeroplane's tail is to stabilize the plane', we are saying something about the intention of the designer. If we look at a bird, it is evident that its tail does much the same thing. If the bird didn't have a tail it would pitch and roll like an aeroplane without a tail. It is natural, therefore, to use the same kind of language: the *purpose* of a bird's tail is to stabilize it in flight; the purpose of a hedgehog's spines is to protect it; the purpose of a rabbit's fur is to keep it warm; and so on. But as we have seen, everything about animals and plants that looks as if it had been designed for a purpose has in fact been shaped by the slow sculpting of natural selection. I shall call this kind of purposiveness 'Purpose Type 1' or 'survival value'. In a way, Purpose Type 1 is not really purpose at all, for there is no need to postulate a designer. Birds have efficient and correctly shaped tails simply because they are built by genes that have come down through generations of successful birds. By definition the *ancestors* of a modern bird were successful: unsuccessful birds left no descendants. A bird was unlikely to be an ancestor unless it had a correctly shaped tail. Therefore modern descendant birds contain the genes that happened to have the effect of building correctly shaped tails. That is why tails, and all other attributes of animals and plants, look as if they had been designed by a clever mind with a purpose in view. But the purpose is pseudopurpose – 'Purpose Type 1'.

Aeroplane tails really *have* been designed by a clever mind with a purpose in view. I call this type of purpose 'Purpose Type 2'. Purpose Type 2 is the kind of purpose we are familiar with from our own designs, schemes and goals. When we say that the purpose of an aeroplane's tail is to stabilize the plane, it is Purpose Type 2 that we are talking about. We are talking about a goal in the mind of a designer.

My thesis is that Purpose Type 2 is, in fact, an evolved adaptation with a survival value (or Purpose Type 1), in the same sense as a feather, an eye or a backbone has a survival value. Brains have evolved with various capacities that assist the survival of the genes that made them. The brain can be seen as a kind of 'on-board computer', which is used to control the body's behaviour in ways that are beneficial to the genes that built it. Among the useful capacities of the brain are the ability to perceive aspects of the outside world; to remember things; to learn the consequences of actions – which ones have good results and which ones have bad results; the ability to set up simulated models in imagination; and – here is the point of the argument – the ability to set up purposes or goals in the sense of Purpose Type 2. The capacity to have a mental goal or purpose (Type 2) is an adaptation with a survival value or purpose (Type 1), in just the same sense as the capacity to run fast, or the capacity to see clearly.

I said that the brain was an on-board computer. This doesn't mean that it works in exactly the same way as a man-made electronic computer. It certainly doesn't. But the brain does the job of an on-board computer, and some of the principles and techniques of computer science apply in brains. Now, why should it be useful for an on-board computer to set up goals, to have purposes (Type 2)? Do man-made electronic machines in fact have purposes (Type 2)?

Yes, they certainly do. This doesn't mean that they are conscious. It is still reasonable and useful to talk of a machine having a goal, even if it is not conscious. Think of a guided missile tracking a moving target like an aeroplane. The missile is controlled by its own on-board computer, which detects the position of the target by radar, by heat, or by some other equivalent of sense organs. The discrepancy between the present positions of plane and missile is measured, and the motors and steering surfaces of the missile are manipulated by the computer in such a way as to reduce the discrepancy. If the target plane takes evasive action, spiralling and twisting and turning, a good missile automatically takes countermeasures. It shows flexible, versatile behaviour to close the gap between itself and the plane. The missile behaves as if its computer contains a mental picture of its target, a Purpose Type 2.

Cannon-balls didn't have this property. They were simply lobbed by the cannon in the direction of the target. Once on their way, they didn't home in on the target; they didn't track the evasive twists and turns of the target. A cannon-ball has no on-board computer, and no Purpose Type 2. It was, of course, designed with a purpose in mind – in fact, much the same purpose as motivates the designer of the guided missile. In their *design* purposes, the two kinds of projectile don't differ much. Where they differ is in how they work. The cannon-ball is just a lump of iron. The guided missile has its own computer on board, and its computer contains within itself a Purpose Type 2. It behaves as if carrying a mental picture of its own goal around with it.

Now, just as human designers have found it expedient to build into their weapons an on-board computer with its own Purpose Type 2, so natural selection has built into some living organisms the same facility. Just as a guided missile is a more effective weapon than a cannon-ball, so an animal with a brain and flexible goal-seeking behaviour is a more effective predator, say, than an animal without a brain, or with a stereotyped and inflexible brain.

Some living things manage without computers at all. Plants don't move, and don't have brains. But most animals move around, and move around in sophisticated ways more reminiscent of guided missiles than of cannon-balls. Many of them are rather simple kinds of guided missiles. Maggots follow a delightfully simple rule in guiding themselves away from light. The maggot swings its head from side to side, while its computer compares the light intensity on the two sides, and instructs the muscles to move the maggot in such a way as to equalize them. Experimenters showed this by switching on a light whenever the maggot turned to the left, and switching it off every time the maggot turned to the right. This caused the maggot to circle to the right indefinitely. In nature, of course, light does not turn itself on and off in such an annoying manner, and the behavioural rule works as an effective guidance system for moving towards darkness. Let's not be snobbish about it either. There is some evidence that new-born human babies use much the same side-to-side swinging technique to find the breast.

Animals employ a range of increasingly sophisticated guidance systems paralleling the techniques developed by human engineers. Dragonflies hunting smaller insects dive and swoop, twist and turn, with all the flexibility of a man-made guided missile. They use their large eyes to detect the position of the moving target; they use their brain to compute the necessary movements of their steering surfaces; and they frequently intercept and catch the target. A sensible way to interpret their behaviour is to say that their brain is set up as if it had a goal or purpose (Type 2). I suspect that the kinds of computation that go on inside a dragonfly's head when it tracks a gnat are probably rather similar to the computations that go on in the guided missile as it tracks a fighter plane. The same is probably true of the parallel between the sonar systems in bats, whales and man-made submarines.

I don't know whether dragonflies and bats are conscious of their prey, or whether they are just wired up like automatic guided missiles, which are, after all, very effective. My suspicion is that dragonflies are probably not conscious, but bats may be, and whales almost certainly are. I know that I myself am conscious of my goals, and I presume that other people are too. I suggest that conscious goal-seeking is the latest advance in the cybernetic technology of nature – an advance, perhaps, over the dragonfly which is about as great as the advance of the rapidly wheeling and turning dragonfly over the maggot swinging its head alternately left and right, and blundering in vaguely the right direction.

Now, one of the main virtues of an advanced goal-seeking machine is its flexibility. It's easy to reprogram it to seek a different goal. A captured enemy missile may be programmed to seek out and destroy its original creators. The very property that makes the missile so effective in achieving its goal – its flexibility and versatility – that very property makes the machinery easy to subvert to a new purpose.

This brings me back to my original problem. Why is it that humans appear to seek goals that have nothing to do with the survival and propagation of their own genes? Why do we set up goals like making money, composing a brilliant cantata, winning a war, or an election, or a game of chess or tennis? Why aren't all our goals related to the one central goal of propagating our genes?

The answer I am giving is this. It is our *capacity to set up goals*, and to reprogram our goal-seeking machinery rapidly and flexibly, that has been built into us by natural selection. This goal-seeking capacity, with its inherent properties of flexibility and reprogrammability, is an immensely useful piece of brain technology. Useful, that is, in propagating genes. That is why it evolved in the first place. But by its very nature it carries the seeds of its own subversion. Precisely because of its flexible reprogrammability, it is highly prone to seeking new goals.

But there is a paradox in this virtue of flexible reprogrammability. If a machine is *too* ready to change its goals it will never achieve any of them. What is required is some mixture of flexibility in setting up new goals, coupled with tenacity and inflexibility in pursuing them. Our brains are flexible enough to be reprogrammed away from goals that are directly concerned with gene survival, and towards adopting a new and arbitrary global purpose, perhaps one inspired by religion, by patriotism, by 'sense of duty', by 'loyalty to the Party'. But they are inflexible enough, once repro-grammed, to spend an entire lifetime seeking the new global goal – and, moreover (yet another paradox) to show great versatility and flexibility in the setting up of new *sub*goals in the service of the inflexibly pursued main goal. This subtle interplay between flexibility and inflexibility is something that we should work hard to understand, for it has vitally important consequences.

To return to my main theme, what natural selection has built into us is the *capacity* to seek, the capacity to strive, the capacity to set up short-term goals in the service of longer-term goals, eventually the capacity for foresight. When natural selection originally built up these capacities, the shorter- and longer-term goals were always in the service of the ultimate long-term goal of gene survival. But it was in the nature of flexible goal-seeking that this original ultimate goal was itself capable of being subverted. From the selfish genes' point of view, their survival machines became too clever by half. An excellent innovation in nervous-system technology, flexible repro-grammability, overreached itself as far as its original Darwinian purpose was concerned.

From our point of view, as the flexible computers involved, our shaking off of thraldom to our original purpose of propagating the selfish genes can be seen as an exhilarating liberation, as exhilarating as Wordsworth found the French Revolution: 'Bliss was it in that dawn to be alive: but to be young was very heaven.' I suspect that our species is indeed still young in its new-found liberation. Although the human brain has been capable of great flexibility for a long time, the takeover by the on-board computers probably ran away with itself in a big way when the rise of language enabled large groups of people to set up *shared* goals, which could be pursued over more than one lifetime. *One* inventor may set himself the task of improv-ing methods of transport, and produce the wheel. *Generations* of inventors, each building on the accumulated achievements of their predecessors who shared the same goal, are capable of producing the supersonic airliner and the space shuttle. This is a new kind of evolution, superficially similar to the old, and producing advances in technology which mirror the old genetic advances, but at a rate which may be a million times faster. The speed of this new kind of evolution, coupled with the ease with which the human brain can be reprogrammed to adopt a new major goal, and the single-minded tenacity with which it can pursue that goal once adopted, are frightening, for they could promise great danger. It's all too easy for rival groups of humans to adopt incompatible goals – for example, patriotic or sectarian claims over disputed territory. Like advanced guided missiles, we are apt to pursue those goals with relentless tenacity and great flexibility in setting up efficient subgoals, the subgoals of war. Finally, the extreme rapidity of cultural evolution, driven by the cumulative pursuit of shared technical goals, makes possible the deployment of devastating technical weapons. We must hope that our species' 'blissful dawn' will not turn as sour as the French Revolution did for Wordsworth. There are *some* grounds for hope. That same flexibility, versatility and foresight, which threaten us by throwing our stately Darwinian evolution into runaway overdrive, could also be our salvation.

WORLDS IN MICROCOSM

In 1888 Adam Lord Gifford, eminent Scottish advocate and judge with an interest in religion and philosophy, endowed an annual series of lectures in Natural Theology at the four great Scottish universities, St Andrews, Edinburgh, Glasgow and Aberdeen. In 1992 I was invited to give one of the Centenary Year Gifford Lectures, in Glasgow, and I felt honoured to accept. The title of the series, published in a volume edited by Neil Spurway in 1993, was *Humanity, Environment and God*. For an atheist lecturer the experience was not as Daniel-in-the-Lion's-Den-ish as it sounds. Modern theologians tend to be pretty science-friendly, and this was borne out by the other centenary lecturers, who included John Habgood, the then Archbishop of York and himself a scientist, the liberal theologian Don Cupitt, the former Jesuit monk Anthony Kenny and the religious physicist John Barrow. I enjoyed the experience.

A living organism is a model of the world in which it lives. That may seem a strange statement. My justification will come later.

The sense in which I am using the word 'model', however, is not strange. It is the ordinary scientific usage: a model resembles the real thing in some important respects, whether or not it looks, to the human eye, like a replica of the real thing. A child's train set is a model, but so also is a railway timetable. I first learned about models in the scientific sense from my grandfather. In the pioneering days of radio, his job was to lecture to young engineers joining

Marconi's company. To illustrate the fact, important in both radio and acoustics, that any complex wave-form can be broken down into summed simple waves of different frequencies, he took wheels of different diameters and attached them with pistons to a clothes-line. When the wheels went round, the clothes-line was jerked up and down, so that waves of movement snaked along it. The wriggling clothes-line was a *model* of a radio wave, giving the students a more vivid picture of wave-summation than the mathematical equations could ever have done.

Today, my grandfather would have used a computer screen rather than a clothes-line. Well, on second thoughts, perhaps he wouldn't. Though he lived on into the age of computers, he was never able to appreciate their beauty and he died in the mistaken belief that they were nothing but crutches for lazy calculators. I should have explained to him that they are really just like his clothes-line. It is, indeed, a significant fact that the clothes-line model, like many other models in mathematics and engineering, is a model of many things simultaneously: not only radio waves and waves in an analogue or digital computer, but sound waves and tidal waves too. This is because the mathematical equations describing the behaviour of all these waves are fundamentally similar. Any of these physical systems can be seen as a 'model' of any other, and mathematicians would use the word 'model' for the set of equations, too.

Grandfather's clothes-line was a teaching aid, but engineers can put models to more practical purposes, to solve problems that defy calculation. Long before a new aeroplane is actually built, models that replicate only its outer shape are exhaustively tested in wind tunnels. Although mathematicians could in theory calculate the turbulence patterns whipped up by any particular shape, these patterns are so intricate and their mathematics so difficult that it is almost infinitely (for once, this is no exaggeration) quicker to put a model of the plane through ordeal by wind tunnel. Once again, by the way, a computer model can stand in for what happens in the wind tunnel – but it must be a very powerful computer.

More complicated turbulence patterns by far come from the winds and ocean currents that eddy about the spinning Earth and bring us our weather. In a perfect world a mathematician, given the present wind directions, wind speeds, temperatures, and rates of change of these and similar measured quantities all round the world, might hope to calculate an infallible forecast of next week's weather – indeed, next century's weather. In practice the very idea of anyone performing such a vast calculation is a joke, and even moderately accurate forecasting is possible for only a few days ahead. The problem is even worse if, as is increasingly believed, weather patterns are 'chaotically' deterministic. Modern weather forecasters in fact make use of a greatly simplified, though still very elaborate, computer *model* of the Earth's weather. This model is not a visible replica like a model train (although it could have a visible 'readout' on the screen – perhaps a simulated picture of what a satellite might 'see', looking down from a great height on the model world). It is a dynamic model, continually updated as new information flows in from weather stations, ships, planes, balloons and satellites all round the world, and calculations are continually being performed on the updated information. To make predictions, forecasters allow the model to free-run into the 'future'.

As a biologist I am endlessly fascinated by the arcade games of this computer age. In one game the player sits in what appears to be the driving seat of a racing car, holding a steering wheel in one hand and a gear lever in the other. On the screen ahead of him (observation suggests that it is seldom her) is a moving coloured image that simulates the road ahead. Trees and other objects appear small in the distance, then rapidly grow as the car approaches and flashes past them. Rival cars loom up on the road ahead and can be overtaken, or crashed into, depending on the skill of the player. Engine noise from a loudspeaker changes in tune with the manoeuvres of the driver. Though the resemblance to the real thing is in truth quite crude, the impact on the senses is surprisingly lifelike. Yet all that is 'really' happening is that cells in the memory of the computer behind the screen are changing their state from (for example) 3 volts to o volts and back again at high speed.

The programmer – the original creator of the game – has had to create a make-believe world that embodies a good dose of reality. He has had to decide where, in his world, to place each tree and hill. In a sense, there is a sort of racing circuit in the computer, with its own geography and landmarks that bear a fixed 'spatial' relationship to one another. The model car, in a mathematical sense, 'moves' through the landscape, all the time obeying the programmed-in laws of 'reality'. If the programmer wanted to he could, of course, make his imaginary cars violate the normal laws of reality – a car might suddenly split into two cars, or turn into a horse – but this would make the game less commercially appealing.

Remember the racing car in the arcade, for we shall see that it has something to tell us about how our own brains perceive reality. But I began by saying, not that an animal's brain contains a simulated model of its world (though it does, and this is a point I shall return to), but that an animal *is* a model of its world. What is the sense of such a statement? One way to approach it is to realize that a good zoologist, presented with an animal and allowed to examine and dissect its body in sufficient detail, should be able to reconstruct almost everything about the world in which the animal lived. To be more precise, she would be reconstructing the worlds in which the animal's *ancestors* lived. That claim, of course, rests upon the Darwinian assumption that animal bodies are largely shaped by natural selection. If Darwin's theory is correct, the animal is the inheritor of the attributes that enabled its ancestors to be ancestors. If they hadn't had those successful attributes they would have been not ancestors but the childless rivals of ancestors.

So, what *are* the attributes that make for success as an ancestor, the attributes that we should expect to find in the body of our animal when we inspect it? The answer is anything that helps the individual animal to survive and reproduce *in its own environment*. If the species happens to live in a desert, individuals will have inherited whatever it takes to survive in arid heat. If the species happens to live in a rainforest, they will have inherited whatever it takes to survive in cloistered humidity. Not just one or two attributes in each case, but hundreds, thousands of them. This is why, if you present an animal's body, even a new species previously unknown to science, to a knowledgeable zoologist, she should be able to 'read' its body and tell you what kind of environment it inhabited: desert, rainforest, arctic tundra, temperate woodland or coral reef. She should be able to tell you, by reading its teeth and its guts, what it fed on. Flat, millstone teeth indicate that it was a herbivore; sharp, shearing teeth that it was a carnivore. Long intestines with complicated blind alleys indicate that it was a herbivore; short, simple guts suggest a carnivore. By reading the animal's feet, and its eyes and other sense organs, the zoologist should be able to tell how it found its food. By reading its stripes or flashes, its horns, antlers or crests, she should be able to tell something about its social and sex life.

But zoological science has a long way to go. By 'reading' the body of a newly discovered species, we could at present come up with only a rough verdict about its probable habitat and way of life – 'rough' in the same way as a pre-computer weather forecast was rough. The zoology of the future will put into the computer many more measurements of the anatomy and chemistry of the animal being 'read'. More importantly, it will not take the teeth, guts and chemistry of the stomach separately. It will perfect techniques of combining sources of information and analysing their interactions, resulting in inferences of enormous power. The computer, incorporating everything that is known about the body of the strange animal, will construct a model of the animal's world, to rival any model of the Earth's weather. This, it seems to me, is tantamount to saying that the animal, any animal, *is* a model of its own world, or the world of its ancestors. Hence my opening sentence.

In a few cases, an animal's body is a model of its world in the literal sense of a doll or a toy train. A stick insect lives in a world of twigs, and its body is a precise replica of a twig. A fawn's pelage is a model of the dappled pattern of sunlight filtered through trees onto the woodland floor. A peppered moth is a model of lichen on the tree bark that is its world when at rest. But models, as we have seen, do not stop at replicas, and if an animal's skin or plumage literally resembles features in its world, this is just the tip of the iceberg. Any animal is a detailed model of its world, whether camouflaged to mimic its background or not.

For the next stage in the argument, we need to make use of the distinction between static and dynamic models. A railway timetable is a static model, while the weather model in the computer is dynamic: it is continually (in advanced systems, continuously) being updated by new readings from around the world. Some aspects of an animal's body are a static model of its world – that millstone slab of a horse's tooth, for instance. Other aspects are dynamic; they change. Sometimes the change is slow. A Dartmoor pony grows a shaggy coat in winter and sheds it in summer. The zoologist presented with a pony's pelt can 'read', not only the kind of place it inhabited, but also the season of the year in which it was caught. Many animals of high northern latitudes, like Arctic foxes, snowshoe hares and ptarmigans, are white in winter and brownish in summer.

In these cases, then, the model of the world that is an animal is dynamic on a slow timescale, a timescale of weeks or months. But animals are dynamic on a much faster timescale as well, a timescale of seconds and fractions of seconds. This is the timescale of behaviour. Behaviour can be seen as high-speed dynamic modelling of the environment. Think of a herring gull adroitly riding a sea cliff's upcurrents. It may not be flapping its wings, but this doesn't mean that its wing muscles are idle. They and the tail muscles are constantly making tiny adjustments, sensitively fine-tuning the bird's flight surfaces to every nuance, every eddy of the air around it. If we fed information about the state of all these muscles into a computer, from moment to moment, the computer could in principle reconstruct every detail of the air currents through which the bird is gliding. It would assume that the bird was well designed to glide, and on that assumption construct a model of the air around the bird. Again, it would be a model in the same sense as the weather forecaster's. Both are continuously revised by new data. Both can be extrapolated to predict the future. The weather model predicts tomorrow's weather; the gull model could 'advise' the bird on the anticipatory adjustments that it should make to its wing and tail muscles, in order to glide on into the next second.

The point we are working towards, of course, is that although no human programmer has yet constructed a computer model to advise gulls on how to adjust their wing and tail muscles, just such a 'computer' model is almost certainly being run continuously in the brain of my gull and of every other bird in flight. Similar models, preprogrammed in outline by genes and past experience, and continuously updated by new sense data from millisecond to millisecond, are running inside the skull of every swimming fish, every galloping horse, every echo-ranging bat.

The Cambridge physiologist Horace Barlow long ago developed an intriguing view of sensory physiology which fits very well with the point I am making. We shall need a digression in order to understand his idea, and its relevance to my theme. Barlow began by pointing out what a formidable problem sensory recognition systems face. They have to respond to a subset of all possible stimulus patterns, while at the same time not responding to the rest. Think of the problem of recognizing a particular person's face: by convention it is assumed to be the face of the distinguished neurophysiologist J. Lettvin's grandmother. Lettvin's recognition mechanism must respond when the image of her face, but not any other image, falls on his retina. It would be easy if we could assume that the face would always fall exactly on a particular part of the retina. There could be a keyhole arrangement, with a grandmother-shaped region of cells on the retina wired up to a grandmother-detecting cell in the central nervous system. Other cells – members of the 'anti-keyhole' – would have to be wired up in inhibitory fashion, otherwise the central nervous cell would respond to a white sheet just as strongly as to Lettvin's grandmother.

But the keyhole arrangement is not feasible. Even if Lettvin needed to recognize nothing but his grandmother, how could he cope with her image falling on different parts of the retina, changing size and shape as she approaches or recedes, as she turns sideways, and so on? If we add up all possible combinations of keyholes and anti-keyholes, the number enters the astronomical range. When you realize that Lettvin can recognize not only his grandmother's face but hundreds of other faces, all the other bits of his grandmother and of other people, all the letters of the alphabet, all the thousands of objects to which a normal person can instantly give a name, the combinatorial explosion gets completely out of control. The psychologist F. Attneave, who independently arrived at the same general idea as Barlow, dramatized the point by the following calculation. If there was just one central nervous cell to cope with each keyhole combination, the volume of the brain would have to be measured in cubic light years.

How, then, with a brain capacity measured only in hundreds of cubic centimetres, do we do it? Barlow's answer is that we exploit the massive redundancy in all sensory information, reducing it with redundancy-detecting circuits (arranged in a hierarchical cascade). These detectors are not scanning for particular objects like Lettvin's grandmother. Instead, they are scanning for statistical redundancy.

Redundancy is information-theory jargon. It refers to messages or parts of messages that are not informative because the receiver already knows the information. Newspapers do not carry headlines saying 'The sun rose this morning'. But if a morning suddenly came when the sun did not rise, headline writers, if any survived, would note the incident. Technically, a message contains redundancy to the extent that, out of the repertoire of possible signals, some of them occur more frequently than others.

Sensory information is full of redundancy. The easiest kind to understand is temporal redundancy. The state of the world at time t is usually not greatly different from the state of the world at time $t - 1$.

The temperature, for instance, changes, but it usually changes slowly. A sense organ that signalled 'It is hot' in any one second would be wasting its time if it gave exactly the same signal in the next second. The central nervous system can usually assume that if it is hot in any one second it will be hot until further notice.

'Further notice' gives the clue to what a well-designed sensory system should do. It should not signal 'It is hot, it is hot, it is hot, it is hot . . . '. It should signal only when there is a change in temperature. The central nervous system can then assume that the status quo is maintained until the next change is signalled. It has long been known to physiologists that most sense organs do exactly that. The phenomenon is called sensory adaptation.

Figure 1 shows a particular example, the dying away of nerve impulses from a single sensory hair of a fly in response to beer. When it first detects the beer the nerve fires rapidly. The beer remains but the rate of firing decreases back to a steady, low, 'no news' value.

Figure 1

There is analogous redundancy in the spatial domain. The top picture in figure 2 (overleaf) is the scene we are actually looking at. If we represent spots on the retina where light falls with '+', and spots where there is darkness with '-', the middle picture shows that there is massive redundancy: most pluses are next to other pluses, minuses next to minuses. In the bottom picture the redundancy has been eliminated, leaving pluses and minuses only round the edges. It is only the edges that need to be signalled. The redundant interiors can be filled in by the central nervous system. This can be done by the mechanism known as lateral inhibition. If every cell in a bank of photocells inhibits its immediate neighbours, maximal

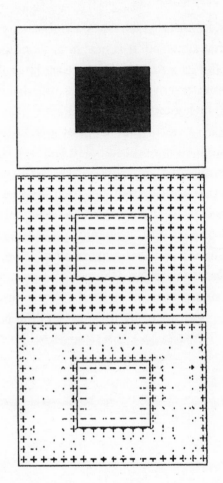

Figure 2

firing comes from cells that lie along edges, for they are inhibited from one side only. Lateral inhibition of this kind is common in both vertebrate and invertebrate eyes.

The straightness of many lines constitutes another kind of redundancy. If the ends of the line are specified, the middle can be filled in. Among the most important neurons that have been discovered in the visual cortex of mammals are the so-called line-detectors, neurons that pick out lines or edges orientated in particular directions. Each cortical line-detector cell has its own preferred direction.

From Barlow's point of view, what is going on here is this. If it were not for the line-detector, all the cells along a straight edge would fire. The nervous system economizes by using a single cell to signal that edge.

Movement over the spatial field is information-rich, in the same kind of way as change over time is information-rich. Just as in the temperature example, a static visual field does not need continuous reporting. But if a part of the visual field suddenly detaches itself from the rest and starts to crawl over the background as a small, black fly, it is news, and should be signalled. Visual physiologists have indeed discovered, again and again, neurons that are silent until something moves in their visual field. They don't respond when the entire field moves – that would correspond to the sort of movement the animal would see when it, itself, moved. They only respond when something moves relative to the rest of the visual field. The most famous of these relative movement detectors are the so-called 'bug-detectors' of Lettvin and Maturana in the retina of the frog.

As for the apparent movements of the world that result when an animal walks around or swivels its eyes, these are of low information content – high redundancy – and should therefore be filtered out before they reach the brain. And indeed, as you walk around, although your retinal image moves your perception is of a steady world. Poke yourself in the eye, however, and you'll see a minor earthquake. Yet the retinal image is moving in the same kind of way in both cases. The difference, according to the well-known reafference theory of von Holst and Mittelstaedt, is this. Whenever the brain sends a message to the eye muscles, telling them to move the eye, it sends a copy of the message to the sensory part of the brain, telling it to expect the apparent position of the world to move accordingly, and to compensate by exactly the right amount. The brain perceives movement only if there is a discrepancy between the expected movement and the observed movement. This is beautifully demonstrated by the following fact. It has long been known that people who have had their eye muscles paralysed see the Earth move when they try to move their own eyes (and in the same direction as the intended eye movement). The brain is told to anticipate movement and allows for it. But the movement never comes, so the model of the world is seen to move.

Returning to information-rich movement relative to the rest of the world, Lettvin and his colleagues found fascinating variations on the basic bug-detector theme in the optic tectum in the brain of frogs. The most interesting from our present point of view are the so-called 'Sameness' neurons. I quote:

> Let us begin with an empty gray hemisphere for the visual field. There is usually no response of the cell to turning on and off the illumination. It is silent. We bring in a small dark object, say 1 to 2 degrees in diameter, and at a certain point in its travel, almost anywhere in the field, the cell suddenly 'notices' it. Thereafter, wherever that object is moved it is tracked by the cell. Every time it moves, with even the faintest jerk, there is a burst of impulses that dies down to a mutter that continues as long as the object is visible. If the object is kept moving, the bursts signal discontinuities in the movement, such as the turning of corners, reversals, and so forth, and these bursts occur against a continuous background mutter that tells us the object is visible to the cell.

It is as if the nervous system is tuned, at successive hierarchical levels, to respond strongly to the unexpected, weakly or not at all to the expected. Switching back to Barlow's language derived from the theory of codes, we could say that the nervous system uses short, economical words for messages that occur frequently and are expected; long, less economical words for messages that occur rarely and are not expected. Another way of putting it – and this brings us back to my main theme – is to say that a well-tuned nervous system embodies, in its dictionary of code symbols, a statistical model of the world in which the animal lives. Barlow puts it like this:

The effect of coding to reduce redundancy is not just the elimination of wasteful neural activity. It constitutes a way of organizing the sensory information so that, on the one hand, an internal model of the environment causing the past sensory inputs is built up, while on the other hand the current sensory situation is represented in a concise way which simplifies the task of the parts of the nervous system responsible for learning and conditioning.

When I look at any object, say a wooden box on the table, I have the strong impression that the thing I am looking at really is the box itself, and that it is outside myself. Ancient Greeks thought that the eyes spout invisible rays that feel the object we are looking at, and this idea has the ring of truth to my subjective consciousness. Although I know that in fact rays of light from the sun are being reflected from the box, and refracted through my lens and humours so that a tiny, upside-down image of the box appears on each of my two retinas, I cannot help feeling that my percept of the box is really 'out there' waiting to be felt by my eyes. The box, after all, looks solid; it has depth. The solid percept of it corresponds to neither of the two flat images on my two retinas. It is a compound, and it looks solid because the brain has made some sophisticated calculations based on the disparity between the images on the two retinas. The box that I actually perceive is neither 'out there' nor on either of the two retinas. If it is anywhere, it is in the brain, a computer model of a box. If somebody now rotates the box, the two-dimensional images on my two retinas change in complicated ways, but the solid, three-dimensional model in my brain remains intact. I seem to see a different face of the box as it turns, and the box may appear larger or smaller if it is moved closer to or further from me. But the overwhelming subjective experience is that it is the same box. Even though it rotates, swells or recedes, it has a stream of sameness. This is because what I am seeing is indeed the same 'thing'; it is the same computer model in the brain that I am 'looking at' as the box rotates, and as the images on my retinas change.

This 'model in the head' analysis helps us to think our way into minds that move through alien worlds. Bats don't see the world; they hear it, often in pitch darkness. But I have suggested elsewhere that the subjective sensations of a bat may be very like those of a visual animal such as a bird flying by day. The reason is that both bat and bird 'see' a model that they construct in the brain. The type of model constructed – by natural selection – will be the type of model needed for navigating at high speed in three dimensions through the air. The fact that the external information being used to update the model comes from light rays in the case of the bird, and from sound echoes in the case of the bat, is irrelevant to the functional properties of the internal model itself. I even conjectured that bats may perceive colour. Our subjective experience of colours has no necessary connection with particular wavelengths of light. Hues are parts of our internally constructed model, used as arbitrary labels for wavelength. There is no reason why bats should not use the same arbitrary labels to stand for some aspect of the echo-reflectance of textures. A male bat may, to a female's ears, appear as gorgeously coloured as a peacock's train does to our eyes.

We have seen that, in one sense, an animal's body *is* a model of its world. We have also seen that, at least in animals with advanced nervous systems, like birds and bats, the brain can contain dynamically updated models – computer simulations – of the animal's world. Unlike the animal's body, which is on the whole a static model, these computer simulations in the brain are capable of changing rapidly in time. The aspect of this fluctuation that I have so far stressed is the continuous updating of the model as nerve impulses flood in from the sense organs, like weather data from distant recording stations. But the weather analogy also reminds us that computer models can run on into the future – forecasting. Can an animal's brain model, too, run on into the future, providing the animal with useful forecasts? The answer is yes. But before we discuss this we need to consider what 'forecasting' means in general, what it might mean to an animal and why it might be useful to it.

The first thing to say is that we are talking about statistical fore-casting, prediction that is not guaranteed to be right. Weather forecasting is, of course, statistical. Certain American radio stations announce the probability of rain (or 'precipitation' as it is pompously called) as, say, 80 per cent. English weather forecasts do not state numerical probabilities, perhaps because the probability of even the most probable weather would be embarrassingly low.* Life insurance companies make a living by statistically predicting the longevity of their clients. A racing tip is a statistical prediction that a particular horse is likely to win. Statistical predictions, imperfect though they must be, are well worth making, neverthe-less. It is statistical predictions that animals can be said to make.

The second general thing I want to say is that all sensible predic-tion of the future (this excludes astrology and other techniques used by charlatans to exploit the gullible) is based upon extrapolation from the past. We assume that the laws that govern the future are the same as the laws that govern the past, and so whatever has happened in the past will probably happen in the future. If a day's weather was random with respect to previous days, forecasting would be impos-sible. If all human deaths were due to the purest accident and bore no relation to age, health or habits, actuaries would be out of a job and we'd all pay the same life insurance premium. There'd be no betting on favourites if horses capriciously flouted past form.

The past, then, can be used for forecasting the future only because the world behaves lawfully rather than randomly. Scientists are sometimes successful in investigating the underlying reasons why the future is related non-randomly to the past. But a statistical prophecy does not require such understanding: we may observe, statistically, that red berries in autumn presage a hard winter, with-out having any understanding of the mediating causes. The working

---

* This is now much less true than when this essay was written, in the early 1990s.

rule of thumb, 'Assume that whatever has happened in the past will continue to happen in the future', is a successful rule. It is the one that every animal and plant uses to ensure that it survives in its own little future.

Any forecasting system has at least the theoretical possibility of exploiting natural cycles. The rule of thumb here is: 'Whatever has happened at regular intervals in the past will probably go on happening at the same regular intervals in the future.' But simple rhythms constitute only one kind of patterning in the world. Patterning of any kind is the same thing as non-randomness, and any kind of non-randomness can potentially be used to predict the future. If, in your world, the booming of a gong is reliably followed by a good meal, you will probably find after a while that, whenever you hear the gong and are hungry, your mouth will start to water. As is well known, this is what Pavlov found with his dogs.

What is less well known is that it is the reliability with which the 'gong' and 'food' – the conditional and unconditional stimulus – are paired that matters. This has been shown in ingenious experiments by R. A. Rescorla and others. If you present two conditional stimuli, say a bell then a light, before food, which will be most effective at eliciting mouthwatering? The answer is not necessarily the first of the two, nor necessarily the second. Neither is it that bells are inherently more effective than lights, or vice versa. What really matters seems to be reliability. The stimulus that counts is whichever of the two can more reliably be used to predict the arrival time of food. If the time interval between bell and food is long but *constant*, while the interval between light and food is short but *variable*, the bell is the more reliable predictor of when the food will come, and the bell is the one that makes the animal's mouth water. Conversely, if the time interval between bell and food is long and variable, while the interval between light and food is short and constant, the light is the more reliable and the more mouthwatering. Pavlovian conditioning is a device used by the brain to predict future events from events that have already happened. It exploits the fact that particular events reliably follow others in the animal's world.

We can see all learning, not just Pavlovian conditioning, as a device whereby brains make use of past events in an individual animal's life to make that animal better at predicting the future and therefore at surviving in that future. There is no need to think of the animal as making conscious predictions, of having any kind of mental picture of the future – though of course it might. Learning can achieve its predictions by mechanical means, every bit as mechanical, indeed, as natural selection itself. It is a consequence of learning that the animal's brain model becomes an ever more accurate model of its world. Many animals behave as if they have a detailed mental picture of the world around their home, in the same kind of way as the arcade computer has a 'mental' picture of its model racetrack's geography; or in the same kind of way as a chess computer has a picture of the chessboard inside it. I should not wish, by using the metaphor of the computer, to imply that brains work like modern digital electronic computers. They probably don't. It is the principle of getting about the real world by simulating it internally that I want to emphasize, and it happens that the digital electronic computer is a familiar and powerful tool for simulation.

I want now to return to the question: Can an animal's mental model of its world free-run into the future and so simulate future events, like the computer model of the world's weather? Suppose we set up the following experiment. Find a steep cliff in a mountainous area of Ethiopia inhabited by Hamadryas baboons, and place a plank so that it sticks out over the edge of the precipice, with a banana on its far tip. The centre of gravity of the plank is just on the safe side of the edge, so that it does not topple into the gorge below, but if a monkey were to venture out to the end of the plank it would be enough to tip the balance. Now we hide and watch what the monkeys do. They are clearly interested in the banana, but they do not venture out along the plank to get it. Why not?

We can imagine three stories, any of which might be true, to account for the baboons' prudence. In all three stories the prudent behaviour results from a kind of trial and error, but of three different kinds. According to the first story, the baboons have an 'instinctive' fear of precipitous heights. This fear has been built into their brains directly by natural selection. Contemporaries of their ancestors that did *not* possess a genetic tendency to fear cliffs failed to become ancestors because they got killed. Consequently, since modern baboons are all descended, by definition, from successful ancestors, they have inherited the genetic predisposition to fear cliffs. There is indeed some evidence that the newly born young of a variety of species have an innate fear of heights. The first story, then, involves trial and error of the crudest and most drastic kind: Darwinian natural selection, dicing with ancestral life and death. We can call this the 'Ancestral Fear' story.

The second story talks about the past experience of the individual baboons. Each young baboon, as it grows up, has experience of falling. If it fell down a huge cliff, of course, the experience would be its last. But it has enough encounters with small cliffs to learn that falls can be painful. Pain, in trial and error learning, is the analogue of death in natural selection. Natural selection has built brains with the capacity to experience as pain those very sensations that, in a stronger dose, would lead to the animal's death. Pain is not only the analogue of death; it is a kind of symbolic substitute for death if we think in terms of an analogy between learning and natural selection. Baboons build up in their brains, through experience of the pain of falling down small cliffs, perhaps through experience that the bigger the cliff the worse the pain, a tendency to avoid cliffs. This is the second story, the 'Painful Experience' story, of how it has come about that the baboons resist their natural tendency to rush out along the plank to seize the banana.

The third story is the one this is all leading up to. According to this story, each baboon has a model of the situation in its head, a computer simulation of the cliff, the plank and the banana, and it can run the simulation program on into the future. A baboon's brain simulates his body walking along the plank. Just as the arcade computer simulates the racing car passing a tree, the baboon's computer simulates his body advancing towards the banana, the model plank teetering, then toppling and crashing into the simulated abyss. The brain simulates it all, and evaluates the results of the computer run. And that, according to our 'Simulated Experience' story, is why the baboon doesn't venture out in actuality. It is a trial and error story, just like the Ancestral Fear and Painful Experience stories, but this time it is trial and error in the head, not in reality. Obviously, trial and error in the head, if you have a powerful enough computer there to do it, is preferable to trial and error in real earnest.

Now, as you read these stories, I have little doubt that you had an imaginary picture of the scene. You 'saw' the cliff, you saw the plank and you saw the baboons. If you have a vivid imagination you may have seen everything in great detail. Your imagined banana may have been bright yellow, perhaps with a precise pattern of black spots on the skin. Or you may have imagined it peeled. You may have seen the chasm in every detail of its rocks and crannies, with stunted bushes clinging to the scarp, whereas I, not having a vivid imagination, saw a rather abstract precipice like a sketch in a mathematics textbook. The details of all of our imaginary pictures are, no doubt, very different. But we all set up a simulation of the scene, which was adequate to the task of predicting a baboon's future. We all know, from the inside, what it is like to run a computer simulation of the world in our heads. We call it imagination, and we use it all the time to steer our decisions in wise and prudent directions.

The experiment with the baboons and the banana has not been done. If it were, could the results tell us which of our three stories was true, or whether the truth were some combination? If the Painful Experience story were true, we should be able to find out by looking at the behaviour of young or inexperienced baboons. One who had been sheltered all his life from falls should prove fearless when eventually confronted with an edge. If such a naive baboon turned out in fact to be fearful, this would still leave both the other two stories open. He might have inherited ancestral fear, or he might have a vivid imagination. We could try to decide the issue by a further experiment. Say we place a heavy rock on the near end of the plank. Now we humans, at least, can see from our own mental simulation that it is safe to venture along the plank: the rock is obviously a secure counterbalance. What would the baboons do? I don't know, and I'm not sure that it would be a very informative experiment in any case. I know that, however certain I was from my mental model that the rock would be a staunch counterweight, I wouldn't go out along the plank, not for a crock of gold. I just can't take heights. The Ancestral Fear story sounds pretty plausible to me. What is more, so powerful is this fear that it enters into my Simulated Experience! When I simulate the scene in imagination, I literally feel a *frisson* of fear up my spine, however vividly I simulate a ten-ton rock firmly clamped down on the plank. Since I know that all three stories are true for me, I could easily believe the same of baboons. Incidentally, what I am now simulating in my mind is an enterprising baboon hauling in the plank and seizing the banana when it has reached safe haven on the cliff top. My imaginary baboon has arrived at this sagacious resolution by a neat piece of simulation of her own, but as for whether this would be true of a real baboon, your guess, your prediction, your simulation, is as good as mine.

The imagination, the capacity to simulate things that are not (yet) in the world, is a natural – emergent – progression from the capacity to simulate things that are in the world. The weather model is continually updated by information from weather ships and weather stations. To this extent it is a simulation of conditions as they really are. Whether or not it was originally designed to run on into the future, its ability to do so, to simulate things not only as they are but as they may turn out to be, is a natural, almost inevitable, consequence of the fact that it is a model at all. An economist's computer model of the economy of Britain is, so far, a model of things as they are and have been. The program hardly needs to be modified at all in order to take that extra step into the simulated future, to project probable future trends in the gross national product, the currency and the balance of payments.

So it was in the evolution of nervous systems. Natural selection built in the capacity to simulate the world as it is, because this was necessary in order to perceive the world. Think of that well-known visual illusion, the Necker Cube (figure 3). The distinguished psychologist Richard Gregory has shown how visual illusions can easily be understood once we realize that there is a sense in which we are looking, not at reality itself, but at a model of reality in the brain. This is not a cube. It is a two-dimensional pattern of ink on paper. Yet a normal human sees it as a solid cube, with one of the two end faces nearer than the other. The brain has made a three-dimensional model based upon the two-dimensional pattern on the paper. This is, indeed, the kind of thing the brain does almost every time we look at a picture. You cannot see that two-dimensional patterns of lines on two retinas amount to a single solid cube unless you simulate, in your brain, a model of the solid cube. Having built in the capacity to simulate models of things as they are, natural selection found that it was but a short step to simulate things as they are not quite yet – to simulate the future. This turned out to have valuable emergent consequences, for it enabled animals to

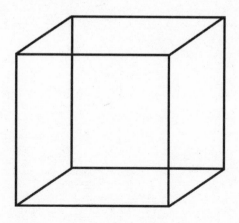

Figure 3

benefit from 'experience': not direct trial and error experience in their own past history, nor the life and death 'experience' of their ancestors, but vicarious experience in the safe interior of the skull.

And once natural selection had built brains capable of simulating slight departures from reality into the imagined future, a further emergent capacity automatically flowered. Now it was but another short step to the wilder reaches of imagination revealed in dreams and in art, an escape from mundane reality that has no obvious limits.

REAL GENES AND
VIRTUAL WORLDS

Leda Cosmides and John Tooby, together with another couple, Martin Daly and Margot Wilson, and Steven Pinker, could fairly be regarded as founders of the subject of evolutionary psychology. Together with Jerome Barkow, Cosmides and Tooby edited a founding theoretical volume, *The Adapted Mind*. David Buss is another leader in the field. I wrote this afterword to the *Handbook of Evolutionary Psychology* edited by him and published in 2005.

At the end of such a compendium – truly a worthy ten-years-on successor to *The Adapted Mind* – what is there left for an afterword to say? Some sort of summing up of all thirty-four chapters? Too repetitious. A prophetic 'Whither evolutionary psychology?' Too presumptuous. An idiosyncratic *jeu d'esprit*, playfully calculated to send the reader diving back into the book to view the whole corpus again but from a different angle of illumination? Nice idea but too ambitious. Reflective musings of a sympathetic observer of the scene? Well, let me start that way, and see what develops.

First, a confession. As an observer of the scene, I had not been a very clear-sighted one. I was one of those who mistakenly thought evolutionary psychology a euphemistic mutation of sociobiology, favoured (like 'behavioral ecology') for its cryptic protection against the yapping ankle-biters from 'science for the people' and their fellow-travellers.* I now think that was a travesty, not even a half-truth, perhaps at most a quarter-truth. For one thing, intellectual pugilists of the calibre of Cosmides, Tooby and other authors of this book need no camouflage. But even that isn't the point. The point is that evolutionary psychology really is quite different. Psychology it is, and psychology is by no means all, or even mostly, about social life, sex, aggression or parental relationships. Evolutionary psychology is about the evolution of so much more than that: about perceptual biases, about language, about errors in information processing. Even within the narrower field of social behaviour, evolutionary psychology distinguishes itself from sociobiology by emphasizing the psychological and information-processing mediation between natural selection and the behaviour itself.

Evolutionary psychology and sociobiology do, however, have one bane in common. Both are subject to a level of implacable hostility which seems far out of proportion to anything sober reason or even common politeness might sanction. E. O. Wilson, struggling to understand the onslaught that engulfed *Sociobiology* at the hands of left-wing ideologues, invoked what Hans Küng in another context had called 'the fury of the theologians'. I have known sweetly reasonable philosophers, with whom I could have an amicable and constructive conversation on literally any other topic, descend to the level of intemperate ranting at the mere mention of evolutionary psychology or even the name of one of its leading practitioners. I have no desire to explore this odd phenomenon in detail. It is well discussed by evolutionary psychologists including contributors to this book, and also by Ullica Segerstråle in *Defenders of the Truth*. I just want to make one additional

---

* This is a reference to the Marxist-inspired thought police, at Harvard and elsewhere, who inspired a savage attack on Edward Wilson's great book *Sociobiology* (see also my review of *Not in Our Genes*, pp. 62–8 above).

remark, about *a priori* scepticism: about the height to which we set the hurdle bar for different sciences.

Sceptical investigators of paranormal claims have a much-quoted maxim: extraordinary claims require extraordinary evidence.* All of us would set the bar very high for, say, a claimed demonstration that two men, sealed in separate soundproof rooms, can reliably transmit information to one another telepathically. We should demand multiple replications under ultra-rigorous double-blind controlled conditions, with a battery of professional illusionists as sceptical scrutineers, and with a statistical *p*-value less than one in a billion. On the other hand, an experimental demonstration that, say, alcohol slows down reflexes would be accepted without a second glance. While nobody would approve poor design or shoddy statistics, we wouldn't go out of our way to scrutinize the alcohol experiment very sceptically before accepting the conclusion. The hurdle in this case would be set so low as almost to escape notice. In the middle, there is a spectrum of scientific claims, of intermediate capacity to arouse *a priori* scepticism. Evolutionary psychology, weirdly, seems to be seen by its critics as way out on the 'telepathy' end of the spectrum, a red rag to critical bulls.

Something similar was true of the earlier controversy over socio-biology. Philip Kitcher's *Vaulting Ambition* is widely touted as a devastating critique of human sociobiology. In reality it is mostly a catalogue of methodological shortcomings of particular studies. The supposed faults range from peccadillo to shoddy, but they are of a type which is in principle remediable by new and improved studies along the same lines. Criticisms like Kitcher's of sociobiology, or like those more recently hurled at evolutionary psychologists such as Daly and Wilson on step-parental abuse, Cosmides and Tooby on social exchange, or Buss on sexual jealousy, are made so strongly

---

* Often attributed to Carl Sagan, but in various forms it goes back a long way.

only because the critics are treating the hypotheses under test as if they were extraordinary claims that demand extraordinary evidence. Evolutionary psychology is seen by its critics as out at the high hurdle end – the 'telepathy' end – of the spectrum, while it is simultaneously seen by its practitioners as down at the plausible end of the spectrum with the alcohol and the reflexes. Who is right?

Without a doubt, the evolutionary psychologists are right in this case. The central claim they are making is not an extraordinary one. It amounts to the exceedingly modest claim that minds are on the same footing as bodies where Darwinian natural selection is concerned. Given that feet, livers, ears, wings, shells, eyes, crests, ligaments, antennae, hearts and feathers are shaped by natural selection as tools for the survival and reproduction of their possessors, in the particular ecological niche of the species, why on earth should the same not be true of brains, minds and psychologies? Put it like that, and the central thesis of evolutionary psychology moves right to the plausible end of the spectrum. The alternative is that psychology is uniquely exempt from the Darwinian imperatives that govern the whole of the rest of life. *That* is the extraordinary claim which, if not downright bonkers, at least demands extraordinary evidence before we should take it seriously. Maybe it is right. But given that we are Darwinians, the onus of proof is on those who would deny the central thesis of evolutionary psychology. It is the critics who lie closer to the 'telepathy' end of the spectrum.

Could it be that the sticking point for critics is that old bugbear, the supposed uniqueness of humans? Is evolutionary psychology permissible for 'animals', but not for *Homo sapiens*? Once again such exceptionalism, though conceivably justifiable, bears the heavy burden of proof. Perhaps ten million species live on this planet at the moment, and as many as a billion species have done so in history. It is, of course, *possible* that our species really is the one in a billion species that, with respect to psychology, has emancipated itself from the purview of evolutionary explanation. But if that is what you think, the onus of demonstration is on you. Don't underestimate the magnitude of the surprisingness of what it is that you purport to believe.

Or could it be 'modularity' that sticks in the craw of critics? Maybe. Maybe they are right, and in any case some evolutionary psychologists are less enamoured of modularity than others. But, yet again, modularity is not an extraordinary claim. It is the *alternative* to modularity that bears the burden of coming up with extraordinary evidence in its favour. Modularity is a universally good design principle which pervades engineering, software and biology, to say nothing of political, military and social institutions. Division of labour among specialist units (experts, organs, parts, subroutines, cells) is such an obvious way to run any complex operation, we should positively expect that the mind would be modularized unless there is good reason to believe the contrary. Yet again, the detailed arguments are to be found in this book. I merely repeat my point about the onus of proof lying on the opponents of evolutionary psychology.

Of course, some individual evolutionary psychologists need to clean up their methodological act. Maybe many do. But that is true of scientists in all fields. Evolutionary psychologists should not be weighed down by abnormal loads of scepticism and *a priori* hostility. On the contrary, they should hold their heads high and go to work with confidence, for the enterprise they are engaged upon is normal science within the neo-Darwinian paradigm.

That's all I want to say about that. I want to end by making the nearest approach I can to the *jeu d'esprit* that I quailed from as too ambitious in my opening paragraph. Sometimes science proceeds not by experiment or observation but by changing the point of view: seeing familiar facts from an unfamiliar angle. I'd like to think that the 'extended phenotype' is an example. Another candidate is 'the genetic book of the dead', and a third is 'continuously updated virtual reality'.

The idea of the genetic book of the dead is that an animal, since it is well adapted to its environment, can actually be seen as a *description* of its environment. A knowledgeable and perceptive zoologist, allowed to examine and dissect a specimen of an unknown species, should be able to reconstruct the habitat and way of life of the animal. To be strict, the reconstructed habitat and way of life is a complicated average of the ancestral habitats and ways of life of the animal's ancestors: its EEA,* to use the evolutionary psychology jargon.

This conceit can be phrased in genetic terms. The animal you are looking at has been constructed by a sampling of genes from the gene pool of the species: genes that have successfully come down through a long sequence of generational filters. These are the genes that survive in the EEA. The genes can therefore be seen as a description of the EEA: hence the phrase 'genetic book of the dead' (for a full account, see the chapter with that title in my book, *Unweaving the Rainbow*).

'Constrained virtual reality' is the idea that every brain constructs a virtual reality model of the world through which the animal is moving.† The virtual reality software is 'continuously updated' in the sense that, although it might theoretically be capable of simulating scenes of pure fantasy (as in dreams), it is in practice constrained by data flowing in from the sense organs.

Visual illusions such as Necker Cubes and other alternating figures are best interpreted in these terms. The data sent to the brain by the retina are equally compatible with two virtual models of a cube. Having no basis on which to choose, the brain alternates between them. The virtual world that our brains construct is, no doubt, very different from the virtual world constructed by the brain of a squirrel, a mole or a whale. Virtual as these constructed

---

* Environment of Evolutionary Adaptation. This would not have needed spelling out to any reader coming to the end of the *Handbook* in which this essay first appeared.

† See the previous essay in this book.

worlds are, there is a profoundly real sense in which an animal lives in, and survives in, its own constructed world.

Each species will construct virtual models that are useful for its particular way of life. A swift and a bat both move at high speed through a three-dimensional world catching insects on the wing. Both therefore need the same kind of virtual model of the world, even though swifts hunt by day using their eyes, bats hunt by night using their ears. Qualia that swifts associate with colour are actually constructions of the virtual reality software. My conjecture could never, I think, be tested, but I have suggested that bats might 'hear in colour' in the sense that their virtual reality software is likely to make use of the same qualia to signify equally salient features of their auditory world. Surface textures presumably reflect echoes in particular ways. Such textures are likely to be as important to bats as colour is to swifts. So the virtual reality software of bats is, I suggest, likely to use the same qualia – red, blue, green etc. – as internal labels for different acoustic textures.

My bat suggestion is just an example of how the idea of continuously updated virtual reality changes our view of animal psychology. (For a more thorough discussion, see the chapter called 'Reweaving the world' in *Unweaving the Rainbow*.) Now I want to unite it with the idea of the genetic book of the dead. If a knowledgeable zoologist can reconstruct an animal's EEA using data from its anatomy and physiology, could a knowledgeable psychologist do something similar for mental worlds? Surely the mental world of a squirrel would, if we could peer into it, be a world of forests, a three-dimensional maze of trunks and twigs, branches and leaves. The mental world of a mole is dark, damp and filled with smells, because the genes that built its brain have survived through a long line of similarly dark and damp ancestral places. The virtual reality software of each species would, if we could reverse engineer it, allow us to reconstruct the environments in which natural selection built up that software.

Nowadays we are accustomed to saying, in a sense which is more literal than metaphorical, that all the genes of a species have survived through a long succession of ancestral worlds. We usually mean physical and social environments when we speak of ancestral worlds. The parting shot of my afterword, my little *jeu d'esprit*, is that the long succession of ancestral worlds in which our genes have survived include the virtual worlds constructed by our ancestors' brains. Real genes have – again in something close to a literal sense – been selected to survive in a *virtual EEA*, constructed by ancestral brains.

NICE GUYS (STILL)
FINISH FIRST

In 1990, I wrote a foreword for the Penguin edition of Robert
Axelrod's *The Evolution of Cooperation*; this is the updated version
I wrote for the new edition published in 2006.

This is a book of optimism. But it is a believable, realistic optimism,
more satisfying than the naive, pie-in-the-sky optimisms of Christianity, Islam or Marxism. Promises of rewards in another life for
privations or martyrdoms in this one are so transparently self-
serving for the promisers that our scepticism should be aroused
before we even notice the lack of evidence. Prophecies of an ultimate, inevitable withering away of the state in the face of universal
proletarian brotherly love are marginally less self-serving but
scarcely more realistic. An optimism, to be believed, must first
acknowledge fundamental reality, including the reality of human
nature. Not just human nature but the nature of all life. Life as we
know it, and probably throughout the universe if there is life elsewhere, means Darwinian life. In a Darwinian world, that which
survives survives, and the world becomes full of whatever qualities
it takes to survive. As Darwinians we start pessimistically by
assuming deep selfishness at the level of natural selection, pitiless
indifference to suffering, ruthless attention to individual success at

the expense of others. And yet, from such warped beginnings,* something that is in effect, if not necessarily in intention, close to amicable brotherhood and sisterhood can come. This is the uplifting message of Robert Axelrod's remarkable book.

My own credentials for writing this foreword have been peripheral but recurrent. In the late 1970s, a few years after publishing my own first book *The Selfish Gene*, which explained the pessimistic principles mentioned above, I received, out of the blue, a typescript from an American political scientist whom I didn't know, Robert Axelrod. It announced a 'computer tournament' to play the game of Iterated Prisoner's Dilemma and invited me to compete. To be more precise – and the distinction is an important one for the very reason that computer programs don't have conscious foresight – it invited me to submit a computer program that would do the competing. I'm afraid I didn't get around to sending in an entry. But I was hugely intrigued by the idea, and I did make one valuable, if rather passive, contribution to the enterprise at that stage. Axelrod was a professor of political science and, in my partisan way, I felt that he needed to collaborate with an evolutionary biologist. I wrote him an introduction to W. D. Hamilton, probably the most distinguished Darwinian of our generation, now sadly dead after an ill-fated expedition to the Congo jungle in 2000. In the 1970s, Hamilton was a colleague of Axelrod in a different department of the University of Michigan, but they didn't know each other. Upon receiving my letter, Axelrod immediately contacted Hamilton, and they collaborated on the paper that was the forerunner of this book and is abridged as chapter 5. It had the same title as the book, was published in *Science* in 1981 and won the Newcomb–Cleveland prize of the American Association for the Advancement of Science.

The first American edition of *The Evolution of Cooperation* was published in 1984. I read it as soon as it appeared, with mounting

---

* The allusion is to a poem by Cecil Day-Lewis, 'The Unwanted'.

excitement, and took to recommending it, with evangelical zeal, to almost everyone I met. Every one of the Oxford undergraduates I tutored in the years following its publication was required to write an essay on Axelrod's book, and it was one of the essays they most enjoyed writing. But the book was not published in Britain; and in any case the written word sadly has a limited constituency compared with television. So I was pleased when in 1985 I was invited by Jeremy Taylor of the BBC to be the presenter of a *Horizon* programme largely based upon Axelrod's work. We called the film *Nice Guys Finish First*. I had to speak my lines from such unaccustomed locations as a football pitch, a school in Britain's industrial midlands, a ruined medieval nunnery, a whooping cough vaccination clinic and a replica of a First World War trench. *Nice Guys Finish First* appeared in the spring of 1986 and it enjoyed some critical success, although it was never shown in America – whether because of my unintelligible British accent I don't know.* It also brought me temporary standing as a public partisan of 'forgiving', 'non-envious' 'nice guys' – a welcome relief, at least, from notoriety as the alleged high priest of selfishness, and salutary testimony to the power of title over content: my book had been *The Selfish Gene* and I was regarded as an advocate of selfishness; my film was called *Nice Guys Finish First* and I was hailed as Mr Nice Guy. Neither accolade was borne out by the content of book or film. Nevertheless, in the weeks after *Nice Guys* was broadcast, I was lunched and consulted on niceness by industrialists and manufacturers. The chairman of Britain's leading chain of clothes shops gave me lunch in order to explain how nice his company was to its employees. A spokeswoman from a leading confectionery company also took me to lunch on a similar mission, in her case to explain that her company's dominant motivation in selling chocolate bars was not to make money but

---

* Almost beyond belief as it seems, at least two of David Attenborough's remarkable BBC documentaries were revoiced using American voices for the American market.

literally to spread sweetness and happiness among the population. I was invited by the world's largest computer company to organize and supervise a whole day's game of strategy among their executives, the purpose of which was to bond them together in amicable cooperation. They were divided into three teams, the reds, the blues and the greens, and the game was a variant on the prisoner's dilemma game which is the central topic of Axelrod's book. Unfortunately, the cooperative bonding which was the company's goal failed to materialize – spectacularly. As Robert Axelrod could have predicted, the fact that the game was known to be coming to an end at exactly 4 p.m. precipitated a massive defection by the reds against the blues, immediately before the appointed hour. The bad feeling generated by this sudden break with the previous day-long goodwill was palpable at the post-mortem session that I conducted, and the executives had to have counselling before they could be persuaded to work together again.

In 1989 I acceded to Oxford University Press's request for a second edition of my own *The Selfish Gene*. It contains two chapters based upon the two books that most excited me during the intervening dozen years. It will come as no surprise that the first of these chapters was an exposition of Axelrod's work, again called 'Nice guys finish first'. But I still felt that Axelrod's own book should be available in my own country. I took the initiative by approaching Penguin Books and was pleased that they accepted my recommendation to publish it here, and that they invited me to write a foreword to their British paperback edition (1990). I am doubly pleased that Robert Axelrod himself has now invited me to update that foreword for this new edition (2006) of his book.

In the years since *The Evolution of Cooperation* was first published, it is no exaggeration to say that it has spawned a whole new research industry. In 1988 Axelrod and a colleague, Douglas Dion, compiled an annotated bibliography of research publications more or less directly inspired by *The Evolution of Cooperation*. They listed more than 250 works up to that date under the following headings: 'politics and law', 'economics', 'sociology and anthropology', 'biological applications', 'theory (including evolutionary theory)', 'automata theory (computer science)', 'new tournaments' and 'miscellaneous'. Axelrod and Dion collaborated on another paper with the title 'The further evolution of cooperation', summarizing the progress of the field in the four years since 1984. Since that review, the growth of research fields inspired by this book has continued apace. But, in contemplating the welter of new research, the main impression I am left with is how little the basic conclusions of the book need to be changed. Ancient Mariner-like, I have continued over the years to press it upon students, colleagues and passing acquaintances. I really do think that the planet would be a better place if everybody studied and understood it. The world's leaders should all be locked up with this book and not released until they have read it. This would be a pleasure to them and might save the rest of us. *The Evolution of Cooperation* deserves to replace the Gideon Bible.

ART, ADVERTISEMENT
AND ATTRACTION

Robin Wight is one of Britain's more imaginative and creative advertising executives. He has long been fascinated by the analogy between animal and human advertising techniques. I wrote this foreword for his 2007 essay entitled *The Peacock's Tail and the Reputation Reflex: the neuroscience of art sponsorship.*

Darwin would have liked this thoughtful essay by Robin Wight. His co-discoverer of natural selection, Alfred Russel Wallace, would have loved it. These two scientific heroes stand at opposite ends of a continuum of opinion, which we can represent as 'art for art's sake' at Darwin's end of the spectrum, and 'art repays sponsorship' at Wallace's. The specific field of their disagreement was Darwin's 'other theory' of sexual selection, epitomized by the peacock – poster boy of nature's advertising industry, the Robin Wight of the bird world.

Natural selection, narrowly understood as a drab utilitarian bean-counter obsessed with survival, was always going to have trouble with peacocks and peacock butterflies, with angel fish and birds of paradise, with the song of a nightingale or the antlers of a stag. Darwin realized that individual survival was only a means to the end of reproduction. As we would put it today, it is not peacocks that survive in the evolutionary long run anyway, it is their genes, and genes survive only if they make it into the next generation, manipulating a succession of short-term bodies to that long-term end. For peacocks and other animals whose biggest hurdle in the way of reproduction is competitors of the same sex, natural selection – or sexual selection as Darwin called it in this case – will tend to favour extravagant attractiveness or formidable weaponry, cost what it might in economic resources or risk to individual survival.

Attractiveness in whose eyes? The eyes of the peahen, of course, and if her tastes happen to coincide with ours so much the better for us. The peacock is a walking advertisement, a colourful hoarding, a neon come-hither, an expensive commercial. Even a work of art? Yes, why not? The case is even clearer for the bower birds of Australia and New Guinea. Not particularly bright or showy in their plumage, male bower birds build an 'external peacock tail', a labour of love which serves the same purpose of attracting females. Fashioned from grass, twigs and leaves woven into the shape of an arch or a maypole, paved with stones, decorated with berries or painted with their juice, adorned with flowers, shells, coloured feathers from other species of birds, fragments of coloured glass, even beer-bottle tops, no two bowers are the same. Females survey the bowers and then choose the male whose architectural offering they find most pleasing.

Males will spend hours titivating and perfecting their bowers. If an experimenter moves an item while the bird is away, he will carefully replace it when he returns. The trope of the 'external peacock tail' is reinforced by a telling observation. Those species of bower bird with the drabbest plumage tend to be the ones with the most elaborate bowers. It is as though in evolutionary time they shifted their advertising budget, bit by bit, from body to bower.

When a male bower bird stands back, cocks his head to one side and surveys his creation, then darts forward to move a blueberry two inches to the right, then steps back to survey again, it is hard not to see an artist delicately touching up his canvas. And that narrative is entirely plausible, for the following reason. The eye and brain that the male seeks to impress are of the same species as his own eye and brain. If he likes his own bower, there's a good chance that a female will too. Whatever turns you on, if you are a male bower bird, will probably turn on a female of the same species.

American song sparrows teach themselves to sing by burbling experimentally, and learning to repeat those fragments or phrases that sound good to the male himself, or, as the experimenters put it, that conform to a 'built-in template', a kind of idealized song genetically lodged in the brain. But there is another way to understand the 'template'. As with the bower, the gradual improvement of the young bird's song can be regarded as the shaping and perfecting of a work of art, in this case a musical composition. And again, since the purpose is to appeal to the nervous system of a (female) member of the same species, what better way to perfect the composition than to try out random phrases on himself? Again, 'Whatever turns me on will probably turn her on too, because we share the song sparrow nervous system.' How else does a composer proceed than to try out fragments of melody or candidate harmonies, in his head or on the piano, varying them and modifying them to suit his own taste, implicitly reasoning that what appeals to him will also play well in the concert hall?

'Works of art' is one way to look at the perfected products of sexual selection. 'Drugs' is another. A nightingale in a cool greenwood sang of summer to John Keats, and a drowsy numbness pained his sense as though of hemlock he had drunk. Keats was not a bird, but he shared its vertebrate nervous system, and I have made the case elsewhere that female nightingale nervous systems might be drugged in the same way. Male canary song is known to cause female canary ovaries to swell, and secrete hormones that affect reproductive behaviour. It is entirely plausible that Keats's experience of being drugged was a (probably reduced, for in his case it was incidental) version of what a female nightingale experiences when she hears the male pour forth his soul abroad in such an ecstasy. Is birdsong the avian equivalent of a date rape drug?

Works of high art? Seductive drugs? Piccadilly Circus neon signs? However we describe them, the products of sexual selection make demands on Darwinian theory that go beyond ordinary utilitarian natural selection. This was where Wallace and Darwin parted company, as Helena Cronin has shown in her beautiful book *The Ant and the Peacock*. Darwin accepted as a given fact of life that females have tastes, aesthetic preferences, inexplicable whims, and these simply dictate how aspirant males must sing, or be adorned, or build their bowers. Wallace, who described himself as more Darwinian than Darwin, hated the arbitrary assumption of feminine whim. Having given up on the hope that peacock tails and other bright beauties might have some hidden usefulness of a practical, utilitarian nature, Wallace fell back on the strong belief that, at the very least, they were demonstrations to the female of the male's practical worth. The gorgeous plumage of the peacock was not just beauty for beauty's sake, as Darwin had it. It was a demonstration, a badge of worth, an unfakeable certificate of a male's quality. A large and elaborate bower, perhaps, is tangible evidence that here is a diligent male, hard-working, skilled, with stamina and perseverance, all desirable qualities in the father of one's children.

Two of Darwin's and Wallace's greatest successors, R. A. Fisher in the first half of the twentieth century and W. D. Hamilton in the second, have championed the Darwinian and the Wallacean view of sexual selection respectively. Fisher showed that beauty for beauty's sake need not depend on arbitrary female whim, but nor need it be a badge of real quality. If we assume that female preference itself evolves alongside male ornamentation, mathematical reasoning can predict a runaway process in which the two advance together 'with ever increasing speed' until the joint evolutionary race is finally brought up short by overwhelming utilitarian pressures. Fisher really did succeed in modelling beauty for beauty's sake, while never departing from rigorous Darwinian standards. Though not denying the ingenuity of Fisher's mathematical reasoning, Hamilton was more drawn to Wallacean models of sexually selected advertisements of genuine male quality. For Darwin, the male's colourful hoarding says (albeit using compulsively seductive arts to do so) nothing more than 'Choose me! Choose me! Choose me!' The Wallacean male, by contrast, says: 'Choose me because I am a healthy male, strong and fit, likely to be a good father, as you can see from the following evidence.' In human terms, the Darwinian/Wallacean distinction is, of course, entirely familiar to Robin Wight and his colleagues in the world of commercial advertising. Both schools of advertising surely have their merits, but in different circumstances.

Hamilton's computer models were particularly concerned with advertisements of male health and resistance to parasites. He was indebted to the more general 'handicap principle' of the Israeli zoologist Amotz Zahavi, which plays so large a role in Robin Wight's 'reputation reflex', as developed in this work. Zahavi was, in turn, indebted to Hamilton's friend and Oxford colleague (and mine) Alan Grafen, who showed, in the teeth of massive scepticism among zoologists,* that Zahavi's seemingly bizarre idea can actually work.[†] Grafen's rigorous mathematical models show us that the handicap principle, maddeningly paradoxical though it at first appeared, is a serious candidate for explaining real facts in the real world. Natural selection can favour advertisements that are costly, extravagant or dangerous, not (as Darwin and Fisher thought) in spite of their costs but precisely *because* they are costly. It is the unavoidable cost of a good advertisement that authenticates it. Males deliberately endanger themselves, or perform feats that are difficult to execute, or demand expensive equipment or scarce and precious resources, because females will not accept cheap substitutes. To put it in Darwinian terms, natural selection would penalize females who did accept cheap substitutes, and would favour rival females who insist on authenticated evidence of

---

* Regrettably including me, in the first edition of *The Selfish Gene*.

[†] See the footnote to my introduction in the thirtieth-anniversary edition of *The Selfish Gene* on pp. 93–4 above.

quality. In Hamilton's version, male advertisements and displays are nicely calculated to show healthy males to their best advantage and, more surprisingly, even paradoxically, are also calculated to expose unhealthy males to female detection. As I have put it before, females evolve to become skilled diagnostic doctors while, at the same time and more surprisingly but in accordance with the logic of the Zahavi/Grafen theory, males evolve to give the game away even if they are unhealthy, with the equivalent of clinical thermometers protruding ostentatiously from every orifice.

FROM AFRICAN EVE TO THE BANDA STRANDLOPERS

Jonathan Kingdon is an extraordinary amalgam of zoologist and world-class artist. The world authority on African mammals, his six-volume opus *Mammals of Africa*, exquisitely illustrated by himself, finally appeared in 2013 after many years' gestation. This review of his theory of human evolution, published under the title *Self-Made Man and his Undoing*, appeared in the *Times Literary Supplement* on 26 March 1993.

> We carry within us the wonders we seek without us. There is all Africa and her prodigies in us.
>
> *Sir Thomas Browne*

A gentle wisdom comes out of Africa, a timeless vision that looks through and beyond the effete faddishness, the forgettable ephemerality of contemporary culture and preoccupation. With the eyes of an artist and the mind of a scientist and polymath, Jonathan Kingdon gazes deep into the past; and if you look deeply enough into our human past, you come inevitably to the home continent of Africa. He was born there and so, as it happens, was I, although we'd both be classed under that ludicrous name (which he rightly ridicules) 'Caucasian'. But, as Kingdon reminds us, wherever we were individually born and whatever our 'race', we are all Africans. And also, whatever our race and tastes, we are all 'Moderns', but this doesn't mean what an art historian or a literary critic might mean by it. With the long perspective of Jonathan Kingdon the Modern era began between a hundred thousand and a quarter of a million years ago, for it was then that skulls lost their massive struts and brow-ridges and assumed the fine-boned grace that all today's races share.

There are two 'out of Africa' stories, one undoubted, the other controversial. I shall return to the alleged recent migration, the disputed one, in a moment. The earlier Great Trekkers out of Africa, one and a half to two million years ago, were the massive, low-browed 'Erects'. Sufficiently different from us to be classified as a different species – *Homo erectus** – their remains are found in Java and China, in Iraq and Israel, as well as, of course, in the mother continent of Africa. Their eyebrows were heavily buttressed and their brains, in long, back-swept skulls, were markedly smaller than ours. They had stone tools and probably wood, bone and horn ones too, and they early discovered the use of fire. Before the Erect exodus, the story of humanity reverts to Africa alone, and the lost genus of 'Southern Apes', *Australopithecus*. This is not seriously doubted. We are African apes.

Jonathan Kingdon is a zoologist and naturalist who is best known as a consummate biological artist. His seven-volume atlas of evolution in Africa, *East African Mammals*, established his pencil drawings as sought-after collectors' items. His *Island Africa*, as well as being an important treatise on biogeography, shows his vivid talents as a painter. He is also a gifted sculptor, both representational and abstract, and it is not fanciful to imagine that he sees

---

* According to some authorities the African Erects should be classified as *Homo ergaster*.

evolution as modelling the solid bones of our ancestors with sculptor's hands.

> In the course of individual development, as well as during evolution, genes control whether bone surfaces get pared away or built up. The technicalities of this remodelling are analogous to a sculptor working with gypsum plaster. Where a surface needs building up, active cells plaster it with bone salts. These cells are called osteoblasts from the Greek *blastos* meaning 'sprout' or 'bud'. By contrast osteoclasts (from the Greek *klastos*, 'to break') are erosive cells that actually nibble surfaces away. Such addition and subtraction is highly selective in where it takes place.

His word-paintings, too, are unmistakably those of a sure-fisted artist.

> In my own case, curiosity about origins did not begin in a classroom, laboratory or library. It began on my own home ground in Africa. The landscapes of my childhood still rumbled with the passage of extinct herds and I could relish the same lake breezes that had cooled thirsty hunters a million years ago. These were not waters to be dredged for scientific fodder or fields to be ploughed for data, but a land where I was only the most recent adolescent inhabitant.

And here is his way of summarizing the sources of genetic diversity.

> If the children of Adam and Eve differ from one another today, it is partly due to recombination and partly to innumerable tiny changes in the chemistry of our genes that have taken place after our ancestors' long-forgotten copulations under an African moon.

He is not shy of unqualified allusion to the Eden myth. In introducing his theme of 'self-made man' itself, he identifies technology with the fruit of the Tree of Knowledge. And he espouses the theory of 'African Eve'. This is the popular label for the second 'out of Africa' theory, the controversial one. I shall spend some time on this interesting theory because it is less secure than Kingdon admits, and we need to decide whether this affects the validity of his book. It is the theory that all the descendants of the original Erect exodus across Asia have died out, and all surviving humans stem from a single female who lived in Africa less than a quarter of a million years ago. This 'African Eve's' descendants spread out of Africa during the last couple of hundred thousand years or so.

How could we verify such a hypothesis? Brief as a hundred thousand years is to geologists, those times are unthinkably remote by ordinary historical standards. Even if language goes back that far (controversial), oral traditions of such antiquity are ruled out by the low fidelity of linguistic communication between the generations. We remember a few anecdotes about the world in which our parents grew up; perhaps even our grandparents too, but family history before that is degraded like Chinese whispers. Written records reach back a little further, but only a little. Writing is at most a few thousand years old and 'Eve' lived, if at all, hundreds of thousands of years ago. DNA 'tradition' is a much better bet. It is far more high-fidelity than linguistic tradition. Some DNA documents survive from the common ancestor of all life, and the fidelity of DNA copying far surpasses the most meticulous scribe. Our DNA is like a family Bible faithfully transcribed into all our cells at birth with only an occasional error, but it has one drawback: sex. Our father's text and our mother's are inextricably shuffled. Only some tens of generations back, your ancestry and mine merge into the same few individuals, and it becomes hard to sort out the migrations and diasporas that we are trying to reconstruct.

But there is a parallel strand of DNA tradition that is blessedly free from sexual contamination: mitochondrial DNA. Mitochondria are tiny, self-reproducing bodies, thousands of them in each of our cells, almost certainly the lineal descendants of free-living bacteria although they have lived intimately, and indispensably, in our cells for probably 2,000 million years and they are vital for every second of our existence. Mitochondria have their own, entirely independent, DNA, and now here is the point. They reproduce asexually inside us and we receive them only from our mothers, not our fathers. Males have mitochondrial DNA but essentially none of it do they pass on to their children.* The ordinary DNA in my cell nuclei is a mixture of the DNA from all my ancestors, one sixteenth part coming from each of my great-great-grandparents, for example. The DNA in my mitochondria, on the other hand, came from my mother's mother's mother's mother. No part of this rich text came from any of the rest of my sixteen great-great-grandparents. Follow your pedigree back to the nth generation and, no matter how large you make n, only one ancestress of that generation gave you all your mitochondrial DNA just as only one ancestor gave you your surname.

As the generations go by, mitochondrial DNA is subject to random mutational change like any other DNA. Actually it changes rather faster than ordinary nuclear DNA because it lacks sophisticated proofreading mechanisms, and it changes at a roughly constant rate, which can be calibrated using lineages with a known fossil history. This means that you can take any two people and work out how long ago they shared a common ancestor along the pure female female female line. When this was first done, by workers in the laboratory of the late Allan Wilson, two rather interesting conclusions seemed to emerge. First, the grand ancestress lived surprisingly recently: one or only a few hundreds of thousands of years ago. This finding is comparatively uncontroversial. The second, more controversial conclusion concerns where in the world 'Eve' lived. To understand it, we need to understand what a cladogram is.

---

*Although still widely accepted, this fact has been called into question (see p. 191). This is my only reason for the qualifier 'essentially'.

A cladogram is a bifurcating tree of cousinship. Here is a very tiny cladogram to show the principle.

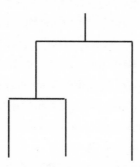

Ingroup cousins

Outgroup. Equally close to both ingroup cousins

Closer cousins are identified as those who share the most characteristics (in this case mitochondrial DNA characteristics). For every pair of cousins, at a given level of distance, there will be an 'outgroup' – a more distant cousin or set of cousins. The outgroup is the next most closely related to the ingroup cousins, and it is equally related to all members of the ingroup because the members of the ingroup are connected via the same, more recent, common ancestor.

My diagram shows a very simple tree with only three terminating twigs. Wilson and his colleagues constructed a tree for modern humans with more than a hundred twigs, each one representing mitochondria from a person of a particular local group. The tree is constructed by a computer attempting to minimize the number of unparsimonious convergences of the mitochondrial text messages. To see the meaning of this, note that it is theoretically possible that humans and rhinos are close cousins, with chimpanzees the outgroup. But this would be a very unparsimonious cladogram and it would be rejected, because all the many resemblances between humans and chimps would have to have been acquired convergently. Wilson and his colleagues looked for the most parsimonious cladogram of human mitochondria and came to the following striking conclusion. All the non-African DNA constitutes a single clade – a single united sub-branch of cousinship. The outgroup of this non-African group was an African group. And the outgroup of *this* larger grouping was another African group, and so on. If this pattern is correct, it strongly suggests that all non-Africans have a more recent common ancestor than the set of all Africans. This, in turn, suggests that the common ancestor of the non-Africans and the Africans lived in Africa.

By the way, don't be misled by the 'Eve' title into thinking that she was lonely! To say that all surviving humans are descended from one female is not to say that she was the only one around. All it means is that the other females' descendants in the purely female line have died out: a lineage of mitochondria can go extinct exactly as easily as a noble surname. Nor does it mean that she was the most recent common ancestor to exist. On the contrary. She is only the most recent common ancestor if we confine our reckoning exclusively to the female line. It is statistically almost certain that our most recent common ancestor, counting male as well as female lines, lived more recently than the mitochondrial common ancestor.

To get the full, startling impact of the African Eve theory in its pure form, consider the following remarkable *sequitur*. The primary divisions among humankind separate Africans from other Africans, and these from yet other Africans. The whole of the rest of humanity put together – Europeans, Australian Aborigines, Chinese, Native Americans and all – are lumped together in one relatively small subdivision, on a par with a small subset of Africans. To put it another way, if you wanted to preserve all the main branches of the human species, you could wipe out the whole world except Africa south of the Sahara and down to the Kalahari. It doesn't, on the face of it, seem plausible. To the superficial and possibly prejudiced gaze, the rest of the world seems to offer more variation among its humans than Africa alone. Certainly, if the theory is true, there must have been considerable, rather rapid, evolutionary modification to have produced the great variety of non-African types since the supposed recent emigration. Jonathan Kingdon is prepared to go along with this. Indeed, the evolutionary specialization of peoples to different parts of the non-African world is one of his principal topics.

The theory of a recent origin in Africa is indeed startling, and it is not surprising that it is controversial. Some anthropologists favour an ancient origin for our racial divergences, going back even to different local populations of *Homo erectus*. A greater number accept the evidence for a recent common ancestor of all Moderns, but dispute whether she necessarily lived in Africa. The principal reason for their doubts is this. I said that the computer is programmed to look for the most parsimonious tree, the tree that minimizes the number of convergences that have to be postulated. Unfortunately it is difficult to home in on the most parsimonious tree, and it is all too easy to think you've found it when you haven't. Here is the problem. The number of possible bifurcating trees is unthinkably large (it would take several lines of print to write out all the digits). All these trees have in theory to be looked at by the computer, but to do so would take the world's fastest supercomputer more time than the universe has existed. In practice the computer has to use some kind of intelligent sampling or 'heuristic' procedure. Unless you are careful, irrelevant trivialities such as the order in which your computer looks at the trees can have a crucial effect. You can come up with what you think is the most parsimonious tree, only to discover subsequently that there are lots more parsimonious trees that your computer didn't get around to trying.

More recent analyses of the mitochondrial data have found several other, equally parsimonious trees. Indeed, there are probably at least a thousand million trees sharing the top rank for parsimony! And many of these equally parsimonious trees are rooted in places other than Africa. So the mitochondrial evidence is at least compatible with an 'Asian Eve' or an 'Australasian Eve'. The issue remains open. The 'out of Africa' theory is still accepted by many biologists and Jonathan Kingdon is among them. They are probably right and there is fossil evidence in their favour.

If the African Eve theory turns out to be wrong, Kingdon would not, I think, have much difficulty in rescuing the main themes of his book. Indeed there are respects in which an Asian or an Indonesian Eve might actually suit him better. At times he almost seems to see the whole Indian Ocean fringe, the Coast of Coromandel, the Andaman Islands and the great south-eastern archipelago through Indonesia to Australia as a kind of honorary extension of Africa. This brings me to his own theory of the 'Banda strandlopers'.

Popular imagination has been caught by what Kingdon calls the 'eccentric theory' of an aquatic phase in our ancestry. As originally suggested by Sir Alister Hardy in apologetic old age and ably championed more recently by Elaine Morgan, the theory refers to an earlier, pre-*Australopithecus* phase in hominid history. In this form Kingdon does not give the idea serious consideration. Nevertheless he feels that it 'had the virtue of drawing attention to lakes and seashores as prime habitats'. He conjures up a haunting vision of 'Banda strandlopers', migrating seashore folk, fishing and beachcombing around the Red Sea and along the Indian Ocean coasts from Africa via India to the Far East. They built rafts and then canoes and became island-hoppers all the way to Australia. They get their name not from Africa but from the Banda Sea off Indonesia. They were opportunists and brilliant technologists. Their technology made it possible for them to migrate and in turn speeded up their evolution. Not only are island archipelagos the classic workshops of evolutionary divergence. For Kingdon, as we shall see, technology had a more focused influence on human genetic evolution.

The Banda strandlopers could be called the heroes of Kingdon's book. They are an exciting creation of a naturalist and artist's imagination. One would have expected that they would have been introduced with a fanfare, and their importance rammed home with the poetic rhetoric of which Kingdon is demonstrably capable. But the Banda don't make a proper entrance; they steal unnoticed into the reader's consciousness. We gradually become aware that some mysterious beings of whom we have not heard, the Banda strandlopers, are among us. Allusions to them increase in frequency, and the bewildered reader begins to fret. Did I miss something? Was there another chapter that dropped out of my copy? Who are these strandlopers? We are never really told. We gradually piece the idea together, in the way that a child assembles its developing conception of the world. When we finally have the picture of Kingdon's strandlopers it is a haunting and imaginative as well as a plausible one, but the reader has to shoulder much of the responsibility for painting it. I can only think that Kingdon himself has been living with the Banda strandlopers in his mind for so long that he no longer realizes the need for others to be properly introduced to them. A good editor could have pointed this out. Literary folk might suspect that Kingdon deliberately intended the strandlopers to tiptoe incognito into the book, their modest entrance being some kind of metaphor for their piecemeal, unheralded arrival in Asia all those centuries ago. I am sure not, for that is the kind of literary tricksterism to which a scientist and an honest man does not stoop.

The reader also has to work pretty hard to descry the book's eponymous theme of 'self-made man'. We get it in the jacket blurb and the preface and the introduction. But in the book itself this important theme is muted almost to vanishing point. It isn't that Kingdon ignores the details of primitive technology. On the contrary, his treatment of fire use, boat-building, stone tools, woomeras and string is detailed and enthralling, like an expert guided tour of the Pitt Rivers Museum. I was absorbed by his imaginative speculations on the biological origins of important inventions – basket-weaving imitated from weaver birds, string developed from spider silk, probably by children. But the theory of technology as an influence and guide for genetic evolution is not forcefully laid out. Here is an inkling of it, on page 37:

In an extraordinary mimicry of natural adaptations, these primates began to manipulate elements and use materials in a way that rapidly multiplied the number of ecological niches they could invade. Each new tool opened possibilities that were formerly the prerogative of very specialised animals. Where diggers had needed heavy nails, now there were stone picks, cats no longer had the monopoly of sharp claws, spears mimicked horns, porcupine quills or canine teeth and so on. Here, for the first time, was an animal that was learning a multiplicity of roles via the invention of technology. An increasing number of animals now had a new competitor that would encroach on at least a part of their former niche.

And here is a relevant passage from the introduction:

Those familiar contours of the face in the mirror or on the street evolved only after tools and technology had become a decisive force in our destiny . . . For instance, it was cooking and processing food that removed the need for massive jaws and teeth and changed the proportions of our faces. It was control of fire that let us invade the cold north and it was building boats that took us to new islands and continents. Such adventures arose from a material culture that was self-made rather than the result of biological adaptation but they precipitated changes in physique that can still be seen in the appearance and complexions of living people . . . As prehistoric humans dispersed out of Africa their obvious adaptations, such as black or white skins, were made not to climate alone but also to the technologies and cultures that pushed humans beyond their biological norms.

Natural selection is differential survival of genes in gene pools. Whether a gene survives through the generations depends upon its effects upon the successive bodies in which it finds itself. Normally this means effects upon the survival skills and reproductive success of those bodies. Skills and success are not measured in a vacuum: they depend upon the environment. Environment may be interpreted conventionally as meaning weather, soil conditions, temperature and so on. It also includes important biological components – predators and prey, parasites and hosts, social rivals, potential mates, offspring and parents. Somewhat less conventionally, I find it important to emphasize that the environment of any one gene includes all the other genes that it is likely to meet in its journey down the generations, and in a sexual species this means all the other genes in the species. Natural selection of genes against the background of other genes in the species gives rise to so-called co-adapted gene complexes. To come finally to Kingdon's point: the environment, in the special case of humans, also includes technology.

Let me carry the theme a little beyond Kingdon's book but not, I think, beyond where he would be prepared to go. Humans live not only in a real world but in a virtual world, created inside our own skulls. It is a simulated, imaginary world, bearing some resemblance to the real world but simplified in the same way as a computer simulation is simplified. It is a dynamic simulation, continually updated by intelligence from the real world. It is an artefact of a kind. Some schools of social science would say that it is a collective artefact, put together by a whole society. If that is going too far, it must at least be true that the virtual worlds of all individuals brought up in a given culture share many features which differ between cultures: religion and superstitions, language, folklore, aesthetic and ethical values, suppositions about the 'natural' way to classify things and about the natural suture lines of perceived reality. What I want to suggest is that this virtual environment is an important member of the set of environments in which genes have to survive; not virtual genes, real ones, winnowed by natural selection in just the same kind of way as genes are winnowed in 'real' environments. There is nothing mystical or fey about this suggestion. At some level the genes must be selected by agents in the real world. But this does not stop it being simultaneously sensible to speak of the virtual reality in our skulls as an environment in which genes are naturally selected and, for what it is worth, I offer the thought as a natural extension of Kingdon's central thesis.*

Kingdon ends his book by showing, with eloquence and passion, that his heart is in the right place. His scorn for the ignorant bigotry of racism is especially moving coming from a white African. As an African and as a lover of wild creatures and places, his loathing for GATT, the World Bank and the unspeakable Japanese whaling industry comes across with stinging conviction.

> Africa's savannas being the primary habitat of mankind, their fate is of obvious concern. Like tropical forest, the savanna is a target for some of the most destructive so-called 'development' that is taking place anywhere on earth. The cowboys and pirate loggers are not picturesque outlaws in remote ranches and log cabins; they are the Great and the Good, ministers, bank presidents and captains of industry . . . Primitive attitudes to the wilderness have been encouraged by banks and businesses, because what else is the tradition of making a fast buck if not primitive? . . . It is popularly promoted that poachers are the main threat to African wildlife, but poachers are generally marginal to the onslaught being made on wildlife habitats by livestock and timber industries that are being encouraged to expand, and expand again, by GATT, the World Bank and national governments. The extinction of some of Africa's remaining ungulates and of the spectacular carnivores that prey upon them could ultimately be the responsibility of international markets that promoted the conversion of viable and very ancient ecosystems into chemical-dependent beef-lots.

There speaks a good man. Listen to him.

---

* See also pp. 226–8.

WE ARE STARDUST

Foreword to *My Name is Stardust* by Bailey and Douglas Harris (2017).

Publishers of children's books do their noble best to second-guess what children like to read. But the best possible judges would surely be children themselves. And when a book is itself written by a highly intelligent child, isn't that best of all? Bailey Harris began work on this book when she was eight, around the age of her target audience. She honed her explaining skills, and grew her knowledge of science, through exposure to bullies in her Utah school who tried to tell her she'd go to hell if she didn't believe in the Book of Mormon. That's why she originally wanted to call her first book the Book of Truth. Now fourteen, she's invited me to introduce the second edition and I'm delighted to do so.

Bailey was the youngest presenter at the 2019 CSICon (Committee for Skeptical Inquiry conference) in Las Vegas, and she was – as everyone agreed – a star. She is also stardust, and so are we all.

'Hello everyone! My name is Stardust and this is my little brother, Vincent.' Did ever a book have a more charming beginning? Actually it's the beginning of another of Bailey's Stardust books, but the charm runs equally through all of them. This one has 375-million-year-old Tiktaalik, one of my favourite animals. Also another of my favourite animals, Lucy, who lived 3 million years ago. Some would say Lucy wasn't an animal but a person. Others, including me, would say she was both, for we too are animals – we are African apes, closer cousins to chimpanzees than chimps are to gorillas. Tiktaalik may not be our direct ancestor but he was pretty much like our gazillion-greats-grandfather, the fish who first crawled out on land. Our several-hundred-thousand-greats-grandmother might have met Lucy, so it's thrilling for us to meet her in Bailey's book – and see her in one of Natalie Malan's beautiful illustrations.

We are stardust: you and me, Lucy and Tiktaalik, dinosaurs, bacteria and redwood trees. Let Stardust Bailey tell you all about it.

THE DESCENT OF
EDWARD WILSON

This review of Edward O. Wilson's *The Social Conquest of Earth*
appeared in *Prospect*, June 2012.

When he received the manuscript of *The Origin of Species*, John
Murray, the publisher, sent it to a referee who suggested that Darwin
should jettison all that evolution stuff and concentrate on pigeons.
It's funny in the same way as the spoof review of *Lady Chatterley's
Lover*, which praised its interesting 'passages on pheasant raising,
the apprehending of poachers, ways of controlling vermin, and
other chores and duties of the professional gamekeeper' but added:

> Unfortunately one is obliged to wade through many pages of
> extraneous material in order to discover and savour these side-
> lights on the management of a Midland shooting estate, and in
> this reviewer's opinion this book can not take the place of J. R.
> Miller's *Practical Gamekeeping*.

I am not being funny when I say of Edward Wilson's latest book
that there are interesting and informative chapters on human evolution, and on the ways of social insects (which he knows better
than any man alive), and it was a good idea to write a book compar-
ing these two pinnacles of social evolution, but unfortunately one is

obliged to wade through many pages of erroneous and downright perverse misunderstandings of evolutionary theory. In particular, Wilson now rejects 'kin selection' (I shall explain this below) and replaces it with a revival of 'group selection' – the poorly defined and incoherent view that evolution is driven by the differential survival of whole groups of organisms.

Nobody doubts that some groups survive better than others. What is controversial is the idea that differential group survival drives evolution, as differential individual survival does. The American grey squirrel is driving our native red squirrel to extinction, no doubt because it happens to have certain advantages. That's differential group survival. But you'd never say of any part of a squirrel that it evolved to promote the welfare of the grey squirrel over the red. Wilson wouldn't say anything so silly about squirrels. He doesn't realize that what he does say, if you examine it carefully, is as implausible and as unsupported by evidence.

I would not venture such strong criticism of a great scientist were I not in good company. The Wilson thesis is based on a 2010 paper that he published jointly with two mathematicians, Martin Nowak and Corina Tarnita. When this paper appeared in *Nature* it provoked very strong criticism from more than 140 evolutionary biologists, including a majority of the most distinguished workers in the field. They include Alan Grafen, David Queller, Jerry Coyne, Richard Michod, Eric Charnov, Nick Barton, Alex Kacelnik, Leda Cosmides, John Tooby, Geoffrey Parker, Steven Pinker, Paul Sherman, Tim Clutton-Brock, Paul Harvey, Mary Jane West-Eberhard, Stephen Emlen, Malte Andersson, Stuart West, Richard Wrangham, Bernard Crespi, Robert Trivers and many others. These may not all be household names but let me assure you they know what they are talking about in the relevant fields.

I'm reminded of the old *Punch* cartoon where a mother beams down on a military parade and proudly exclaims, 'There's my boy, he's the only one in step!' Is Wilson the only evolutionary biologist in step? Scientists dislike arguing from authority, so perhaps I shouldn't have mentioned the 140 dissenting authorities. But one can make a good case that the 2010 paper would never have been published in *Nature* had it been submitted anonymously and subjected to ordinary peer review, bereft of the massively authoritative name of Edward O. Wilson. If it was authority that got the paper published, there is poetic justice in deploying authority in reply.

Then there's the patrician hauteur with which Wilson ignores the very serious drubbing his *Nature* paper received. His book doesn't even mention those many critics: not a single, solitary sentence. Does he think his authority justifies going over the heads of experts and appealing directly to a popular audience, as if the professional controversy didn't exist – as if acceptance of his (tiny) minority view were a done deal? 'The beautiful theory [kin selection, see below] never worked well anyway, and now it has collapsed.' Yes it did and does work, and no it hasn't collapsed. For Wilson not to acknowledge that he speaks for himself against the great majority of his professional colleagues is – it pains me to say this of a lifelong hero – an act of wanton arrogance.

The argument from authority, then, cuts both ways, so let me now set it aside and talk about evolution itself. At stake is the level at which Darwinian selection acts: 'survival of the fittest' but, to quote Wilson's fellow entomologist-turned-anthropologist R. D. Alexander, the fittest what? The fittest gene, individual, group, species, ecosystem? Just as a child may enjoy addressing an envelope: Oxford, England, Europe, Earth, Solar System, Milky Way Galaxy, Local Group, Universe, so biologists with non-analytical minds warm to multi-level selection: a bland, unfocused ecumenicalism of the sort promoted by (the association may not delight Wilson) the late Stephen Jay Gould. Let a thousand flowers bloom and let Darwinian selection choose among all levels in the hierarchy of life. But it doesn't stand up to serious scrutiny. Darwinian selection is a very particular process, which demands rigorous understanding.

The essential point to grasp is that the gene doesn't belong in the hierarchy I listed. It is on its own as a 'replicator', with its own unique status as a unit of Darwinian selection. Genes, but no other units in life's hierarchy, make exact copies of themselves in a pool of such copies. It therefore makes a long-term difference which genes are good at surviving and which ones bad. You cannot say the same of individual organisms (they die after passing on their genes and never make copies of themselves). Nor does it apply to groups or species or ecosystems. None make copies of themselves. None are replicators. Genes have that unique status.

Evolution, then, results from the differential survival of genes in gene pools. 'Good' genes become numerous at the expense of 'bad'. But what is a gene 'good' at? Here's where the organism enters the stage. Genes flourish or fail in gene pools, but they don't float freely in the pool like molecules of water. They are locked up in the bodies of individual organisms. The pool is stirred by the process of sexual reproduction, which changes a gene's partners in every generation. A gene's success depends on the survival and reproduction of the bodies in which it sits, and which it influences via 'phenotypic' effects. This is why I have called the organism a 'survival machine' or 'vehicle' for the genes that ride inside it. Genes that happen to cause slight improvements in squirrel eyes or tails or behaviour patterns are passed on because individual squirrels bearing those improving genes survive at the expense of individuals lacking them.* To say that genes improve the survival of groups of squirrels is a mighty stretch.

With the exception of one anomalous passage in *The Descent of Man*, Darwin consistently saw natural selection as choosing between individual organisms. When he adopted Herbert Spencer's phrase 'survival of the fittest' at the urging of A. R. Wallace, 'fittest' meant something close to its everyday meaning, and Darwin applied it strictly to organisms: the strongest, swiftest, sharpest

---

* That is to say, rival individuals within the same species – red squirrel rivalling red squirrel, grey squirrel rivalling grey.

of tooth and claw, keenest of ear and eye. Darwin well understood that survival was only a means to the end of reproduction, so the 'fittest' should include the most sexually attractive, and the most diligent and devoted parents.

Later, when twentieth-century leaders of what Julian Huxley called the 'Modern Synthesis' deployed mathematics to unite Darwinism with Mendelian genetics, they co-opted 'fitness' to serve as a variable in their equations. 'Fitness' became 'that which is maximized in natural selection'. 'Survival of the fittest' thus became a tautology, but it didn't matter for the equations. The 'fitness' of an individual lion, say, or cassowary, became a mathematical expression of its capacity to leave surviving children, or grandchildren, or descendants into the indefinite future. Parental care and grandparental care contribute to individual fitness because an individual's descendants are vehicles in which ride copies of the genes that engineer the caring.

But lineal descendants are not the only such vehicles. In the early 1960s, W. D. Hamilton, arguably the most distinguished Darwinian since R. A. Fisher, formalized an idea that had been knocking around since Fisher and Haldane. If a gene happens to arise which works for the benefit of a sibling, say, or a niece, that gene can survive in the same kind of way as a gene that works for the benefit of offspring or grandchildren. A gene for sibling care, under the right conditions, has the same chance of surviving in the gene pool as a gene for parental care. A copy is a copy is a copy, whether it sits in a lineal or a collateral relative.

But the conditions have to be right, and in practice they often aren't. Full siblings are usually harder to identify than offspring, and usually less obviously dependent. For practical reasons, therefore, sibling care is rarer in nature than parental care. But as far as Darwinian principle is concerned, sibling care and parental care are favoured for the same reason: the cared-for individual contains copies of the genes that programme the caring behaviour.

Half-siblings, nephews, nieces and grandchildren are half as likely as full siblings or offspring to share a caring gene. First cousins are half as likely again, and are harder to identify. Hamilton summarized all this in the form of a simple equation, which has become known as Hamilton's Rule. A gene for altruism towards kin will be favoured if the cost to the altruist C is outweighed by the benefit to the recipient B devalued by r, which is a subtle but computable index of probability of sharing genes.* For example, r for full siblings and parents and offspring is ½; r for grandchildren, half-siblings, nephews and nieces is ¼; r for first cousins is ⅛, and so on. A gene for altruistic care will spread through the population if $rB > C$. It is extremely important not to forget B and C and conclude that only r matters in evaluating the success of the theory in particular cases. I am sorry to say that Wilson, in his allegation that Hamilton's ideas don't apply to particular cases, comes perilously close to doing just that. It is as though r is so interesting and novel that B and C are overshadowed.

Hamilton replaced 'classical fitness' (which took account only of

---

* The exact meaning of r and how to compute it is complicated and was beyond the scope of this review. It confuses students who have read that we share more than 90 per cent of our genes with chimpanzees, 60 per cent with fruit flies and 40 per cent with bananas. How do we reconcile that with '50 per cent shared by full siblings'? The problem disappears for the case of rare genes (for example, recent mutations), where siblings, but not chimps or bananas, really do have an approximately 50 per cent probability of sharing genes. But this is unsatisfactory, because one of Hamilton's achievements was to show that his Rule applies to common genes as well as rare ones. Hamilton himself referred to proportions of genes 'identical by descent'. This is correct, but not the easiest way to understand the problem. Alan Grafen resolved the paradox in an elegant geometric visualization published in *Oxford Surveys in Evolutionary Biology* in 1985. My own preferred verbal explanation is that siblings share 50 per cent of their genes *over and above* a baseline shared with all members of the population.

lineal descendants) by 'inclusive fitness', which is a carefully weighted sum embracing collateral as well as lineal kin. I have informally (and a touch facetiously but with Hamilton's blessing) defined inclusive fitness as 'that quantity which an individual will appear to maximize, when what is really being maximized is gene survival'. In his previous books, Wilson was a supporter of Hamilton's ideas, but he has now turned against them in a way that suggests to me that he never really understood them in the first place.

'Inclusive fitness' was coined as a mathematical device to allow us to keep treating the individual organism ('vehicle') as the level of agency, when we could equivalently have switched to the gene ('replicator'). You can say that natural selection maximizes individual inclusive fitness, or that it maximizes gene survival. The two are equivalent, by definition. On the face of it, gene survival is simpler to deal with, so why bother with individual inclusive fitness? Because the organism has the appearance of a purpose-driven agent in a way that the gene does not. Genes lack legs to pursue goals, sense organs to perceive the world, hands to manipulate it. Gene survival is what ultimately counts in natural selection, and the world becomes full of genes that are good at surviving. But they do it vicariously, by embryologically programming 'phenotypes': programming the development of individual bodies, their brains, limbs and sense organs, in such a way as to maximize their own survival. Genes program the embryonic development of their vehicles, then ride inside them to share their fate and, if successful, get passed on to future generations.

So, 'replicators' and 'vehicles' constitute two meanings of 'unit of natural selection'. Replicators are the units that survive (or fail to survive) through the generations. Vehicles are the agents that replicators program as devices to help them survive. Genes are the primary replicators, organisms the obvious vehicles. But what about groups? As with organisms, they are certainly not replicators, but are they vehicles? If so, might we make a plausible case for 'group selection'?

It is important not to confuse this question – as Wilson regrettably does – with the question of whether individuals benefit from living in groups. Of course they do. Penguins huddle for warmth. That's not group selection: every individual benefits. Lionesses hunting in groups catch more and larger prey than a lone hunter could: enough to make it worthwhile for everyone. Again, every individual benefits: group welfare is strictly incidental. Birds in flocks and fish in schools achieve safety in numbers, and may also conserve energy by riding each other's slipstreams – the same effect as racing cyclists sometimes exploit.

Such individual advantages in group living are important but they have nothing to do with group selection. Group selection would imply that a group does something equivalent to surviving or dying, something equivalent to reproducing itself, and that it has something you could call a group phenotype, such that genes might influence its development, and hence their own survival.

Do groups have phenotypes, which might qualify them to count as gene vehicles? Convincing examples are vanishingly hard to find. The classic promoter of group selection, the Scottish ecologist V. C. Wynne-Edwards, suggested that territoriality and dominance hierarchies ('peck orders') might be group phenotypes. Territorial species are more spaced out, and species with peck orders show less overt aggression. But both phenomena are more parsimoniously treated as emergent manifestations of individual phenotypes, and it is individual phenotypes that are directly influenced by genes. You may choose to treat a dominance hierarchy as a group phenotype if you insist, but it is better seen as emerging from each hen, say, being genetically programmed to learn which other hens she can beat in a fight and which normally beat her.

But what about the social insects, Wilson's area of expertise? Hamilton's, too, and indeed the social insects were an early, stunningly successful showcase for his theory.

Female bees, ants and wasps are genetically capable of developing into fertile queens or sterile workers. Each individual is switched into either the queen pathway or the worker pathway (one of several worker pathways in ants) by an environmental switch, and the point is utterly crucial. No gene for outright sterility could survive. But a gene for sterility under some environmental conditions but not others could easily be favoured, and it was. A female bee larva fed on royal jelly and housed in a large queen cell will develop into a fertile queen. Otherwise she will develop into a sterile worker. Genes that find themselves in sterile bodies programme them to work for copies of the same genes in fertile bodies – either the old queen (their mother), or young queens (their sisters*) or young males. The result is that queens evolve to become more efficient, full-time specialist egg-layers, with all their needs taken care of by their sterile daughters or sisters.

Because of how the B, C and r values in Hamilton's Rule turn out for bees, genes for sterility are favoured under some conditions, hyper-fertility under others. The same is true for ants and wasps; and termites but with differences of detail (for example, termites have male as well as female workers – alas I have no space to expound Hamilton's elegant explanation of this difference and many other intriguing facts). With more differences of detail, the same is true for some non-insect species such as naked mole rats and a few crustaceans.

It truly is a beautiful theory. Everything fits, exactly as it should. Darwin himself, with characteristic prescience but using the pre-genetic language of his time, got the point. As so often, he drew inspiration from domestication:

> Thus, a well-flavoured vegetable is cooked, and the individual is destroyed; but the horticulturist sows seeds of the same stock, and confidently expects to get nearly the same variety; breeders of cattle wish the flesh and fat to be well marbled together; the

---

* Or nieces.

animal has been slaughtered, but the breeder goes with confidence to the same family. I have such faith in the powers of selection, that I do not doubt that a breed of cattle, always yielding [sterile] oxen with extraordinarily long horns, could be slowly formed by carefully watching which individual bulls and cows, when matched, produced oxen with the longest horns; and yet no one ox could ever have propagated its kind.

In modern, Hamiltonian terms we would interpret Darwin's 'seeds of the same stock' as sharing genes with the vegetable that has been cooked. The sterile ox with the long horns shares genes with the same stock from which we breed. Darwin, lacking the concept of the discrete, Mendelian gene, spoke of going with confidence to the 'same family' rather than the same genes. Wilson now interprets this as a form of 'group selection', the 'group' in this case being the family. But what a staggeringly unpenetrating – even perverse – use of language. Kin share genes, that is the point, and Darwin would have loved it. The fact that a family can also be seen as a 'group' is entirely beside the point and an unhelpful distraction from it.

When Hamilton's twin papers on inclusive fitness were first published in 1964, John Maynard Smith, who was the referee chiefly responsible for recommending them, published a short paper in *Nature* in which he called attention to Hamilton's brilliant innovation. Maynard Smith coined the phrase 'kin selection' specifically in order to distinguish it from group selection, then in the process of being discredited by him, and others such as the ecologist David Lack. Soon after this, Wilson, in *The Insect Societies* (1971), enthusiastically adopted Hamilton's ideas. He continued to press them in *Sociobiology* (1975), but in an oddly misleading way which indicates that he was already flirting with a watered-down version of his current folly. He treated kin selection as a special case of group selection, an error which I was later to highlight in my paper on 'Twelve Misunderstandings of Kin Selection' as Misunderstanding

Number Two.* Kin may or may not cling together in a group. Kin selection works whether they do or not.

Misunderstanding Number One, which is also perpetrated by Wilson, is the fallacy that 'kin selection is a special, complex kind of natural selection, to be invoked only when the allegedly more parsimonious "standard Darwinian theory" proves inadequate'. I hope I have made it clear that kin selection is logically entailed by standard Darwinian theory, even if the B and C terms work out in such a way that collateral kin are not cared for in practice. Natural selection without kin selection would be like Euclid without Pythagoras. Wilson is, in effect, striding around with a ruler, measuring triangles to see whether Pythagoras got it right. Kin selection was always logically implied by the neo-Darwinian synthesis. It just needed somebody to point it out – Hamilton did it.

Edward Wilson has made important discoveries of his own. His place in history is assured, and so is Hamilton's. Please do read Wilson's earlier books, including the monumental *The Ants*, written jointly with Bert Hölldobler (yet another world expert who will have no truck with group selection). As for the book under review, the theoretical errors I have explained are important, pervasive, and integral to its thesis in a way that renders it impossible to recommend. To borrow from Dorothy Parker, this is not a book to be tossed lightly aside. It should be thrown with great force. And sincere regret.

---

* The article is reproduced in *Science in the Soul.*

IV

THE MINER'S CANARY: SUPPORTING SCEPTICISM

IN CONVERSATION WITH CHRISTOPHER HITCHENS

IS AMERICA HEADING FOR THEOCRACY?

I was invited to guest-edit the 2011 Christmas issue of *New Statesman*, and in this role spoke to Christopher Hitchens in what was to be his final interview before his much-lamented death. The text reproduced here is the second part of the interview, which focused on the central theme of this section, 'supporting scepticism'. We met in the garden of the beautiful house that he and his wife had been lent during his treatment for cancer in Houston, Texas. Our long conversation, which I recorded with no fewer than three devices for fear of missing anything, was followed by a delicious dinner which he was too ill to eat but not too ill to enliven with his sparkling conversation. The Atheist Alliance International had elected him the 2011 winner of the Richard Dawkins Award, and it was my honour to present the award to him at the Texas Freethought Convention the day after our interview and dinner (https://www.youtube.com/watch?v=Bo6tiTwAuvg). It was the last time I saw him.

If Christianity loses ground in the West, will Islam replace it? Why is America, founded as a secular republic, so much more religious today than much of western Europe? Should children be taught about religion in schools?

RD: Do you think America is in danger of becoming a theocracy?

CH: No, I don't. The people who we mean when we talk about that – maybe the extreme Protestant evangelicals, who do want a God-run America and believe it was founded on essentially fundamentalist Protestant principles – I think they may be the most overrated threat in the country.

RD: Oh, good.

CH: They've been defeated everywhere. Why is this? In the 1920s, they had a string of victories. They banned the sale, manufacture and distribution and consumption of alcohol. They made it the constitution. They more or less managed to ban immigration from countries that had non-Protestant, non-white majorities. From these victories, they have never recovered. They'll never recover from [the failure of] Prohibition. It was their biggest defeat. They'll never recover from the Scopes trial. Every time they've tried [to introduce the teaching of creationism], the local school board or the parents or the courts have thrown it out and it's usually because of the work of people like you, who have shown that it's nonsense. They try to make a free speech question out of it but they will fail with that, also. People don't want to come from the town or the state or the county that gets laughed at.

In all my tours around the South, it's amazing how many people – Christians as well – want to disprove the idea that they're all in thrall to people like [the fundamentalist preacher Jerry] Falwell. They don't want to be a laughing stock.

And if they passed an ordinance saying there will be prayer in school every morning from now on, one of two things would happen: it would be overthrown in no time by all the courts, with barrels of laughter heaped over it, or people would say: 'Very well, we're starting with Hindu prayer on Monday.' They would regret it so bitterly that there are days when I wish they would have their own way for a short time.

RD: Oh, that's very cheering.

CH: I'm a bit more worried about the extreme, reactionary nature of the papacy now. But that again doesn't seem to command very big allegiance among the American congregation. They are disobedient on contraception, flagrantly; on divorce; on gay marriage, to an extraordinary degree that I wouldn't have predicted; and they're only holding firm on abortion, which, in my opinion, is actually a very strong moral question and shouldn't be decided lightly. I feel very squeamish about it. I believe that the unborn child is a real concept, in other words. We needn't go there,* but I'm not a complete abortion-on-demand fanatic. I think it requires a bit of reflection. But anyway, even on that, the Catholic Communion is very agonized. And also, [when] you go and debate with them, very few of them could tell you very much about what the catechism really is. It's increasingly cultural Catholicism.

RD: That is true, of course.

CH: So, really, the only threat from religious force in America is the same as it is, I'm afraid, in many other countries – from outside. And it's jihadism, some of it home-grown, but some of that is so weak and so self-discrediting.

---

* We didn't go there. It's one of the two areas where I disagreed with Christopher (the other being the Iraq War), but I didn't pursue it because I didn't think it strongly relevant to the point he was making about the decline of Christianity. If I had pursued it I'd have quoted Jeremy Bentham: Can they *suffer*? Before the development of a uniquely human nervous system, a human embryo is obviously less equipped to suffer than an adult cow or pig.

RD: It's more of a problem in Britain.

CH: And many other European countries, where its alleged root causes are being allowed slightly too friendly an interrogation, I think. Make that much too friendly.

RD: Some of our friends are so worried about Islam that they're prepared to lend support to Christianity as a kind of bulwark against it.

CH: I know many Muslims who, in leaving the faith, have opted to go . . . to Christianity or via it to non-belief. Some of them say it's the personality of Jesus of Nazareth. The mild and meek one, as compared to the rather farouche, physical, martial, rather greedy . . .

RD: Warlord.

CH: . . . Muhammad. I can see that that might have an effect.

RD: Do you ever worry that if we win and, so to speak, destroy Christianity, that vacuum would be filled by Islam?

CH: No, in a funny way, I don't worry that we'll win. All that we can do is make absolutely sure that people know there's a much more wonderful and interesting and beautiful alternative. No, I don't think that Europe would fill up with Muslims as it emptied of Christians. Christianity has defeated itself in that it has become a cultural thing. There really aren't believing Christians in the way there were generations ago.

RD: Certainly in Europe that's true – but in America?

CH: There are revivals, of course, and among Jews as well. But I think there's a very long-running tendency in the developed world and in large areas elsewhere for people to see the virtue of secularism, the separation of church and state, because they've tried the alternatives . . . Every time something like a jihad or a sharia movement has taken over any country – admittedly they've only been able to do it in very primitive cases – it's a smouldering wreck with no productivity.

RD: Total failure. If you look at religiosity across countries of the world and, indeed, across the states of the US, you find that religiosity tends to correlate with poverty and with various other indices of social deprivation.

CH: Yes. That's also what it feeds on. But I don't want to condescend about that. I know a lot of very educated, very prosperous, very thoughtful people who believe.

RD: Do you think [Thomas] Jefferson and [James] Madison were deists, as is often said?

CH: I think they fluctuated, one by one. Jefferson is the one I'm more happy to pronounce on. The furthest he would go in public was to incline to a theistic enlightened view but, in his private correspondence, he goes much further. He says he wishes we could return to the wisdom of more than two thousand years ago. That's in his discussion of his own Jefferson Bible, where he cuts out everything supernatural relating to Jesus. But also, very importantly, he says to his nephew Peter Carr in a private letter [on the subject of belief]: 'Do not be frightened from this inquiry by any fear of its consequences. If it ends in a belief that there is no God, you will find incitements to virtue in the comfort and pleasantness you feel in its exercise and the love of others which it will procure you.' Now, that can only be written by someone who's had that experience.

RD: It's very good, isn't it?

CH: In my judgement, it's an internal reading, but I think it's a close one. There was certainly no priest at his bedside. But he did violate a rule of C. S. Lewis's and here I'm on Lewis's side. Lewis says it is a cop-out to say Jesus was a great moralist. He said it's the one thing we must not say; it is a wicked thing to say. If he wasn't the Son of God, he was a very evil impostor and his teachings were vain and fraudulent. You may not take the easy route here and say:

'He may not have been the Son of God and he may not have been the Redeemer, but he was a wonderful moralist.' Lewis is more honest than Jefferson on this point. I admire Lewis for saying that. Rick Perry said it the other day.

RD: Jesus could just have been mistaken.

CH: He could. It's not unknown for people to have the illusion that they're God or the Son. It's a common delusion but, again, I don't think we need to condescend. Rick Perry once said: 'Not only do I believe that Jesus is my personal saviour but I believe that those who don't are going to eternal punishment.' He was challenged at least on the last bit and he said, 'I don't have the right to alter the doctrine. I can't say it's fine for me and not for others.'

RD: So we ought to be on the side of these fundamentalists?

CH: Not 'on the side', but I think we should say that there's something about their honesty that we wish we could find.

RD: Which we don't get in bishops . . .

CH: Our soft-centred bishops at Oxford, and other people, yes.

RD: I'm often asked why it is that this republic [of America], founded in secularism, is so much more religious than those western European countries that have an official state religion, like Scandinavia and Britain.

CH: [Alexis] de Tocqueville has it exactly right. If you want a church in America, you have to build it by the sweat of your own brow and many have. That's why they're attached to them.

RD: Yes.

CH: [Look at] the Greek Orthodox community in Brooklyn. What's the first thing it will do? It will build itself a little shrine. The Jews – not all of them – remarkably abandoned their religion very soon after arriving from the shtetl.

RD: Are you saying that most Jews have abandoned their religion?

CH: Increasingly in America. When you came to escape religious persecution and you didn't want to replicate it, that's a strong memory. The Jews very quickly secularized when they came. American Jews must be the most secular force on the planet now, as a collective. If they are a collective – which they're not, really.

RD: While not being religious, they often still observe the Sabbath and that kind of thing.

CH: There's got to be something cultural. I go to Passover every year. Sometimes, even I have a seder, because I want my child to know that she does come very distantly from another tradition. It would explain if she met her great-grandfather why he spoke Yiddish. It's cultural, but the Passover seder is also the Socratic forum. It's dialectical. It's accompanied by wine. It's got the bones of quite a good discussion in it. And then there is manifest destiny. People feel America is just so lucky. It's between two oceans, filled with minerals, wealth, beauty. It does seem providential to many people.

RD: Promised land, city on a hill.

CH: All that and the desire for another Eden. Some secular utopians came here with the same idea. Thomas Paine and others all thought of America as a great new start for the species.

RD: But that was all secular.

CH: A lot of it was, but you can't get away from the liturgy: it's too powerful. You will end up saying things like 'promised land' and it can be mobilized for sinister purposes. But in a lot of cases, it's a mild belief. It's just: 'We should share our good luck.'

[. . .]

CH: The reason why most of my friends are non-believers is not particularly that they were engaged in the arguments you and I have been having, but they were made indifferent by compulsory religion at school.

RD: They got bored by it.

CH: They'd had enough of it. They took from it occasionally whatever they needed – if you needed to get married, you knew where to go. Some of them, of course, are religious and some of them like the music but, generally speaking, the British people are benignly indifferent to religion.

RD: And the fact that there is an established church increases that effect. Churches should not be tax-free the way that they are. Not automatically, anyway.

CH: No, certainly not. If the Church has demanded that equal time be given to creationist or pseudo-creationist speculations . . . any Church that teaches that in its school and is in receipt of federal money from the faith-based initiative must, by law, also teach Darwinism and alternative teachings, in order that the debate is being taught. I don't think they want this.

RD: No.

CH: Tell them if they want equal time, we'll jolly well have it. That's why they've always been against comparative religion.

RD: Comparative religion would be one of the best weapons, I suspect.

CH: It's got so insipid in parts of America now that a lot of children are brought up – as their parents aren't doing it and leave it to the schools and the schools are afraid of it – with no knowledge of any religion of any kind. I would like children to know what religion is about because [otherwise] some guru or cult or revivalists will sweep them up.

RD: They're vulnerable. I also would like them to know the Bible for literary reasons.

CH: Precisely. We both, I was pleased to see, have written pieces about the King James Bible. The AV [Authorized Version], as it was called in my boyhood. A huge amount of English literature would be opaque if people didn't know it.

RD: Absolutely, yes. Have you read some of the modern translations? 'Futile, said the priest. Utterly futile.'

CH: He doesn't!

RD: He does, honestly. 'Futile, futile said the priest. It's all futile.'

CH: That's Lamentations.

RD: No, it's Ecclesiastes. 'Vanity, vanity.'

CH: 'Vanity, vanity.' Good God. That's the least religious book in the Bible. That's the one that Orwell wanted at his funeral.

RD: I bet he did. I sometimes think the poetry comes from the intriguing obscurity of mistranslation. 'When the sound of the grinding is low, the grasshopper is heard in the land . . . The grasshopper shall be a burden.' What the hell?

CH: The Book of Job is the other great non-religious one, I always feel. 'Man is born to trouble as the sparks fly upward.' Try to do without that. No, I'm glad we're on the same page there. People tell me that the recitation of the Qur'an can have the same effect if you understand the original language. I wish I did. Some of the Catholic liturgy is attractive.

RD: I don't know enough Latin to judge that.

CH: Sometimes one has just enough to be irritated.

RD: Yes [laughs]. Can you say anything about Christmas?

CH: Yes. There was going to be a winter solstice holiday for sure. The dominant religion was going to take it over and that would have happened without Dickens and without others.

RD: The Christmas tree comes from Prince Albert; the shepherds and the wise men are all made up.

CH: Cyrenius wasn't governor of Syria, all of that. Increasingly, it's secularized itself. This 'Happy Holidays' – I don't particularly like that, either.

RD: Horrible, isn't it? 'Happy holiday season.'

CH: I prefer our stuff about the cosmos.

*

When I presented Christopher with his award the day after this interview, the large audience gave him a standing ovation, first as he entered the hall and again at the end of his deeply moving speech. My own presentation speech ended with a tribute,* in which I said that every day he demonstrates the falsehood of the lie that there are no atheists in foxholes: 'Hitch is in a foxhole, and he is dealing with it with a courage, an honesty and a dignity that any of us would be, and should be, proud to muster.'

---

* Reprinted in *Science in the Soul*.

WITNESS OF INTERNAL DELUSION

Foreword to the 2008 paperback edition of Dan Barker's *Godless*.

It isn't difficult to work out that religious fundamentalists are deluded – those people who think the entire universe began after the agricultural revolution; people who believe literally that a snake, presumably in fluent Hebrew, beguiled into sin a man fashioned from clay and a woman grown from him as a cutting: people who find it self-evident that the origin myth that happened to dominate their own childhood trumps the thousands of alternative myths sprung from all the dreamtimes of the world. It is one thing to know that these faith-heads are wrong. My mistake has been naively to think I can remove their delusion simply by talking to them in a quiet, sensible voice and laying out the evidence, clear for all to see. It isn't as easy as that. Before we can talk to them, we must struggle to understand them; struggle to enter their seized minds and empathize. What is it really like to be so indoctrinated that you can honestly and sincerely believe obvious nonsense – believe it with every fibre of your being?

Just as Helen Keller was able to tell us from the inside what it was like to be blind and deaf, so there are rare individuals who have broken the bonds of fundamentalist indoctrination and are also gifted with the articulate intelligence to tell the rest of us what it was like. Some of these memoirs promise much but end up disappointing.

Ed Husain's *The Islamist* gives a good picture of what it is like to be a decent young man gradually sucked in, step by step, to the mental snake-pit of radical Islamism. But Husain doesn't give us a feeling for what it is really like, on the inside, to believe passionately in arbitrary nonsense. And even at the end of the story, when he has escaped from jihadism, it is only the political extremism that he abjures: he seems even now not to have shaken off his childhood belief in Islam itself. Faith still lurks, and one fears for the author that he remains vulnerable. Ayaan Hirsi Ali's *Infidel* is a fascinating and moving account of her escape from the singular oppression that is a woman's lot under Islam (and 'under' is the right word), including the unspeakable barbarism of genital mutilation. But even at her most devout, she was never the kind of zealot who goes around preaching and actively seeking victims to convert. Again, her book doesn't really help the reader to understand mental possession by religious delusion. The most eloquent witness of internal delusion that I know – a triumphantly smiling refugee from the zany, surreal world of American fundamentalist Protestantism – is Dan Barker.

Barker is now one of American secularism's most talented and effective spokespeople – together with his delightful partner (in all senses of the word) Annie Laurie Gaylor. Dan, to put it mildly, was not always thus. He has a truly remarkable tale to tell of his personal history and breakout from the badlands of religious fundamentalism. He was in it right from the start, up to his ears. Dan was not just a preacher, he was the kind of preacher that 'you would not want to sit next to on a bus'. He was the kind of preacher who would march up to perfect strangers in the street and ask them if they were saved: the kind of doorstepper on whom you might be tempted to set the dogs. Dan knows deeply what it is like to be a wingnut, a faith-head, a fully paid-up nutjob, an all-singing, all-glossolaling religious fruit bat. He can take us – simultaneously laughing and appalled – into that bat-belfry world and even make us sympathize. But he also knows what it is like to stumble upon the unaccustomed pleasure of thinking for oneself, without help from anybody else, right in the teeth of opposition from what was then his entire social world. The socially unacceptable habit of *thinking* led him directly to realize that his entire life so far had been a time-wasting delusion.* All by himself, he came to his senses – in a big way. Unusually, he has the verbal skills and the intelligence and the sensitivity to tell us the whole story, step by painful – and yet exhilarating – step.

His account of his early indoctrination into his parents' fundamentalist sect, his unquestioning faith in the literal truth of every word of it, his disturbingly easy facility in 'saving' souls, his successful career as a preacher and musician and composer for Jesus, is riveting. Even more fascinating is the process by which the doubts set in and gradually multiplied in that intelligent but naive young mind. Then there is the pathos with which he tells of the interregnum period when he was already a convinced atheist, but could not quite bring himself to leave the church of which he was a minister – mostly because this was the only life he knew, and he found it hard to face the world outside, or confront his family with the truth. As so often where there is pathos, comedy is not far behind, and there is a sort of dismal comedy in the responses of Dan's religious friends when he finally announced his atheism. In all the many letters he received, not one offered any kind of *reason* why atheistic beliefs might actually be wrong. Perhaps the funniest example is that of the Rev. Milton Barfoot, who said to Dan's brother, in apparently honest bafflement, 'But isn't Dan afraid of hell?' No, Reverend, Dan doesn't believe in hell any more, that's one of the things about being an atheist, you see.

Dan's delay, after he became an atheist and before he resigned from the ministry, carries the likely implication that there are lots

---

* Fortunately he was quite young when it happened. Dan Dennett's observation on the same theme, 'There's simply no polite way to tell people they've dedicated their lives to an illusion', has led me to zip my lip when talking to religiously deluded people towards the end of their lives.

more clergymen out there who have ridden the same course as Dan
but shied at the final fence: reverend atheists who dare not jump
from the only way they know to make a living, dare not lose their
ticket to respect in the limited society in which they move – their
big-fish-hood in a very small pond. How hard must that be to give
up? Fascinatingly, since the publication of his previous book, *Losing Faith in Faith*, Dan Barker himself has become a kind of secret
rallying post for large numbers of now faith-free but still pusillan-
imous clerics. Like a kind of atheistic father confessor, Dan is a
magnet for the disillusioned clergy.* As a good confessor, he will
not betray their confidence as individuals, but there is nothing to
stop him from telling their story in a general way and, once again,
there's comedy mixed in with the pathos.

Dan Barker's own confessor is each reader of this book, and it is
hard not to revel in the role. It is hard to disavow the exultation as
Dan breaks the shackles, and even more so when he is later joined
in unbelief by the parents who had been responsible for his original
religious fervour, and by one of his two brothers. It wasn't that he
turned his preaching skills on his family in reverse, and worked
hard to deconvert them. Rather, it simply had never *occurred* to any
of the family that being an atheist was even an option. As soon as
they had the example of Dan before them, to show that a decent
and good person could be a non-believer, they started thinking for
themselves about the real issues and it didn't take them long to
reach the obvious conclusion. In his mother's case, it only took her
a few weeks to conclude that 'religion is a bunch of baloney' and a
little later she was able to add, happily, 'I don't have to hate anymore.' Dan's father and one of his two brothers followed a similar
course. The other brother remains a born-again Christian. Perhaps
one day he too will see the light.

---

* This magnet effect was one of the factors leading to the Clergy Project,
  of which more in the next two pieces in this collection.

Deconversion stories occupy only the opening chapters of this book. Just as Dan's religiously doped youth gave way to a more ful-filled maturity, so later chapters of his book move on, and give us the generous reflections of a mature atheist. Dan Barker's road from Damascus will, I predict, become well trodden by many others in the future, and this book is destined to become a classic of its kind.

KICKING THE HABIT

Foreword to *Caught in the Pulpit* by Daniel C. Dennett and Linda LaScola (2015).

What must it be like to be frightened of your own opinions? If asked 'What do you think about X?' most of us enjoy the luxury of reflecting: X, well yes, what do I think? What is the evidence bearing on X? What are the good things about X, what are the bad? Maybe I should read up on X before formulating my opinion. Maybe I should discuss X with friends. X is interesting, I'd like to think about X.

Imagine being in a job which forbids any of that: a profession to which you have vowed your life with solemn commitments; and where the terms of the job are that you must hold certain rigid beliefs about the world, the universe, morality and the human condition. You are allowed opinions about football or chimney pots,* but when it comes to the deep questions of existence, origins, much of science, everything about ethics, you are

---

* An obscure tribute to my late godfather, the Reverend Valentine Fletcher, who, among other accomplishments, published the standard work on the British Domestic Chimney Pot and possessed an unrivalled collection of at least two hundred of the things. To become expert in some improbable field is one of many endearing characteristics I associate with the best sort of Anglican clergyman. Darwin himself aspired to become an ordained authority on beetles.

told what to think; or you have to parrot your thoughts from a book written by unknown authors in ancient deserts. If your reading, your thinking, your conversations, lead you to change your opinions you can never divulge your secret. If you breathe a hint of your doubts you will lose your job, your livelihood, the respect of your community, your friends, perhaps even your family. At the same time, the job demands the highest standards of moral rectitude, so the double life you are leading torments you with a wasting sense of shameful hypocrisy. Such is the predicament of those priests, rabbis and pastors who have lost their faith but remain in post. They exist in surprisingly large numbers, and they are the subjects of this fascinating, sometimes harrowing, sometimes uplifting book.

Other jobs might make demands on your skills, but if you are deficient you can do something about it. A lumberjack or a musician can improve with practice: try harder. But lumberjacks and musicians are not required to believe. You can't change your beliefs as an act of will, in the way you can decide to improve your skills with chainsaw or keyboard. If the evidence before your eyes doesn't support a belief, you cannot will yourself to believe it anyway. This, incidentally, is one of the many arguments against Pascal's Wager: no matter how you rate the odds of immortal salvation, you can't decide to believe in God, as if it were a decision to bet on Red or Manchester United.

It is hard not to feel sympathy for those men and women caught in the pulpit. It is for this reason that a group of us set up the Clergy Project. I originally even thought of trying to raise money for scholarships – retraining clergy as carpenters, say, or secular counsellors. That proved too expensive. But we did provide apostate clergy with a website where they can secretly go to talk to each other, in confidence, protected from parishioners and even from family by pseudonyms and passwords: a safe haven where they can share experiences, advice, even metaphorical shoulders to cry on. Dan Dennett and Linda LaScola were involved in the Clergy Project from the start: indeed it was largely inspired by their pilot study, together with the experiences of the pioneer apostate Dan Barker.* LaScola and Dennett's pilot study involved five Protestant ministers. Four of them are founder members of the Clergy Project. Subsequently the Clergy Project grew and grew, and it now numbers more than five hundred,† including Catholics, Protestants, Jews, Muslims and Mormons. The sheer numbers are heartwarming to those of us who see religion as unfavourable to human welfare and positively hostile to educational values. But the figures also imply a level of human misery on the part of the unfortunate clergy themselves, and the gentle kindness of Dan Dennett and Linda LaScola shines from every page of this book.

Their expanded study looks in detail at thirty individual cases. Some of these men and women believed passionately for many years before losing their faith. Others seem to have been already sceptical while still in seminary, but went ahead with their priestly career for reasons that need to be explored – and are. These are human beings, every one different, and the victims are allowed the space to tell their own stories, woven together with intelligent and insightful commentary by the two authors.

The book is a collaboration between a patient, sympathetic social worker and one of the world's great philosophers. It will surprise as it fascinates. If, as I hope and expect, the five hundred‡ apostates now in the Clergy Project turn out to be the thin end of a very large wedge, tip of a reassuringly large iceberg, harbingers of a coming and very welcome tipping point, this book will be seen as – to mix metaphors yet again with pardonable glee – the miner's canary. It will help us to understand what is happening as the floodgates open, as I hope they soon will. I also hope it will be widely read by still-believing clergy, and that it will give them the courage to join their colleagues who have already seen the light and walked away from the dark shade of the pulpit.

---

* See the previous essay in this volume.

† In 2019 the number reached a thousand.

‡ Now over one thousand.

THE UNBURDENING
LIGHTNESS OF RELIEF

This is a slightly abbreviated version of my foreword to Catherine Dunphy's *Apostle to Apostate* (2015).

Catherine Dunphy was a founding member of the Clergy Project* and one of the first to come out. She bears moving witness to the difficulties that face all her colleagues, now more than six hundred strong,[†] as they struggle, in their different ways, to come to terms with their loss of faith, and start to emerge into the light. The struggle is different in every case, of course. Catherine testifies to the particular difficulties of being both a woman and a Roman Catholic, cradle votary of that singularly misogynistic and unbendingly hidebound institution. Her reluctance to leave, born of relentless brainwashing in childhood, led her to seek last-ditch reconciliation with the church via 'liberation theology' and 'feminist theology'. Such liberal apologists seem to have provided her with temporary respite before she finally realized that the problem was with theology itself. She finally gave up her faith altogether and experienced

---

* On the Clergy Project, see the previous essay in this volume.
[†] As noted above, by 2019 this figure had risen to over one thousand.

287

the unburdening lightness of relief that so many others, in their different ways, have encountered.

Her book includes revealing interviews with other members of the Clergy Project, some named, some still anonymous, all of whom have their own versions of a similar story. One observes that coming out as atheist was far harder, and led to far more hurtful ill-treatment, than coming out as gay. Through all these testimonials I, as a scientist who lost faith because of the conspicuous failure of religion to explain the natural world, was struck by how many members of the Clergy Project travelled a different route. For them, disillusionment with the morals of their respective churches, and the personal shortcomings of their priests, was paramount. They internally repudiated their church before they stopped believing in God. Catherine herself was especially disgusted by the behaviour of her bishop, covering up and making light of the sexual abuse of children, while pulling outraged rank over those who questioned him – both particular vices of the Roman Catholic Church.

My own, probably over-hasty response to those people of faith who experience the hollowness of disillusionment is, 'Why don't you just leave? After all there's no positive reason, not even the smallest grounds, for believing in the factual claims of your church. Now that you've perceived its moral feet of clay, why don't you just walk?' Catherine Dunphy has taught me that this is too dispassionately scientific, too briskly cold. Her intelligence shines a painfully revealing light on the tragedy of entrapment. Great is the power of childhood imprinting. Those who inflict it can be forgiven only because they themselves were the innocent victims of it in their own early and impressionable years. Correspondingly great is the courage and strength of character of those who rise above it and break the cycle, refusing to hand the scourge on to the next generation. Those who were unfortunate enough to take the additional step of commitment to a life career in the clergy deserve special sympathy and praise when they have the courage to leave.

Catherine Dunphy's book will serve to stiffen the resolve of those still in the closet – and the six hundred who have so far joined the Clergy Project must surely be the tip of a large and still swelling iceberg. It will also serve as a warning to those who might otherwise have been lured – 'called' as they would perhaps put it – into the blind alley of commitment to the career that she has thankfully renounced. The good things that clergy sometimes do – fostering good fellowship and community, charitable work, education of a non-religious character – can be achieved in secular ways without supernatural nonsense and without dictatorial hierarchies. Here, as elsewhere, Catherine Dunphy serves as a role model, and her book as a beacon of hope.

A PUBLIC AND
POLITICAL ATHEIST

Herb Silverman might be the most genial man I know. A stalwart of
the American atheist scene, if he is one of many audience members
to stand up at a conference's question time, the chairman will
reliably be bombarded with a chorus of 'Choose Herb!' from all
corners of the room. His affable wit never disappoints. I was
delighted to be asked to write the foreword to his autobiography,
Candidate without a Prayer (2012).

If a man is going to publish his life story, he had best take the pre-
caution of leading an interesting life first. Or at least of being a very
funny writer or of lacing his pages with wittily unconventional
wisdom. Or even of being just an exceptionally nice person. Fortunately, Herb Silverman ticks all these boxes, and more.

Not every autobiographer can begin his life with an amusing
childhood supervised by amusing parents but, by Silverman's hilar-
ious account, his mother was the mother of all Jewish mother jokes.
And his story just goes on getting better, through adolescent encounters with girls to his career as an academic mathematician, then
secular activist, and his gentle and courteous puncturing of hypocrisy and illogicality whenever he finds it – which is pretty much
every day in the life of a sensitive atheist. Silverman has the endearing capacity to laugh at himself and poke fun at his shortcomings.

Boswell to his own Johnson, he quotes his own past sayings and writings, but with a conspicuous lack of the irritating self-regard that this might, in others, suggest.

Endearing pleasantries adorn every page. When his schoolfellows, asked to write an essay on a chosen US President, selected the obvious ones like Washington and Lincoln, the young Herb chose John Adams. Why? For the sufficient reason that his family could afford only two volumes of the encyclopedia: A and B. 'Were it not for the Adams family,' Herb added, 'it would have been considerably more difficult to justify why my favorite president was Chester A. Arthur or James Buchanan.'

Later in life, he visited Israel and was standing by the River Jordan at John the Baptist's reputed *Stammtisch* when a young man approached and asked Herb to baptize him. Herb's 'spiritual' demeanour had impressed him, and the beard and sandals reminded him of Jesus. The genial atheist unhesitatingly obliged, and no doubt did it beautifully.

Back in America, he has undertaken various political campaigns, losing them with his own distinctive panache as a means to winning a more timeless battle. The constitution of South Carolina stipulated that no person could be eligible for the office of governor who denied the existence of 'the' Supreme Being. That Herb's sole motive in running for governor was to test the legality of that prohibition is attested by the answer he gave when asked what would be his first action, if elected: 'Demand a recount.' I'm reminded of the paradoxical maxim that anybody who actively wants high office should be disqualified from holding it.

I once publicly criticized American atheists for tokenism (defacing banknotes, for example, in protest against the 1957 addition of 'In God We Trust') when they should be going after what I saw as more important issues (like tax exemptions for fat-cat televangelists). I now realize that that particular criticism was misplaced (because ignorant people demonstrably *use* the banknote slogan as alleged *evidence* that the United States is a Christian foundation).

One might still criticize token gestures like refusing to bow the head in prayer at university prizegiving ceremonies. But this criticism, too, receives a beautifully Silvermanian response. At one gathering when most eyes were closed and heads bowed in prayer, Herb reflected that his erect, open-eyed posture was the perfect dissenting gesture. It couldn't give offence because the sincerely devout wouldn't see it, while those not offended could catch each other's eyes and take reassurance from the company.

This last is an important point, as I have discovered when lecturing to surprisingly large but beleaguered audiences around the so-called (though overrated) Bible Belt.* When people tell Herb Silverman he is the only atheist they know, he says, 'No I'm not. You know hundreds. I'm the only one who has been public about it.'

Silverman enjoys arguing – he might say it is a Jewish trait – and he takes a gentle delight in teasing his opponents. Persuaded to attend a Billy Graham rally, he characteristically went forward to be 'saved'. He was received by one of Billy Graham's underlings (vicars in the literal sense, I suppose), Pastor A. Pastor A discovered Silverman's Jewish background and handed him on to Pastor B, who had converted from Judaism. On hearing that Pastor B's parents were dead, Herb asked whether he relished the thought that, as Jews, they were roasting in hell. When Pastor B demurred, Herb simply summoned Pastor A over, and happily left the two of them to fight it out.

In an effort to convert him, Christian apologists might quote a verse like 'I am the Way, the Truth, and the Light'. Did they expect him to slap his forehead and say, 'Gee, I never knew that. Now I'm a believer'? How many times have we all wanted to say something like that? Equally familiar, the media often refer to Silverman as an 'admitted atheist' or a 'self-confessed atheist'. How do they feel

---

* The Belt's exact geographic location is hard to pin down. I've lost count of the number of apologetic hosts who have described their home town to me as 'the Buckle of the Bible Belt'. One feels the fabled Belt has almost as many buckles as there are fragments of the True Cross.

when described as an 'admitted Baptist' or a 'self-confessed Catholic'? And the following is vintage Silverman:

> However, the oddest comments came from those who thought my
> not believing in a judging God meant I must feel free to rape, murder,
> and commit any atrocity I can get away with. I'd respond, 'With an
> attitude like that, I hope you continue to believe in God.'*

He regularly horrifies 'Bible-believing Christians' by showing them what is actually in the Bible.† He happily accepts invitations to debate with religious apologists, usually Christian but, on one notable occasion, Jewish. Silverman's Orthodox opponent expressed religious objections to medical research on dead human bodies. He conceded to Herb that many lives had been saved by such medical research, but argued, 'There are lots of *goyim* and animals available for such things.' Wow, just wow, as young people say.

On another occasion Herb was pitted against a Christian apologist, a 'philosopher' from an unknown 'university', who seems to do nothing but travel the country from one debate to another. This full-time debater fatuously asserted that the resurrection of Jesus must be a historical fact because the disciples were prepared to die for their beliefs. Herb's answer was devastatingly succinct: '9/11.'

---

* The celebrated conjuror Penn Jillette had another good response to the
  same point: 'The question I get asked by religious people all the time
  is, without God, what's to stop me from raping all I want? And my
  answer is: I do rape all I want. And the amount I want is zero. And I
  do murder all I want, and the amount I want is zero. The fact that
  these people think that if they didn't have this person watching over
  them that they would go on killing, raping rampages is the most
  self-damning thing I can imagine.'
† As Dan Barker has also done, notably in his book *God: the most
  unpleasant character in all fiction*, to which I wrote the foreword
  reproduced on pp. 298–301 below.

Another moment to savour took place in the Oxford Union, in my own university, a debate for which Herb took the unprecedented step of hiring a (too large) tuxedo. The motion was that 'American religion undermines American values'. Herb well deserved his applause for the following:

> In the melting pot called America, we are one nation under the Constitution . . . but not one nation under God. Given how the religious right opposes the teaching of evolution or any scientific and social view that conflicts with a literal interpretation of the Bible, we are really becoming one nation *under-educated*. And this is not an American value to be proud of.

Once, when debating with a Pastor Brown, Herb asked the pastor what he would do if God commanded him to kill a member of his family, as God had commanded Abraham: 'Depending on your answer, I might move a bit farther away from you.' He doubtless said it with such good humour that the pastor could not take offence – but was consequently all the more stuck for an answer.

Pastor Brown generalized the question to one of whether he was ever tempted to disobey God: 'I'm sometimes tempted by women to cheat on my wife, but I resist because I know how much it would hurt Jesus.' Herb Silverman's retort was almost too easy: 'I'm sometimes tempted by women to cheat on my wife, Sharon, but I resist because I know how much it would hurt Sharon.'

Herb and Sharon married late, and their love story is moving because it flies above mawkish sentimentality. By Herb's account the mystery is how she puts up with him, the answer being that he makes her laugh every day. It is a story both humorous and touching.

Herb Silverman is such a legendarily nice guy that he is the perfect mediator – albeit in his unambassadorial shorts and T-shirt (saying something like 'Smile, there is no hell'). He loves fraternizing with those who wish to argue with him, perhaps because he wins the argument. He supported the Moonies when they were denied access to his campus, on the grounds that a university should hear all points of view (and in any case the Moonies are no more bonkers than other branches of Christianity, just more recently founded).

If a religious person says to an atheist (I can confirm that they often do), 'I'll pray for you,' Herb Silverman is too nice to use the reply that first occurs to him, 'OK, I'll think for both of us.' Instead, he says, 'Thank you.' He knows how to disagree without being disagreeable. Nowhere is this gift more necessary than when reconciling rival groups of atheists, agnostics, humanists and freethinkers. 'Herding cats' may be a cliché, but clichés can be true as well as tiresome. Herb Silverman is the cat herder beyond compare: quite possibly the only person in America who could amicably unite all factions of the non-believing community.

In this capacity he is the founder and president of the Secular Coalition for America, a union of ten member organizations: the American Atheists, American Ethical Union, American Humanist Association, Atheist Alliance of America, Camp Quest, Council for Secular Humanism, Institute for Humanist Studies, Military Association of Atheists and Freethinkers, Secular Student Alliance and Society for Humanistic Judaism. Endorsed by an even larger number of organizations, and with an advisory board on which I have the honour to serve, the Secular Coalition runs the only lobby in Washington dedicated to secular causes, and its officers coordinate activities countrywide. But the driving force and guiding spirit of the Secular Coalition is the gentleman – in the best sense of the word – who is the author of this splendid and idiosyncratic book. Let me end with one of his most characteristic aphorisms: 'Changing minds is one of my favorite things, including my own when the evidence warrants it.'*

---

* John Maynard Keynes said something similar: 'When the facts change, I change my mind. What do you do, sir?'

THE GREAT ESCAPE

Foreword to Seth Andrews' *Deconverted: a journey from religion to reason* (2019).

Imagine what it must be like to really believe there's a place called Hell, red hot and real, a post-mortem torture chamber of unspeakable horror where you will certainly, without any doubt, spend all eternity if you don't accept Jesus. I mean really, really, really believe it: believe it in the core of your being, believe it with as much total conviction as we all believe there's a place called the South Pole where it's very cold.

So, God is a petulant, vengeful sadist, then? A celestial narcissist so ignominiously vain that he'd spit-roast you for all eternity for the petty slight of not believing in him? Why would you credit anything so ridiculous? Really, really believe something so implausible? The reason is childishly simple. Literally childish. Your parents told you about it when you were a helpless child. Your beloved parents, your teachers and other respected adults, fed it to you when you were too young to know better: when you were vulnerable and gullible, credulous and trusting. That's all it takes.

Seth Andrews is spot-on right to call it psychological abuse: it is using the weapon of real, palpable terror, literally believed terror, against children. But we should excuse the parents on the grounds that they themselves believe the very same nonsense with the utmost sincerity, and for exactly the same reason. Their own parents fed it to them. And so on back through the generations. And so it will go on forward, unless we find some way to break the tragic cycle and 'save the children'. To achieve that, we need lots more courageous breakaways like Seth Andrews.

Seth did more than break the cycle for himself. He is a gifted communicator, equipped to guide others out of the pit into which so many were thrown as children by well-meaning parents and grandparents, teachers and preachers. A charismatic broadcaster, podcaster and producer, this book shows him to be a gripping writer too. His is a great escape story, a lighthouse beam of hope for others. There's charm here too, not least in Seth's gentle sympathy towards his own parents. He's careful not to shame them. He doesn't blame them for stunting his education, shielding him from the glories of science and the true poetry of reality. Magnanimous, he forgives their thankfully failed attempt to impoverish his intellectual development and scare him witless with high Hadean threats. They meant well. They sincerely believed – still believe – that they must do all in their power to save him from the terrifying fate with which they in their turn were threatened by their own parents. What must it feel like to believe your much-loved son is hell-bent on hell and damnation? It is a horrible position they are in – one from which they themselves, and countless other loving parents in America's benighted Bible Belt, need rescuing.

Is it too much to hope (I fear it is) that parents like these will read Seth's book and repent? It's surely too much to hope they'll listen to me. Perhaps the fanatically religious Oliver Cromwell will strike a chord: 'I beseech you, in the bowels of Christ, think it possible you may be mistaken.'

Oh and, by the way, you actually are mistaken. Read Seth's book with anything remotely approaching a receptive mind and you'll see why.

IN HIS OWN WORDS:
A PORTRAIT OF GOD

Foreword to Dan Barker's *God: the most unpleasant character in all fiction* (2016).

Next time you find yourself in a hotel, look in the bedside drawer. You know what you'll find – it's so predictable, you need scarcely bother to check. Yes, it's the 'Gideon Bible'. According to the Gideons' own website,* they have distributed nearly two billion bibles, free of charge, around the world since their foundation in 1899. They claim to hand out bibles at a rate of two every second. Using a minimal estimate of the cost of production and distribution of a typical hardback book, the 81 million Gideon Bibles distributed in 2013 must have cost about $300 million, or about $1,000 for each of the three hundred thousand members of the Gideons organization. Tax-deductible, no doubt, but still we have here a very considerable outlay in cash and dedication. Why do they do it?

They do it, in their own words, 'to make it possible for others to learn about the love of God by giving them access to his word'. Yet if that really is their aim, I can't help wondering how many of these loyal Gideons have actually read the Bible. Even a cursory look

---

* http://www.gideons.org/about.

should be enough to convince a reasonable person that it's the very last document you should thrust in front of someone if you want to convince them of the love of God. If you happen to be a member of the Gideons, I challenge you to read Dan Barker's book and then ask yourself why you don't instantly resign. Indeed, I can't help wondering whether a good many of those three hundred thousand Gideon subscribers are actually undercover atheists who have calculated that the best possible way to turn people against Abrahamic religion would be to expose them to the Bible. Fortunately, from the point of view of the Gideons' stated aim, I suspect that very few of their bibles are ever actually opened.

Dan Barker's book had its origin, as his introduction explains, in a single sentence of my own *The God Delusion*: the opening sentence of chapter 2. It says that the God of the Old Testament is 'arguably the most unpleasant character in all fiction' and goes on to list nineteen character traits which, if they were all combined in a single fictional villain, would strain the reader's credulity to the point of ridicule. Certifiable psychopaths apart, no real human individual is quite so irredeemably nasty as to combine all of the following: 'jealous and proud of it; a petty, unjust, unforgiving control-freak; a vindictive, bloodthirsty ethnic cleanser; a misogynistic, homophobic, racist, infanticidal, genocidal, filicidal, pestilential, megalomaniacal, sadomasochistic, capriciously malevolent bully'.

The sentence has been controversial. I suspect that it is single-handedly responsible for *The God Delusion*'s reputation for 'stridency'. The rest of the book is far from strident. This is the sentence that led the British Chief Rabbi to accuse me of anti-semitism – an accusation he graciously withdrew on further reflection. Apparently he had focused on the phrase 'God of the Old Testament' and took it to mean 'as opposed to the God of the New Testament'. He took this to mean the Jewish God as opposed to the Christian God, a comparison which revived cultural memories of pogroms and persecutions. Of course that was not my intention. Indeed I gave reasons, elsewhere in *The God Delusion*, for regarding the God of the New Testament as almost equally bad. There's little in the Old Testament to match the horror of St Paul's version of the ancient principle of the scapegoat: the Creator of the Universe and Inventor of the Laws of Physics couldn't think of a better way to forgive our sins (especially the sin of Adam, who never existed and never sinned) than to have himself hideously tortured and executed in human form as vicarious punishment. As Paul's Epistle to the Hebrews (9: 22) puts it, 'without the shedding of blood there is no forgiveness'. To be fair, Paul's authorship of this epistle is disputable, but it is fully in the spirit of his often-expressed doctrine of atonement.

From the day I wrote it, I always knew that every one of my list of nineteen nasty attributes could be fully substantiated in the Bible, but I didn't know quite how richly and thoroughly every one of them could be documented. It occurred to me that it would be an amusing exercise to make a picture with nineteen hyperlinks, each clicking through to a verse, or set of verses, from the Bible. I soon realized that my biblical knowledge was inadequate to the task of assembling all the verses, and indeed that they were numerous enough to fill a book. A book! Now, there was an idea. Not a book that I could write. But I knew just the man who could. Dan Barker, of course. Fortunately he jumped at the idea, and this volume is the splendid result.

Dan knows his Bible inside out. He's preached from it, thrust it in the face of countless victims of the doorstepping young pastor he once was. He has now seen the light in a big way, and I know nobody better qualified to assemble this remarkable anthology of sheer, unadulterated nastiness.

What will the apologists say in response? 'The Bible was never meant to be taken literally.' Wasn't it? Did the generations upon generations of scribes faithfully reproducing the story of Adam and Eve and the talking snake not take it literally? Really? Were the pious generations of churchgoers, through the Dark and Middle Ages, so 'theologically' sophisticated as to realize it was all a metaphor? Really? When the peasants mustered to defy Richard II and sang 'When Adam delved and Evé span / Who was then the gentle-man?', did they mutter apophatically under their breath that, of course, it was never meant to be taken literally? Don't be so ridiculous. Even today, more than 40 per cent of Americans think the world began exactly in the way Genesis describes, less than ten thousand years ago. The pretence by 'sophisticated theologians' that the typical Christian in the pew doesn't take the Bible literally is nothing short of dishonest.

But anyway, if not literally, how else might our apologists wish to interpret the Bible? As a set of moral tales? A handy guide to what's right and wrong? What? Moral tales? You cannot be serious! Read this book.

Is it, perhaps, unfair of Dan and me to pick and choose illustrations of the unpleasant characteristics of God? Couldn't an apologist come up with a similar list of virtues, backed up by an even larger list of supporting verses? Couldn't the apologist compile a counter to my damning litany? 'The God of the Old Testament is . . . magnanimous, generous, encouraging, forgiving, charitable, loving, friendly, good-humoured, supportive of women, of homosexuals, of children, freedom-loving, open-minded, broad-minded, non-violent.' Go on. Try and find those nice verses. Do your best. You think you'll succeed? Want a bet?

We have all become acculturated to the idea that criticizing religion is somehow not done; it's bad taste, you just don't do it. The result is that even mild criticism sounds a lot stronger than it really is. My chapter 2, first sentence, is not mild. But it is nothing less than the truth, truth about both the letter and the spirit of the Bible. The evidence? Read on.

LIBERATION FROM THEOLOGY

This is an edited version of my foreword to *The New Encyclopedia of Unbelief*, edited by Tom Flynn and published in 2007.

'Unbelief' sounds negative. How can you have an encyclopedia of anything defined as an absence? There are encyclopedias of music, but who would buy an encyclopedia of tone-deafness? I have seen and enjoyed an encyclopedia of food, but an encyclopedia of hunger?

The first thing wrong with these comparisons is that music and food have positive associations. Tone-deafness is an absence of something valued. So is hunger. Among the many things this encyclopedia demonstrates is that religious unbelievers are not similarly deprived. For many – and I shall return to this – unbelief comes as a liberating gateway to a more fulfilled life.

The second thing that might give us pause about unbelief is that there is an all but infinite number of things in which we don't believe, but we don't go out of our way to say so. I am an unbeliever in fairies, unicorns, werewolves, Elvis on Mars, spoon-bending by mental energy, the Easter Bunny, green kangaroos on Uranus . . . the list could be expanded trivially and without end. We don't bother to declare ourselves unbelievers in all the millions of things that nobody else believes in. It is only worth bothering to declare unbelief if there is a default assumption that we all must believe in some particular hypothesis unless we positively state the contrary.* Manifestly, and in spades, that is the common assumption over the hypothesis of divine intelligence.

Divine intelligences are not the only things that are both widely believed and widely doubted. The word 'sceptic' rather than 'unbeliever' is commonly applied to those who doubt the widespread claims of astrology, homeopathy, telepathy, water divining, clairvoyance, alien sexual abduction and communication beyond the grave. Sceptics in this sense do not necessarily deny the validity of these claims. Instead they demand evidence, and sometimes go out of their way to set up the rather stringent conditions – much more stringent than supporters usually realize – that proper evidence requires. Supporters of homeopathic medicine, for example, or dowsing, seldom understand the need for statistically analysed double-blind control experiments to guard against chance effects, placebo effects and the unwitting suggestibility of the believing mind. It is a matter of convention that 'sceptic' has come to be associated with those matters, while the superficially synonymous 'unbeliever' specifies *religious* unbelief. The two kinds of scepticism/unbelief often go together, but you can get into trouble if you simply assume that they do.

An American student asked her professor whether he had a view about me.

> 'Sure,' he replied. 'He's positive science is incompatible with religion, but he waxes ecstatic about nature and the universe. To me, that *is* religion!'

---

* A young woman of my acquaintance had been admitted to hospital, and a nurse came round to fill in her personal details form. To the nurse's question 'Religion?' my friend answered 'None.' Later she overheard a pair of nurses gossiping about her: 'She doesn't *look* like a nun!' Isn't it bizarre that such official forms always ask our religion, and assume that we have one? Why not note down our political persuasion, our taste in music or our favourite colour?

But is religion the right word to use? Words are our servants, not our masters, but there are many people out there, passionate believers in supernatural religion, who are only too eager to misunderstand.

There is a tactical, political point to be made here. Maybe Einsteinian 'religion' – the view that God does not exist as a personal intelligence at all, but that the word may be used as a poetic metaphor for the deep laws of the universe which we don't yet understand – provides a useful way for atheists to euphemize their way into American society and lessen the grip of fundamentalist theocracy. In the twenty years since the original *Encyclopedia of Unbelief* was published, the expected decline in religiosity has continued apace in western Europe, but the reverse has happened in North America* and the Islamic world. The hapless European sometimes feels cornered in a nightmarish pincer movement between Islamic and Christian jihadists in holy alliance. The United States of America is now suffering an epidemic of religiosity that seems almost medieval in its intensity and positively sinister in its political ascendancy. At various times in history it has been impossible for a woman, a Jew, a homosexual, a Roman Catholic or an African American to gain high political office. Today this negative privilege is pretty much restricted to atheists and criminals. The actress Julia Sweeney, in her beautiful theatrical monologue 'Letting Go of God', recounts with black humour her parents' response to her own gentle atheism:

My first call from my mother was more of a scream. 'Atheist? ATHEIST?!?!'
My dad called and said, 'You have betrayed your family, your school, your city.' It was like I had sold secrets to the Russians. They both said they weren't going to talk to me anymore. My dad

---

* Actually, polls show that even the United States is moving in the right direction. It just has further to go than western Europe.

said, 'I don't even want you to come to my funeral.' After I hung up, I thought, 'Just try and stop me.'

I think that my parents had been mildly disappointed when I'd said I didn't believe in God anymore, but being an *atheist* was another thing altogether.

One can argue that the problem lies in the ludicrously demonized word 'atheist' itself. If you look at the detail of what Einstein said, for example, notwithstanding his God-encumbered language he was just as much of an atheist as, say, Bertrand Russell or Robert Ingersoll or any typical Fellow of the Royal Society or the National Academy of Sciences.* Einsteinian euphemisms have probably enabled more than one intelligent, thinking person to achieve election to high office.

Some see this as a tactical argument for ditching the word 'atheist' altogether, and calling ourselves 'Brights', by analogy with the way homosexuals positively rebranded themselves as 'gays'.† Various more or less attractive attempts are made to turn Einstein-style pantheism into organized quasi-religions with names like Ursula Goodenough's 'Religious Naturalism', or World Pantheism; and perhaps this, or Universism, is the politically expedient way to go. Others prefer to tough it out and call a spade a spade, while making consciousness-raising efforts to rehabilitate the word 'atheist' itself.

Maybe 'politically expedient' is too cynical. Carl Sagan's ringing declaration in *Pale Blue Dot* can be read as positively inspiring:

---

* Surveys of the members of both these august bodies agree that the percentage of religious believers in both of them is approximately 10 per cent (https://link.springer.com/article/10.1186/1936-6434-6-33).
† See http://www.the-brights.net/. This move is widely unpopular because it seems to imply that religious believers should be called the Dims. Daniel Dennett counters thus. Since the Brights are defined as lacking supernatural beliefs, religious believers could be called the Supers. Surely nobody could take offence at that?

How is it that hardly any major religion has looked at science and concluded, 'This is better than we thought! The Universe is much bigger than our prophets said, grander, more subtle, more elegant'? Instead they say, 'No, no, no! My god is a little god, and I want him to stay that way.' A religion, old or new, that stressed the magnificence of the Universe as revealed by modern science might be able to draw forth reserves of reverence and awe hardly tapped by the conventional faiths. Sooner or later, such a religion will emerge.

Back to liberation: we experience liberation only by comparison with captivity. Nobody today feels liberated by unbelief in Thor's hammer or Zeus's thunderbolts, though our ancestors might have done so. Today, huge numbers of people are brought up to treat belief in – depending on an accident of birth – Christianity, Islam, Judaism or Hinduism as an expected norm, departure from which is a grievous and onerous decision (to put it no more strongly: the official Islamic penalty for apostasy is death). Women, in many parts of the world, have additional reasons to regard the end of religion as a liberation.

I not infrequently receive letters from readers of my books, thanking me for liberating them from the bondage of religion. I'll quote just one example, renaming the author Jerry because – reasonably enough, given Julia Sweeney's experience – he is deeply worried lest his parents discover his new-gained unbelief. Jerry's youthful indoctrination as an evangelical Christian was depressingly successful. He recounts how, in his last year at school,

the headmaster chose a small group of brighter boys to study philosophy with him. He probably regretted selecting me, as I made it perfectly clear during class debates that there really was no need for such discussions – the answer to all life's problems was simple, and it was Jesus.

Jerry's liberation had to wait until his postgraduate years:

My postgraduate studies, however, opened my mind to a world of ideas I barely knew existed. I met highly intelligent fellow students who had applied their rationality to all aspects of their lives, and come to the conclusion that there was no God. And, amazingly, they were happy, they enjoyed life, they didn't feel the 'God-shaped hole' that I had warned people about so often [Jerry had done a stint as a missionary, financed by well-meaning donations from his home church] . . .

For the first time in my life, I was willing to be challenged. 'Bliss was it in that dawn to be alive.' I craved more intellectual meat . . . and so I spent the summer of 1998 devouring book upon book.

Among the books he devoured were two of mine, which explains his writing to me. He goes on to expound his reaction to these and other books:

The dawning realization that I could safely jettison my increasingly tenuous faith, and that a world without God wouldn't be the joyless hell I had always imagined it would be, was overwhelming. I felt liberated. All the benefits of Christianity that I had promoted – that faith brings you freedom, meaning, purpose, joy etc. – I now discovered for real, but on the wrong side!

How are we liberated when we forsake religion? Let me count the ways! Morally we are freed – especially if we were brought up Catholic as I, thank goodness, was not – from an ominous burden of guilt and fear. The awful notion of private 'sin' is never far from the minds of the pious, and we cannot but feel joyous release when we shake it off and replace it by the open good sense of moral philosophic reasoning. In place of private 'sin', we choose to behave in such a way as to avoid causing suffering to others and increase their happiness.

Practically we are freed, depending on the details of the particular faith in which we were raised, from the fatuity and inconvenience of time-wasting rituals: freed from the necessity to pray five times a day; freed from the duty to confess our 'sins' to a priest; freed from having to buy two refrigerators, lest meat and milk should meet; freed from enforced laziness on Saturdays to the point of being unable to move a light switch or lift a telephone; freed from having to wear uncomfortable and unflattering clothes lest a flash of forbidden skin should be exposed; freed from the obligation to mutilate children too young to defend themselves.

Intellectually, we are freed to pursue evidence and scholarship wherever it might lead, without constantly looking over our spiritual shoulder to check whether we are straying from the party line. 'Party line' is right, even when it is not dictated by living priests, elders or ayatollahs but fossilized in a book (actually a motley collection of arbitrarily stitched-together fragments whose anonymous authors were writing in different times and to different audiences, reflecting different local and long-dead issues).

Personally, we are freed to direct our lives towards a worthwhile fulfilment, in the full knowledge and understanding that this is the one life we shall ever have. We are freed to exult in the privilege – the extraordinary good fortune – that we, each individual one of us, enjoys through the astronomically improbable accident of being born. The measure of the privilege you and I enjoy through existing is the number of possible people divided by the number of actual people. That ratio – too large for decent computation – should be the measure of our gratitude for life and our resolve to live it to the full. Well might the newly liberated unbeliever quote Wordsworth and his blissful dawn.

THE GOD TEMPTATION

This is an abbreviated version of the piece I wrote as a new introduction for the tenth-anniversary edition of *The God Delusion*, published in 2016.

The fact that you exist should brim you over with astonishment. You and I, and every other living creature, are machines of ineffable complexity, complexity of a magnitude to challenge credulity. Complexity here means statistical improbability in a non-random direction, the direction of seeming designed for a purpose. The ultimate purpose (gene survival) hides behind a more up-front 'design', details of which vary from species to species. Whatever its specialism – wings for flying, tails for swimming, hands for climbing or digging, galloping legs for prey-catching or predator-escaping – every animal embodies a statistically improbable complexity of detail which approaches (but revealingly falls short of) perfection as an engineer might judge it. 'Statistically improbable' means 'unlikely to have come about by chance'. The God Temptation here is the temptation to evade, by invoking a designer, the responsibility to explain. The point is that the designer himself, in order to be capable of designing, would have to be another complex entity of the kind that, in his turn, needs the same kind of explanation. It's an evasion of responsibility because it invokes the very thing it is supposed to be explaining.

I'm a biologist, so I speak first of the biological version of the God Temptation, the false argument destroyed by Darwin. There is also a cosmological version, which lies outside the Darwinian domain and precedes it by ten billion years. The cosmos may not look so obviously designed as a peacock or its eye. But the laws and constants of physics are fine-tuned in such a way as to set up the conditions under which, in the fullness of time, eyes and peacocks, humans and their brains, will come into existence. The God Temptation here is to invoke an Intelligent Knob-Twiddler who adjusts the dials of the physical constants so that they have the exquisitely precise values required to bring evolution, and eventually us, into being.

To succumb to the God Temptation in either of those guises, biological or cosmological, is an act of intellectual capitulation. If you are trying to explain something improbable, it can never suffice to invoke an entity that is, in itself, at least as improbable. If you'll stoop to magicking into existence an unexplained peacock-designer, you might as well magic an unexplained peacock and cut out the middleman.

Nevertheless, it's hard not to feel sympathy for such capitulation. The complexity of a living body, indeed of every one of its trillion cells, is so mind-shattering to anyone who truly grasps it (not all do) that the temptation to buckle at the knees and succumb to a non-explanation is almost overwhelming. Even a magic trick can draw the same reaction. There's an old card trick where the conjuror invites a member of the audience to pick a card and show it to the audience. He then burns the card, grinds the ash to powder and rubs it on his forearm. The image of the card appears on his arm, picked out in ash. A conjuror recently told me he performed the trick to a band of Arabs round a camp fire. The tribesmen's reaction made him fear for his life. They sprang up and reached for their guns, thinking he was a djinn. You can see why. You have to smack yourself and shout, 'No! However loudly my senses and my instincts are screaming "Miracle!", it really isn't. There really is a rational explanation. The conjuror prepared the ground in some unknown way before the trick started, and then did some clever prestidigitation while he cunningly distracted my attention.' It's almost as though you have to have 'faith' that it really is only a trick. Faith that nothing supernatural has happened. The laws of physics have not been suspended.

In the case of conjurors we know this to be the case because the best and most honest ones, like Jamy Ian Swiss, or James Randi, or Penn and Teller, or Derren Brown (as opposed to spoon-bending charlatans) assure us it is so.* Even if they didn't, the rational thinker falls back on the elegant parsimony of the eighteenth-century philosopher David Hume. Which should surprise you more – that you have been fooled by a trick, or that the laws of physics really have been violated?

When we contemplate the vertebrate eye, or the fine structure of a cell, once again our instincts scream 'Miracle!' and once again we need to smack ourselves. Darwin plays a role akin to the honest conjuror – but he goes further. The honest conjuror tells us it is only a trick but risks expulsion from the Magic Circle if he reveals how it's done. Darwin patiently tells us exactly how the Trick of Life works: cumulative natural selection.

Admittedly that isn't (or probably isn't) how the Cosmological Trick is done. Natural selection explains the miracle of life but it doesn't explain the apparent fine-tuning of the laws and constants of physics – unless you count as a version of natural selection the multiverse theory: there are billions of universes having different laws and constants; with anthropic hindsight we could only find ourselves in one of the minority of universes whose laws and constants happen to be propitious to our evolution. There is a weak sense in which you could regard that as a kind of Darwinism: anthropic *post hoc* selection among universes. The physicist Lee Smolin has provocatively suggested a stronger analogy in which

---

* Jamy Ian Swiss ends his emails with a quotation from the celebrated illusionist Karl Germain: 'Conjuring is the only absolutely honest profession – the conjuror promises to deceive, and does.'

universes give birth to daughter universes with mutated laws and constants.

In any case, Darwin can fairly be said to have done the heavy lifting. Before he came along, any impartial judge would have agreed with Archdeacon William Paley (1743–1805) that the apparent design of physics would be a doddle to explain compared with almost any biological organ, let alone the whole magnificent diversity of purpose-ridden life. Both these versions of the God Temptation are logically fallacious but one of them – the biological one – was so eloquently strong before Darwin, it would tempt one to defy even logic itself. The fact that Darwin solved it so convincingly should now stiffen our confidence to reject the much weaker cosmological version too. Darwin is a role model to inspire all who follow the logical and courageous compulsion to explain complex things in the only legitimate way, which is in terms of simpler things and their interactions.

THE INTELLECTUAL AND
MORAL COURAGE OF ATHEISM

This is a slightly edited version of the essay written to accompany
the transcript of the conversation between myself, Daniel Dennett,
Sam Harris and the late Christopher Hitchens, recorded in
Christopher's flat in Washington DC in September 2007 and
published in 2019 as *The Four Horsemen*.

Among the many topics the 'four horsemen' discussed in 2007 was
how religion and science compared in respect of humility and
hubris. Religion, for its part, stands accused of conspicuous overconfidence and sensational lack of humility. The expanding
universe, the laws of physics, the fine-tuned physical constants, the
laws of chemistry, the slow grind of evolution's mills – all were set in
motion so that, in the fourteen-billion-year fullness of time, we
should come into existence. Even the constantly reiterated insist-
ence that we are miserable offenders, born in sin, is a kind of inverted
arrogance: such vanity, to presume that our moral conduct has some
sort of cosmic significance, as though the Creator of the Universe
wouldn't have better things to do than tot up our black marks and
our brownie points. The universe is all concerned with me. Is that
not the arrogance that passeth all understanding?

Carl Sagan, in *Pale Blue Dot*, makes the exculpatory point that our
distant ancestors could scarcely escape such cosmic narcissism. With

no roof over their heads and no artificial light, they nightly watched the stars wheeling overhead. And what was at the centre of the wheel? The exact location of the observer, of course. No wonder they thought the universe was 'all about me'. In the other sense of 'about', it did indeed revolve 'about me'. 'I' was the epicentre of the cosmos. But that excuse, if it is one, evaporated with Copernicus and Galileo.

Turning, then, to theologians' overconfidence, admittedly few quite reach the heights scaled by the seventeenth-century archbishop James Ussher, who was so sure of his chronology that he gave the origin of the universe a precise date: 22 October, 4004 BC. Not 21 or 23 October but precisely on the evening of 22 October. Not September or November but definitely, with the immense authority of the Church, October. Not 4003 or 4005, not 'somewhere around the fourth or fifth millennium BC' but, no doubt about it, 4004 BC. Others, as I said, are not quite so precise about it, but it is characteristic of theologians that they just make stuff up. Make it up with liberal abandon and force it, with a presumed limitless authority, upon others, sometimes – at least in former times and still today in Islamic theocracies – on pain of torture and death.

Such arbitrary precision shows itself, too, in the bossy rules for living that religious leaders impose on their followers. And when it comes to control-freakery, Islam is way out ahead, in a class of its own. Here are some choice examples from the *Concise Commandments of Islam* handed down by Ayatollah Ozma Sayyed Mohammad Reda Musavi Golpaygani, a respected Iranian 'scholar'. Concerning the wet-nursing of babies, alone, there are no fewer than twenty-three minutely specified rules, translated as 'Issues'. Here's the first of them, Issue 547. The rest are equally precise, equally bossy, and equally devoid of apparent rationale:

If a woman wet-nurses a child, in accordance to the conditions to be stated in Issue 560, the father of that child cannot marry the woman's daughters, nor can he marry the daughters of the husband whom the milk belongs to, even his wet-nurse daughters, but it is permissible for him to marry the wet-nurse daughters of the woman . . . [and it goes on].

Here's another example from the wet-nursing department, Issue 553:

If the wife of a man's father wet-nurses a girl with his father's milk, then the man cannot marry that girl.

'Father's milk'? What? I suppose in a culture where a woman is the property of her husband, 'father's milk' is not as weird as it sounds to us.

Issue 555 is similarly puzzling, this time about 'brother's milk':

A man cannot marry a girl who has been wet-nursed by his sister or his brother's wife with his brother's milk.

I don't know the origin of this creepy obsession with wet-nursing, but it is not without its scriptural basis:

When the Qur'aan was first revealed, the number of breast-feedings that would make a child a relative (mahram) was ten, then this was abrogated and replaced with the number of five which is well-known.*

That was part of the reply from another 'scholar' to the following recent *cri de coeur* from a (pardonably) confused woman on social media:

I breastfed my brother-in-law's son for a month, and my son was breastfed by my brother-in-law's wife. I have a daughter and a son

---

* https://islamqa.info/en/27280.

who are older than the child who was breastfed by my brother-in-law's wife, and she also had two children before the child of hers whom I breastfed.

I hope that you can describe the kind of breastfeeding that makes the child a mahram and the rulings that apply to the rest of the siblings? Thank you very much.

The precision of 'five' breast feedings is typical of this kind of religious control-freakery. It surfaced bizarrely in a 2007 *fatwa* issued by Dr Izzat Atiyya, a lecturer at Al-Azhar University in Cairo, who was concerned about the prohibition on male and female colleagues being alone together and came up with an ingenious solution. The female colleague should feed her male colleague 'directly from her breast' at least five times. This would make them 'relatives' and thereby enable them to be alone together at work. Note that four times would not suffice. He apparently wasn't joking at the time, although he did retract his *fatwa* after the outcry it provoked. How can people bear to live their lives bound by such insanely specific yet manifestly pointless rules?

With some relief, perhaps, we turn to science. Science is often accused of arrogantly claiming to know everything, but the barb is capaciously wide of the mark. Scientists love not knowing the answer, because it gives us something to do, something to think about. We loudly assert ignorance, in a gleeful proclamation of what needs to be done.

How did life begin? I don't know, nobody knows, we wish we did, and we eagerly exchange hypotheses, together with suggestions for how to investigate them. What caused the apocalyptic mass extinction at the end of the Permian period, a quarter of a billion years ago? We don't know, but we have some interesting hypotheses to think about. What did the common ancestor of humans and chimpanzees look like? We don't know, but we do know a bit about it. We know the continent on which it lived (Africa, as Darwin guessed), and molecular evidence tells us roughly when (between six million and eight million years ago). What is dark matter? We don't know, and a substantial fraction of the physics community would dearly like to.

Ignorance, to a scientist, is an itch that begs to be pleasurably scratched. Ignorance, if you are a theologian, is something to be washed away by shamelessly making something up. If you are an authority figure like the Pope, you might do it by thinking privately to yourself and waiting for an answer to pop into your head – which you then proclaim as a 'revelation'. Or you might do it by 'interpreting' a Bronze Age text whose author was even more ignorant than you are.

Popes can promulgate their private opinions as 'dogma', but only if those opinions have the backing of a substantial number of Catholics through history: long tradition of belief in a proposition is, somewhat mysteriously to a scientific mind, regarded as evidence for the truth of that proposition. In 1950, Pope Pius XII (unkindly known as 'Hitler's Pope') promulgated the dogma that Jesus' mother Mary, on her death, was bodily – i.e. not merely spiritually – lifted up into heaven. 'Bodily' means that if you'd looked in her grave, you'd have found it empty. The Pope's reasoning had absolutely nothing to do with evidence. He cited 1 Corinthians 15: 54: 'then shall be brought to pass the saying that is written, Death is swallowed up in victory'. The saying makes no mention of Mary. There is not the smallest reason to suppose the author of the epistle had Mary in mind. We see again the typical theological trick of taking a text and 'interpreting' it in a way that just might have some vague, symbolic, hand-waving connection with something else. Presumably, too, like so many religious beliefs, Pius XII's dogma was at least partly based on a feeling of what would be *fitting* for one so holy as Mary. But the Pope's main motivation, according to Dr Kenneth Howell, director of the John Henry Cardinal Newman Institute of Catholic Thought, University of Illinois, came from a different meaning of what was fitting. The world of 1950 was recovering from the devastation of the Second World War and desperately needed the balm of a healing message. Howell quotes the Pope's words, then gives his own interpretation:

Pius XII clearly expresses his hope that meditation on Mary's assumption will lead the faithful to a greater awareness of our common dignity as the human family ... What would impel human beings to keep their eyes fixed on their supernatural end and to desire the salvation of their fellow human beings? Mary's assumption was a reminder of, and impetus toward, greater respect for humanity because the Assumption cannot be separated from the rest of Mary's earthly life.

It's fascinating to see how the theological mind works: in particular, the lack of interest in – indeed, the contempt for – factual evidence. Never mind whether there's any evidence that Mary was assumed bodily into heaven; it would be good for people to believe she was. It isn't that theologians deliberately tell untruths. It's as though they just don't care about truth; aren't interested in truth; don't know what truth even means; demote truth to negligible status compared with other considerations, such as symbolic or mythic significance. And yet at the same time, Catholics are compelled to believe these made-up 'truths' – compelled in no uncertain terms. Even before Pius XII promulgated the Assumption as a dogma, the eighteenth-century Pope Benedict XIV declared the Assumption of Mary to be 'a probable opinion which to deny were impious and blasphemous'. If to deny a 'probable opinion' is 'impious and blasphemous', you can imagine the penalty for denying an infallible dogma! Once again, note the brazen confidence with which religious leaders assert 'facts' which even they admit are supported by no historical evidence at all.

The *Catholic Encyclopedia* is a treasury of overconfident sophistry. Purgatory is a sort of celestial waiting room in which the dead are punished for their sins ('purged') before eventually being admitted to heaven. The *Encyclopedia*'s entry on purgatory has a long section on 'Errors', listing the mistaken views of heretics such as the Albigenses, Waldenses, Hussites and Apostolici, unsurprisingly joined by Martin Luther and John Calvin.*

The biblical evidence for the existence of purgatory is, shall we say, 'creative', again employing the common theological trick of vague, hand-waving analogy. For example, the *Encyclopedia* notes that 'God forgave the incredulity of Moses and Aaron, but as punishment kept them from the "land of promise"'. That banishment is viewed as a kind of metaphor for purgatory. More gruesomely, when David had Uriah the Hittite killed so that he could marry Uriah's beautiful wife, the Lord forgave him – but didn't let him off scot-free: God killed the child of the marriage (2 Samuel 12: 13–14). Hard on the innocent child, you might think. But apparently a useful metaphor for the partial punishment that is purgatory, and one not overlooked by the *Encyclopedia*'s authors.

The section of the purgatory entry called 'Proofs' is interesting because it purports to use a form of logic. Here's how the argument goes. If the dead went straight to heaven, there'd be no point in our praying for their souls. And we do pray for their souls, don't we? Therefore it must follow that they don't go straight to heaven. Therefore there must be purgatory. QED. Are professors of theology really paid to do this kind of thing?

Enough; let's turn again to science. Scientists know when they don't know the answer. But they also know when they do, and they shouldn't be coy about proclaiming it. It's not hubristic to state known facts when the evidence is secure. Yes, yes, philosophers of science tell us a fact is no more than a hypothesis which may one day be falsified but which has so far withstood strenuous attempts to do so. Let us by all means pay lip service to that incantation, while muttering, in homage to Galileo's muttered *eppur si muove*, the sensible words of Stephen Jay Gould:

---

* http://www.catholic.org/encyclopedia/view.php?id=9745.

In science, 'fact' can only mean 'confirmed to such a degree that it would be perverse to withhold provisional assent.' I suppose that apples might start to rise tomorrow, but the possibility does not merit equal time in physics classrooms.*

Facts in this sense include the following, and not one of them owes anything whatsoever to the many millions of hours devoted to theological ratiocination. The universe began between 13 billion and 14 billion years ago. The sun, and the planets orbiting it, including ours, condensed out of a rotating disk of gas, dust and debris about 4.5 billion years ago. The map of the world changes as the tens of millions of years go by. We know the approximate shape of the continents and where they were at any named time in geological history. And we can project ahead and draw the map of the world as it will change in the future. We know how different the constellations in the sky would have appeared to our ancestors and how they will appear to our descendants.

Matter in the universe is non-randomly distributed in discrete bodies, many of them rotating, each on its own axis, and many of them in elliptical orbit around other such bodies according to mathematical laws which enable us to predict, to the exact second, when notable events such as eclipses and transits will occur. These bodies – stars, planets, planetesimals, knobbly chunks of rock, etc. – are themselves clustered in galaxies, many billions of them, separated by distances orders of magnitude larger than the (already very large) spacing of (again, many billions of) stars within galaxies.

Matter is composed of atoms, and there is a finite number of types of atoms – the hundred or so elements. We know the mass of each of these elemental atoms, and we know why any one element can have more than one isotope with slightly different mass. Chemists have a huge body of knowledge about how and why the elements

---

* 'Evolution as fact and theory'.

combine in molecules. In living cells, molecules can be extremely large, constructed of thousands of atoms in precise, and exactly known, spatial relation to one another. The methods by which the exact structures of these macromolecules are discovered are wonderfully ingenious, involving meticulous measurements on the scattering of X-rays beamed through crystals. Among the macromolecules fathomed by this method is DNA, the universal genetic molecule. The strictly digital code by which DNA influences the shape and nature of proteins – another family of macromolecules which are the elegantly honed machine-tools of life – is exactly known in every detail. The ways in which those proteins influence the behaviour of cells in developing embryos, and hence influence the form and functioning of all living things, is work in progress: a great deal is known; much challengingly remains to be learned.

For any particular gene in any individual animal, we can write down the exact sequence of DNA code letters in the gene. This means we can count, with total precision, the number of single-letter discrepancies between two individuals. This is a serviceable measure of how long ago their common ancestor lived. This works for comparisons within a species – between you and Barack Obama, for instance. And it works for comparisons of different species – between you and an aardvark, say. Again, you can count the discrepancies exactly. There are just more discrepancies the further back in time the shared ancestor lived. Such precision lifts the spirit and justifies pride in our species, *Homo sapiens*. For once, and without hubris, Linnaeus's specific name seems warranted.

Hubris is unjustified pride. Pride can be justified, and science does so in spades. So does Beethoven, so do Shakespeare, Michelangelo, Christopher Wren. So do the engineers who built the giant telescopes in Hawaii and in the Canary Islands, the giant radio telescopes and very large arrays that stare sightless into the southern sky; or the Hubble orbiting telescope and the spacecraft that launched it. The engineering feats deep underground at CERN, combining monumental size with minutely accurate tolerances of measurement, literally moved me to tears when I was shown around. The engineering, the mathematics, the physics, in the Rosetta mission that successfully soft-landed a robot vehicle on the tiny target of a comet also made me proud to be human. Modified versions of the same technology may one day save our planet by enabling us to divert a dangerous comet like the one that killed the dinosaurs.

Who does not feel a swelling of human pride when they hear about the LIGO instruments which, synchronously in Louisiana and Washington State, detected gravitation waves whose amplitude would be dwarfed by a single proton? This feat of measurement, with its profound significance for cosmology, is equivalent to measuring the distance from Earth to the star Proxima Centauri to an accuracy of one human hairsbreadth.

Comparable accuracy is achieved in experimental tests of quantum theory. And here there is a revealing mismatch between our human capacity to demonstrate, with invincible conviction, the predictions of a theory experimentally and our capacity to visualize the theory itself. Our brains evolved to understand the movement of buffalo-sized objects at lion speeds in the moderately scaled spaces afforded by the African savannah. Evolution didn't equip us to deal intuitively with what happens to objects when they move at Einsteinian speeds through Einsteinian spaces, or with the sheer weirdness of objects too small to deserve the name 'object' at all. Yet somehow the emergent power of our evolved brains has enabled us to develop the crystalline edifice of mathematics by which we accurately predict the behaviour of entities that lie under the radar of our intuitive comprehension. This, too, makes me proud to be human, although to my regret I am not among the mathematically gifted of my species.

Less rarefied but still proud-making is the advanced, and continually advancing, technology that surrounds us in our everyday lives. Your smartphone, your laptop computer, the satnav in your car and the satellites that feed it, your car itself, the giant airliner that can loft not just its own weight plus passengers and cargo but also the 120 tons of fuel it ekes out over a thirteen-hour journey of seven thousand miles.

Less familiar, but destined to become more so, is 3D printing. A computer 'prints' a solid object, say a chess bishop, by depositing a sequence of layers, a process radically and interestingly different from the biological version of '3D printing' which is embryology. A 3D printer can make an exact copy of an existing object. One technique is to feed the computer a series of photographs of the object to be copied, taken from all different angles. The computer does the formidably complicated mathematics to synthesize the specification of the solid shape by integrating the angular views. There may be life forms in the universe that make their children in this body-scanning kind of way, but our own reproduction is instructively different. This, incidentally, is why almost all biology textbooks are seriously wrong when they describe DNA as a 'blueprint' for life. DNA may be a blueprint for protein, but it is not a blueprint for a baby. It's more like a recipe or a computer program.

We are not arrogant, not hubristic, to celebrate the sheer bulk and detail of what we know through science. We are simply telling the honest and irrefutable truth. Also honest is the frank admission of how much we don't yet know – how much more work remains to be done. That is the very antithesis of hubristic arrogance. Science combines a massive contribution, in volume and detail, of what we do know with humility in proclaiming what we don't. Religion, by embarrassing contrast, has contributed literally zero to what we know, combined with huge hubristic confidence in the alleged facts it has simply made up.

But I want to suggest a further and less obvious point about the contrast of religion with atheism. I want to argue that the atheistic world-view has an unsung virtue of intellectual courage. Why is there something rather than nothing? Our physicist colleague Lawrence Krauss, in his book *A Universe from Nothing*,* controversially suggests that, for quantum-theoretic reasons, Nothing (the

* For which I wrote an afterword; see pp. 353–7 below.

capital letter is deliberate) is unstable. Just as matter and antimatter annihilate each other to make Nothing, so the reverse can happen. A random quantum fluctuation causes matter and antimatter to spring spontaneously out of Nothing. Krauss's critics largely focus on the definition of Nothing. His version may not be what everybody understands by nothing, but at least it is supremely simple – as simple it must be, if it is to satisfy us as the base of a 'crane' explanation (Dan Dennett's phrase), such as cosmic inflation or evolution. It is simple compared to the world that followed from it by largely understood processes: the Big Bang, inflation, galaxy formation, star formation, element formation in the interior of stars, supernova explosions blasting the elements into space, condensation of element-rich dust clouds into rocky planets such as Earth, the laws of chemistry by which, on this planet at least, the first self-replicating molecule arose, then evolution by natural selection and the whole of biology which is now, at least in principle, understood.

Why did I speak of intellectual courage? Because the human mind, including my own, rebels emotionally against the idea that something as complex as life, and the rest of the expanding universe, could have 'just happened'. It takes intellectual courage to kick yourself out of your emotional incredulity and persuade yourself that there is no other rational choice. Emotion screams: 'No, it's too much to believe! You are trying to tell me the entire universe, including me and the trees and the Great Barrier Reef and the Andromeda Galaxy and a tardigrade's finger, all came about by mindless atomic collisions, no supervisor, no architect? You cannot be serious. All this complexity and glory stemmed from Nothing and a random quantum fluctuation? Give me a break.' Reason quietly and soberly replies: 'Yes. Most of the steps in the chain are well understood, although until recently they weren't. In the case of the biological steps, they've been understood since 1859. But more important, even if we never understand all the steps, nothing can change the principle that, however improbable the entity you are trying to explain, postulating a creator god doesn't help you, because the god would itself need exactly the same kind of explanation.' However difficult it may be to explain the origin of simplicity, the spontaneous arising of complexity is, by definition, more improbable. And a creative intelligence capable of designing a universe would have to be supremely improbable and supremely in need of explanation in its own right. However improbable the naturalistic answer to the riddle of existence, the theistic alternative is even more so. But it needs a courageous leap of reason to accept the conclusion.

This is what I meant when I said the atheistic world-view requires intellectual courage. It requires moral courage, too. As an atheist, you abandon your imaginary friend, you forgo the comforting props of a celestial father figure to bail you out of trouble. You are going to die, and you'll never see your dead loved ones again. There's no holy book to tell you what to do, tell you what's right or wrong. You are an intellectual adult. You must face up to life, to moral decisions. But there is dignity in that grown-up courage. You stand tall and face into the keen wind of reality. You have company: warm, human arms around you, and a legacy of culture which has built up not only scientific knowledge and the material comforts that applied science brings but also art, music, the rule of law, and civilized discourse on morals. Morality and standards for life can be built up by intelligent design – design by real, intelligent humans who actually exist. Atheists have the intellectual courage to accept reality for what it is: wonderfully and shockingly explicable. As an atheist, you have the moral courage to live to the full the only life you're ever going to get: to fully inhabit reality, rejoice in it, and do your best finally to leave it better than you found it.

V

COUNSEL FOR THE PROSECUTION: INTERROGATING FAITH

IN CONVERSATION WITH LAWRENCE KRAUSS

SHOULD SCIENCE SPEAK TO FAITH?

After a talk I gave at a conference in San Diego, I was publicly attacked by an audience member whose articulate cogency marked him out as no ordinary audience member. Though obviously not religious himself, he thought I was too aggressive and not conciliatory enough towards the religious. It was my first encounter with Lawrence Krauss.* Afterwards he sought me out for a drink and we found that we agreed more than we disagreed. This led to a dialogue, commissioned by *Scientific American* and published in the magazine in 2007, from which the following is an extract.

What is the best way to teach science to religious people? Will religious faith ever be eradicated? Is religion inherently bad?

LK: I would like to discuss with you what a scientist's primary goals

---

* By no means the last. We have since collaborated on a number of occasions, for example in the documentary film *The Unbelievers*. My afterword to his book *A Universe from Nothing* is reprinted below (pp. 353–7).

should be when talking or writing about religion. Both you and I have devoted a substantial fraction of our time to trying to get people excited about science, while also attempting to explain the bases of our current respective scientific understandings of the universe. So it seems appropriate to ask which is more important: using the contrast between science and religion to teach about science or trying to put religion in its place? I suspect that I want to concentrate more on the first issue, and you want to concentrate more on the second.

I say this because if one is looking to teach people, then it seems clear to me that one needs to reach out to them, to understand where they are coming from, if one is going to seduce them into thinking about science. I often tell teachers, for example, that the biggest mistake any of them can make is to assume that their students are interested in what they are about to say. Teaching *is* seduction. Telling people, on the other hand, that their deepest beliefs are simply silly – even if they are – and that they should therefore listen to us to learn the truth ultimately defeats subsequent pedagogy. Having said that, if instead the primary purpose in discussing this subject is to put religion in its proper context, then perhaps it is useful to shock people into questioning their beliefs.

RD: The fact that I think religion is bad science, whereas you think it is ancillary to science, is bound to bias us in at least slightly different directions. I agree with you that teaching is seduction, and it could well be bad strategy to alienate your audience before you even start. Maybe I could improve my seduction technique. But nobody admires a dishonest seducer, and I wonder how far you are prepared to go in 'reaching out'. Presumably you wouldn't reach out to a Flat Earther. Nor, perhaps, to a Young Earth Creationist who thinks the entire universe began after the Middle Stone Age. But perhaps you would reach out to an Old Earth Creationist who thinks God started the whole thing off and then intervened from time to time to help evolution over the difficult jumps. The difference between us is quantitative, only. You are prepared to reach out a little further than I am, but I suspect not all that much further.

LK: Let me make clearer what I mean by reaching out. I do not mean capitulating to misconceptions but rather finding a seductive way to demonstrate to people that these are indeed misconceptions. Let me give you one example. I have, on occasion, debated both creationists and alien abduction zealots. Both groups have similar misconceptions about the nature of explanation: they feel that unless you understand everything, you understand nothing. In debates, they pick some obscure claim, say, that in 1962 some set of people in Outer Mongolia all saw a flying saucer hovering above a church. Then they ask if I am familiar with this particular episode, and if I say no, they invariably say, 'If you have not studied every such episode, then you cannot argue that alien abduction is unlikely to be happening.'

I have found that I can get each group to think about what they are saying by using the other group as a foil. Namely, of the creationists I ask, 'Do you believe in flying saucers?' They inevitably say 'no'. Then I ask, 'Why? Have you studied all of the claims?' Similarly, to the alien abduction people I ask, 'Do you believe in Young Earth Creationism?' and they say 'no', wanting to appear scientific. Then I ask, 'Why? Have you studied every single counterclaim?' The point I try to make for each group is that it is quite sensible to base theoretical expectations on a huge quantity of existing evidence, without having studied absolutely every single obscure counterclaim. This 'teaching' technique has worked in most cases, except those rare times when it has turned out that I was debating an alien abduction believer who was also a creationist!

RD: I like your clarification of what you mean by reaching out. But let me warn you of how easy it is to be misunderstood. I once wrote in a *New York Times* book review, 'It is absolutely safe to say that if you meet somebody who claims not to believe in evolution, that person is ignorant, stupid or insane (or wicked, but I'd rather not consider that).'* That sentence has been quoted again and again in support of the view that I am a bigoted, intolerant, closed-minded, intemperate ranter. But just look at my sentence. It may not be crafted to seduce, but you, Lawrence, know in your heart that it is a simple and sober statement of fact. Ignorance is no crime. To call somebody ignorant is no insult. All of us are ignorant of most of what there is to know. I am completely ignorant of baseball, and I dare say that you are as completely ignorant of cricket. If I tell somebody who believes the world is six thousand years old that he is ignorant, I am paying him the compliment of assuming that he is not stupid, insane or wicked.

LK: I have to say that I agree completely with you about this. To me, ignorance is often the problem, and, happily, ignorance is most easily addressed. It is not pejorative to suggest that someone is ignorant if they misunderstand scientific issues.

RD: In exchange, I am happy to agree with you that I could, and probably should, have put it more tactfully. I should have reached out more seductively. But there are limits. You would stop short of the following extreme: 'Dear Young Earth Creationist, I deeply respect your belief that the world is six thousand years old. Nevertheless, I humbly and gently suggest that if you were to read a book on geology, or radioisotope dating, or cosmology, or archaeology, or history, or zoology, you might find it fascinating (along with the Bible of course), and you might begin to see why almost all educated people, including theologians, think the world's age is measured in billions of years, not thousands.'

Let me propose an alternative seduction strategy. Instead of

---

* The review – of Maitland Edey and Donald Johanson's *Blueprints: solving the mystery of evolution* – is reproduced below (pp. 401–6).

pretending to respect dopey opinions, how about a little tough love? Dramatize to the Young Earth Creationist the sheer magnitude of the discrepancy between his beliefs and those of scientists: 'Six thousand years is not just a little bit different from 4.6 billion years. It is so different that, dear Young Earth Creationist, it is as though you were to claim that the distance from New York to San Francisco is not 3,400 miles but 7.8 yards. Of course, I respect your right to disagree with scientists, but perhaps it wouldn't hurt and offend you too much to be told – as a matter of deductive and indisputable arithmetic – the actual magnitude of the disagreement you've taken on.'

LK: I don't think your suggestion is 'tough love'. In fact, it is precisely what I was advocating, namely, a creative and seductive way of driving home the magnitude and nature of such misconceptions. Some people will always remain deluded, in spite of facts, but surely those are not the ones we are trying to reach. Rather, it is the vast bulk of the public who may have open minds about science but simply don't know much about it or have never been exposed to scientific evidence. In this regard, let me pose another question, about which you may feel even more strongly: Can science enrich faith, or must it always destroy it?

The question came to me because I was recently asked to speak at a Catholic college at a symposium on science and religion. I guess I was viewed as someone interested in reconciling the two. After agreeing to lecture, I discovered that I had been assigned the title 'Science Enriching Faith'. In spite of my initial qualms, the more I thought about the title, the more rationale I could see for it. The need to believe in a divine intelligence without direct evidence is, for better or worse, a fundamental component of many people's psyches. I do not think we will rid humanity of religious faith any more than we will rid humanity of romantic love or many of the irrational but fundamental aspects of human cognition. While orthogonal from the scientific rational components, they are no less real and perhaps no less worthy of some celebration when we consider our humanity.

RD: As an aside, such pessimism about humanity is popular among rationalists to the point of outright masochism. It is almost as though you and others at the conference where this dialogue began positively relish the idea that humanity is perpetually doomed to unreason. But I think irrationality has nothing to do with romantic love or poetry or the emotions that lie so close to what makes life worth living. Those are not orthogonal to rationality. Perhaps they are tangential to it. In any case, I am all for them, as are you. Positively irrational beliefs and superstitions are a different matter entirely. To accept that we can never be rid of them – that they are an irrevocable part of human nature – is manifestly untrue of you and, I would guess, most of your colleagues and friends. Isn't it therefore rather condescending to assume that humans at large are constitutionally incapable of breaking free of them?

LK: I am not so confident that I am rid of irrational beliefs, at least irrational beliefs about myself. But if religious faith is a central part of the life experience of many people, the question, it seems to me, is not how we can rid the world of God but to what extent can science at least moderate this belief and cut out the most irrational and harmful aspects of religious fundamentalism. That is certainly one way science might enrich faith.

In my lecture to the Catholic group, for instance, I took guidance from your latest book and described how scientific principles, including the requirement not to be selective in choosing data, dictate that one cannot pick and choose in one's fundamentalism. If one believes that homosexuality is an abomination because it says so in the Bible, one has to accept the other things that are said in the Bible, including the allowance to kill your children if they are disobedient or validation of the right to sleep with your father if you need to have a child and there are no other men around, and so forth.

Moreover, science can directly debunk many such destructive literal interpretations of scripture, including, for example, the notion that women are simple chattels, which stands counter to what biology tells us about the generic biological roles of females and the intellectual capabilities of women and men in particular. In the same sense that Galileo argued, when he suggested that God would not have given humans brains if 'he' did not intend people to use them to study nature, science definitely can thus enrich faith.

Still another benefit science has to offer was presented most cogently by [Carl] Sagan, who, like you and me, was not a person of faith. Nevertheless, in a posthumous compilation of his 1985 Gifford Lectures in Scotland on science and religion, he makes the point that standard religious wonder is in fact too myopic, too limited.* A single world is too puny for a real God. The vast scope of our universe, revealed to us by science, is far grander. Moreover, one might now add, in light of the current vogue in theoretical physics, that a single universe may be too puny and that one might want to start thinking in terms of a host of universes. I hasten to add, however, that enriching faith is far different from providing supporting evidence for faith, which is something that I believe science certainly does not do.

RD: Yes, I love that sentiment of Sagan's, and I'm so glad you picked it out. I summed it up for the publishers of those lectures on the book jacket: 'Was Carl Sagan a religious man? He was so much more. He left behind the petty, parochial, medieval world of the conventionally religious; left the theologians, priests and mullahs wallowing in their small-minded spiritual poverty. He left them behind, because he had so much more to be religious about. They have their Bronze Age myths, medieval superstitions and childish wishful thinking. He had the universe.' I don't think there is

---

* For a similar passage from Sagan's *Pale Blue Dot*, see p. 306 above. On the Gifford Lectures, see the opening note to my own contribution on p. 199.

anything I can add in answering your question about whether science can enrich faith. It can, in the sense you and Sagan mean. But I'd hate to be misunderstood as endorsing faith.

LK: I want to close with an issue that I think is central to much of the current debate going on among scientists regarding religion: Is religion inherently bad? I confess here that my own views have evolved over the years, although you might argue that I have simply gone soft. There is certainly ample evidence that religion has been responsible for many atrocities, and I have often said, as have you, that no one would fly planes into tall buildings on purpose if it were not for a belief that God was on their side.

As a scientist, I feel that my role is to object when religious belief causes people to teach lies about the world. In this regard, I would argue that one should respect religious sensibilities no more or less than any other metaphysical inclinations, but in particular they should not be respected when they are wrong. By wrong, I mean beliefs that are manifestly in disagreement with empirical evidence. The Earth is not six thousand years old. The sun did not stand still in the sky. The Kennewick Man was not a Umatilla Indian. What we need to try to eradicate is not religious belief, or faith, it is ignorance. Only when faith is threatened by knowledge does it become the enemy.

RD: I think we pretty much agree here. And although 'lie' is too strong a word because it implies intention to deceive, I am not one of those who elevate moral arguments above the question of whether religious beliefs are true. I recently had a televised encounter with the veteran British politician Tony Benn, a former minister of technology who calls himself a Christian. It became very clear in the course of our discussion that he had not the slightest interest in whether Christian beliefs are true or not; his only concern was whether they are moral. He objected to science on the grounds that it gave no moral guidance. When I protested that moral guidance is not what science is about, he came close to asking what, then, was the use of science. A classic example of a syndrome the philosopher Daniel Dennett has called 'belief in belief'.

Other examples include those people who think that whether religious beliefs are true or false is less important than the power of religion to comfort and to give a purpose to life. I imagine you would agree with me that we have no objection to people drawing comfort from wherever they choose and no objection to strong moral compasses. But the question of the moral or consolation value of religion – one way or the other – must be kept separate in our minds from the truth value of religion. I regularly encounter difficulties in persuading religious people of this distinction, which suggests to me that we scientific seducers have an uphill struggle on our hands.

LK: Having found another place where we definitely agree, it is perhaps a good one to end the discussion for now.

DEFENDING THE WALL
OF SEPARATION

This is a slightly edited version of my foreword to Sean Faircloth's
Attack of the Theocrats (2012).

The United States Founding Fathers, giants of the eighteenth-century Enlightenment, were far-seeing in their plans because wise in history. They knew the European past from which the Pilgrim Fathers had escaped, and they crafted a document of immunization against any such future. 'Congress shall make no law respecting an establishment of religion, or prohibiting the free exercise thereof.' In other words, *the United States shall never be a theocracy.*

That first clause of the Bill of Rights, precious First Amendment to the greatest constitutional document ever enacted, is – or ought to be – the envy of the world. My own country is still nominally a theocracy, with the head of state synonymous with head of the Church of England and constitutionally forbidden to be a Roman Catholic (let alone a Muslim or a Jew*). To this day, the Catholic/

---

* To be a little (though not entirely) frivolous, it might be easier for the
  monarch to be a Muslim or a Jew than a Roman Catholic, simply
  because either would have been inconceivable in historical times when
  such laws were drawn up. There's a persistent but probably untrue story
  that the reason lesbianism was exempt from the English law against

Protestant divide poisons Northern Ireland and, in miniature, Glasgow on a soccer Saturday. We still have twenty-six bishops sitting, unelected and *ex officio*, in the Upper House of Parliament.

None of that would have surprised James Madison and his colleagues. It is exactly the kind of social disease they were trying to forestall in their amended constitution. But even they could not have foreseen the brainless horror of our twenty-first century theocrats . . . in Saudi Arabia, for instance, where Mustafa Ibrahim was judicially executed for practising 'sorcery' (he was a pharmacist): the same Saudi Arabia, our ally and oil-provider, where a woman can be arrested for driving a car,* for showing an arm or an ankle, or for being seen in public without a male relative (who may, as a generous concession, be a child). Or Uganda, where homosexual acts are punishable by fourteen years in jail,[†] and where David Kato, a schoolteacher, was bludgeoned to death by Christian zealots inflamed by Christian newspapers inspired by American missionaries. Or Israel where, in 2006, Tove Johansson, a Swedish human rights worker attempting to escort Palestinian children to school, was violently assaulted, under the callously indifferent gaze of Israeli soldiers, by a Jewish mob chanting: 'We killed Jesus, we'll kill you too.' Or Somalia, where, in 2008, a thirteen-year-old girl, Aisho Ibrahim Dhuhulow, was sentenced to death by stoning in front of a large crowd in a football stadium. Her crime of 'adultery' was actually the crime – under sharia law – of being gang-raped.

The United States is officially not a theocracy. Thomas Jefferson's wall of separation still stands – but precariously, enduring a ceaseless buffeting, a hammering and insidious chipping away by (mainly Christian) saboteurs, who either ignorantly misread the Founders'

---

homosexuality was that Queen Victoria, when asked to sign the bill into law, flatly refused to believe that it was physically possible.

* That particular ridiculous law was repealed in 2017.

[†] Under the Uganda Anti-Homosexuality Act 2014, the maximum penalty is life imprisonment.

intentions or wilfully oppose them. And this is where Sean Faircloth rides in as a latter-day hero of the constitution and of reason. His book is a timely – poignantly timely – manifesto of secularism (not atheism). Faircloth's message is secularist and *conservative* in the true meaning of the term: conserving the original secularist principles of the constitution – unlike the so-called conservatives of the 'Tea Party', whose aim, where religion is concerned, is unashamedly to undermine the core principle of the First Amendment. Sean Faircloth quotes Barry Goldwater, 'I don't have any respect for the Religious Right', as a contrast to Michele Bachmann's 'God called me to run for Congress'. Though Sean Faircloth was a liberal Democrat in the Maine State Congress, the following words of the arch-conservative Senator Goldwater might have inspired Faircloth's book.

There is no position on which people are so immovable as their religious beliefs. There is no more powerful ally one can claim in a debate than Jesus Christ, or God, or Allah, or whatever one calls this supreme being. But like any powerful weapon, the use of God's name on one's behalf should be used sparingly. The religious factions that are growing throughout our land are not using their religious clout with wisdom. They are trying to force government leaders into following their position 100 percent. If you disagree with these religious groups on a particular moral issue, they complain, they threaten you with a loss of money or votes or both. I'm frankly sick and tired of the political preachers across this country telling me as a citizen that if I want to be a moral person, I must believe in A, B, C, and D. Just who do they think they are? And from where do they presume to claim the right to dictate their moral beliefs to me? And I am even more angry as a legislator who must endure the threats of every religious group who thinks it has some God-granted right to control my vote on every roll call in the Senate. I am warning them today: I will fight them every step of the way if they try to dictate their moral convictions to all Americans in the name of conservatism.

Sean Faircloth was trained as a lawyer, and again and again his book uncovers the harm done to today's Americans by religious bias and privileging in law. The least fortunate suffer physical injury, torture and even death. Putting a face on the faceless, giving a voice to the voiceless, Faircloth champions these innocent victims of religious privilege. They include two-year-old Amiyah White, who died unattended in the van of a Christian child-care centre. Why mention that it was 'Christian'? Because the tragedy followed directly from the centre's religious exemption from state child-safety laws. In Tennessee in 2002, Jessica Crank died of cancer, aged fifteen, after her mother decided to have her malignancy treated by 'faith healing' rather than modern medicine. This useless 'treatment' was administered under cover of a religious exemption from a state child-protection law.

Those are just two of many tragic stories. The countless other unfortunate people who have suffered in the same way are lost footnotes in the broader narrative of religion's privileged exemption from civilized law. In an attempt to restore the human element to this narrative, Sean Faircloth calls on all who follow America's historic secular project to share accounts of martyred children and other godly harm. Personal stories serve as entry points to his far-reaching examination of the assault by America's theocratic politicians and hucksters. Readers will marvel at the insidious ways in which theocratic laws exact harm on American citizens – financially, socially, militarily, physically, emotionally and educationally. But the 'us all' to which his subtitle refers is not restricted to Americans. The 'theocratic attack' that has been under way in the United States for more than three decades spills out into the world at large.

Faircloth examines the death of fourteen-year-old Saron Samta from a botched back-alley abortion in Ethiopia, and links it to the global 'gag rule', initiated under Ronald Reagan, which restricts women's access to basic health information and services. He recalls George W. Bush's infamous phone call to Jacques Chirac before the Iraq War, when Bush reportedly warned the French president that 'Gog and Magog are at work in the Middle East' and that 'the biblical prophecies are being fulfilled'. Such crass evangelical certitude, embarrassingly relayed by a sitting US President to another head of state, is chilling. One can only imagine what biblical chestnuts were served and swallowed in the prayer sessions that Bush held with the equally devout Tony Blair in advance of the war.

These are just stray examples of how the 'theocratic attack' by America's religious right has the capacity to overstep US borders. The fundamentalists' undermining of American science has international knock-on effects. The biblical notion of man's 'dominion' resonates in America's environmental policies and informs the religious public's views on climate change, thus contributing to the degradation of the whole planet.

Faircloth pays just attention to one of the great iniquities of the American taxation system, one which is shared with many other countries including my own. Religious institutions, churches, even obscenely wealthy televangelists, are tax-exempt, and exempt from many of the rigours of even *declaring* money for taxation. As he writes:

- Unlike non-profits, churches don't have to file 990 forms (a basic financial disclosure). Thus, their finances are the most secretive of any so-called charitable organization. For-profit businesses, of course, must file detailed tax documents. So must 501(c)(3) non-profits. Because the finances of religious organizations are akin to the proverbial black box, it is difficult to even find out whether something improper has occurred.

- Only a 'high-level' IRS official can even authorize an audit of a religious organization. Meanwhile, the rest of us – whether individuals, for-profit businesses, or secular non-profits – can be audited by any old IRS bureaucrat.

- Religious groups can legally give tax-free housing allowances to so-called clergy (some of whom just might be family), allowances that are not counted as income, exempting the housing from taxation.

To return to my beginning, Faircloth brings out the ugly contrast between today's theocratic attacks and the secular intentions of America's Founding Fathers. They established a nation that should forever separate state from church. Every American child knows this, or at least used to know it (the Texas Board of Education's 2010 decision to excise Jefferson's name from the social studies curriculum awaits a more charitable interpretation than I can muster). By design, the United States was to be kept free of religion's suffocating foot so as to give breath to individual conscience. By putting into practice this cherished ideal, the United States made a great leap forward for civilization and humankind. Other countries followed suit with their own secular constitutions, beginning with France and including, notably, India and Turkey. If America, the world's standard-bearer of secular governance, allows semi-literate Christian know-nothings to erode the wall between church and state – the very foundation of United States exceptionalism – whither the world?

Faircloth paints a sobering picture but fortunately, as anyone who has heard his speeches knows, he also has an inspiring and invigorating vision to offer. The stated intent of this volume is not just to wake people up to how the religious right harms us all, but also to lay out a much-needed plan for action. As a shrewd former politician and a spellbinding orator, Sean Faircloth is well qualified to play a role in returning America to its secular foundation. His manifesto is the optimistic flipside to the dark evidence that his book presents. While the title and parts of the book itself evoke divinely sanctioned barbarism, readers may finish it, like me, exercised, energized, and eager to participate in his bold return to the secular dream of the European Enlightenment and America's enlightened founders.

A MORAL AND INTELLECTUAL EMERGENCY

Foreword to the UK edition of Sam Harris's *Letter to a Christian Nation* (2007). This was Sam's second book on atheism, following his remarkably powerful *The End of Faith* (2005).

Sam Harris doesn't mess about. He writes directly to his Christian reader as 'you', and he pays 'you' the compliment of taking your beliefs seriously: '... if one of us is right, the other is wrong ... in the fullness of time, one side is really going to win this argument, and the other is really going to lose.' But you don't (as I can personally understate) have to fit the 'you' profile in order to enjoy this marvellous little book. Every word zings like an elegantly fletched arrow from a taut bow-string and flies in a gracefully swift arc to the target, where it thuds satisfyingly into the bullseye.

If you are part of the target, I dare you to read this book. It will be a salutary test of your faith. Survive Sam Harris's barrage, and you can take on the world with equanimity. But forgive my scepticism: Harris never misses, not with a single sentence, which is why his short book is so disproportionately devastating. If you already share Harris's and my doubts about religious faith and are not part of his target, this book will powerfully arm you to argue against those who are. Or you may be Christian and still not part of the target. This book freely admits that there are Christians who take, as they would see it, a more nuanced view:

> Liberal and moderate Christians will not always recognize them-
> selves in the 'Christian' I address. They should, however, recognize
> many of their neighbors – and more than one hundred and fifty
> million Americans.

And that's the point. It was the menace of those hundred and fifty millions that provoked this book. If your religious beliefs are so vague and nebulous that even well-aimed arrows bounce off unnoticed, Harris is not writing for you directly. But you should still care about the emergency that concerns him – and me. Where I, as a scientific educator, am dismayed by the 50 per cent of the American population who believe the world is six thousand years old (an error equivalent to believing that the distance from New York to San Francisco is shorter than a cricket pitch), Sam Harris is at least as urgently concerned with other beliefs held by roughly the same 50 per cent:

> It is, therefore, not an exaggeration to say that if London, Sydney,
> or New York were suddenly replaced by a ball of fire, some signifi-
> cant percentage of the American population would see a silver
> lining in the subsequent mushroom cloud, as it would suggest to
> them that the best thing that is ever going to happen was about to
> happen: the return of Christ. It should be blindingly obvious that
> beliefs of this sort will do little to help humanity create a durable
> future for itself – socially, economically, environmentally, or geo-
> politically. Imagine the consequences if any significant component
> of the US government actually believed that the world was about
> to end and that its ending would be *glorious*. The fact that nearly
> half of the American population apparently believes this, purely
> on the basis of religious dogma, should be considered a moral and
> intellectual emergency.

The 'Christian Nation' for whom the book was originally written is, of course, the United States. But it would be complacent folly for us to dismiss it as a purely American problem. The USA, at least, is protected by Jefferson's enlightened wall of separation between church and state. Religion is part of Britain's historic establishment, while at this moment our most pious political leadership since Gladstone is hell bent on supporting 'faith schools'. And not just the traditional Christian schools, be it noticed, for our government, egged on by an heir to the throne who wishes to be known as 'Defender of Faith', is actively sympathetic towards the 'us-too' bleatings of other 'faith communities', eager for state-subsidized indoctrination of their children. Would it be possible to design a more divisive educational formula? More importantly, the world's only superpower is close to domination by electors who believe the entire universe began after the domestication of the dog, and believe that they will be personally 'raptured' up to heaven within their own lifetime, followed by an Armageddon welcomed as harbinger of the Second Coming. Even from this side of the Atlantic, Sam Harris's phrase 'moral and intellectual emergency' begins to look like an understatement.

I began by saying that Sam Harris doesn't mess about. One of his points is that none of us can afford to. *Letter to a Christian Nation* will stir you. Whether it stirs you to defensive or offensive action, it will not leave you unchanged. Read it if it is the last thing you do. And hope that it won't be.

UNMASKING THE
DESIGN ILLUSION

Foreword to Niall Shanks's *God, the Devil and Darwin* (2004).

Who owns the argument from improbability? Statistical improbability is the old standby, the creaking warhorse of all creationists from naive Bible-jocks who don't know better to comparatively well-educated intelligent design 'theorists',* who should. There is no other creationist argument (if you discount falsehoods like 'There aren't any intermediate fossils' and ignorant absurdities like 'Evolution violates the second law of thermodynamics'). However superficially different they may appear, under the surface the deep structure of creationist advocacy is always the same. Something in nature – an eye, a biochemical pathway or a cosmic constant – is too *improbable* to have come about by *chance*. Therefore it must have been *designed*. A watch demands a watchmaker. As a gratuitous bonus, the watchmaker conveniently turns out to be the Christian God (or Yahweh, or Allah, or whichever deity pervaded our particular childhood).

That this is a lousy argument has been clear ever since Hume's time, but we had to wait for Darwin to give us a satisfying replacement. Less often realized is that the argument from improbability,

---

* The emphasis is on 'comparatively'.

properly understood, backfires fatally against its main devotees. Conscientiously pursued, the statistical improbability argument leads us to a conclusion diametrically opposite to the fond hopes of the creationists. There may be good reasons for believing in a supernatural being (admittedly, I can't think of any) but the argument from design is emphatically not one of them. The argument from improbability firmly belongs to the evolutionists. Darwinian natural selection, which, contrary to a deplorably widespread misconception, is the very antithesis of a chance process, is the only known mechanism that is ultimately capable of generating improbable complexity out of simplicity. Yet it is amazing how intuitively appealing the design inference remains to huge numbers of people. Until we think it through . . . which is where Niall Shanks comes in.

Combining historical erudition with up-to-date scientific knowledge, Professor Shanks casts a clear philosopher's eye on the murky underworld inhabited by the 'intelligent design' gang and their 'wedge' strategy (which is every bit as creepy as it sounds) and explains, simply and logically, why they are wrong and evolution is right. Chapter follows chapter in logical sequence, moving from history through biology to cosmology, and ending with a cogent and perceptive analysis of the underlying motivations and social manipulation techniques of modern creationists, including especially the 'intelligent design' sub-species of creationists.

Intelligent design 'theory' (ID) has none of the innocent charm of old-style, revival-tent creationism. Sophistry dresses the venerable watchmaker up in two cloaks of ersatz novelty: 'irreducible complexity' and 'specified complexity', both wrongly attributed to recent ID authors but both much older. 'Irreducible complexity' is nothing more than the familiar 'What is the use of half an eye?' argument, even if it is now applied at the biochemical or the cellular level. And 'specified complexity' just takes care of the point that any old haphazard pattern is as improbable as any other, *with hindsight*. A heap of detached watch parts tossed in a box is, with

hindsight, as improbable as a fully functioning, genuinely complicated watch. As I put it in *The Blind Watchmaker*,

> complicated things have some quality, *specifiable in advance*, that
> is highly unlikely to have been acquired by random chance alone.
> In the case of living things, the quality that is specified in advance
> is, in some sense, 'proficiency'; either proficiency in a particular
> ability such as flying, as an aero-engineer might admire it; or pro-
> ficiency in something more general, such as the ability to stave off
> death.

Darwinism and design are both, on the face of it, candidate
explanations for specified complexity. But design is fatally wounded
by infinite regress. Darwinism comes through unscathed. Designers must be statistically improbable like their creations, and they
therefore cannot provide an ultimate explanation. Specified complexity is the phenomenon we seek to explain. It is obviously futile
to try to explain it simply by specifying even greater complexity.
Darwinism really does explain it in terms of something simpler –
which in turn is explained in terms of something simpler still and
so on back to primeval simplicity. Design may be the temporarily
correct explanation for some particular manifestation of specified
complexity such as a car or a washing machine. But it can never be
the ultimate explanation. Only Darwinian natural selection (as far
as anyone has ever been able to discover or even credibly suggest) is
even a *candidate* as an ultimate explanation.

It could conceivably turn out, as Francis Crick and Leslie Orgel
once facetiously suggested, that evolution on this planet was seeded
by deliberate design, in the form of bacteria sent from some distant
planet in the nose cone of a space ship. But the intelligent life form
on that distant planet then demands its own explanation. Sooner or
later, we are going to need something better than actual design in
order to explain the illusion of design. Design itself can never be
an ultimate explanation. And the more statistically improbable the

specified complexity under discussion, the more unlikely does any kind of design theory become, while evolution becomes correspondingly more powerfully indispensable. So all those calculations with which creationists love to browbeat their naive audiences – the mega-astronomical odds against an entity spontaneously coming into existence by chance – are actually exercises in eloquently shooting themselves in the foot.

Worse, ID is lazy science. It poses a problem (statistical improbability) and, having recognized that the problem is difficult, it lies down under the difficulty without even *trying* to solve it. It leaps straight from the difficulty – 'I can't see any solution to the problem' – to the cop-out – 'Therefore a Higher Power must have done it.' This would be deplorable for its idle defeatism, even if we didn't have the additional difficulty of infinite regress. To see how lazy and defeatist it is, imagine a fictional conversation between two scientists working on a hard problem, say A. L. Hodgkin and A. F. Huxley who, in real life, won the Nobel Prize for their brilliant model of the nerve impulse.

'I say, Huxley, this is a terribly difficult problem. I can't see how the nerve impulse works, can you?'

'No, Hodgkin, I can't, and these differential equations are fiendishly hard to solve. Why don't we just give up and say that the nerve impulse propagates by Nervous Energy?'

'Excellent idea, Huxley, let's write the Letter to *Nature* now, it'll only take one line, then we can turn to something easier.'

Huxley's elder brother Julian made a similar point when, long ago, he satirized vitalism as tantamount to explaining that a railway engine was propelled by *élan locomotif*.

With the best will in the world, I can see no difference at all between *élan locomotif*, or my hypothetically lazy version of Hodgkin and Huxley, and the really lazy luminaries of ID. Yet, so successful is their 'wedge strategy', they are coming close to subverting the schooling of young Americans in state after state, and they are even invited to testify before congressional committees: all this, while ignominiously failing to come up with a single research paper worthy of publication in a peer-reviewed journal.

Intelligent design 'theory' is pernicious nonsense which needs to be neutralized before irreparable damage is done to American education. Niall Shanks's book is a shrewd broadside in what will, I fear, be a lengthy campaign. It will not change the minds of the wedgies themselves. Nothing will do that, especially in cases where, as Shanks astutely realizes, the perceived moral, social and political implications of a theory are judged more important than the truth of that theory. But this book will sway readers who are genuinely undecided and honestly curious. And, perhaps more importantly, it should stiffen the resolve of demoralized biology teachers, struggling to do their duty by the children in their care but threatened and intimidated by aggressive parents and school boards. Evolution should not be slipped into the curriculum timidly, apologetically or furtively. Nor should it appear late in the cycle of a child's education. For rather odd historical reasons, evolution has become a battlefield on which the forces of enlightenment confront the dark powers of ignorance and regression. Biology teachers are front-line troops, who need all the support we can give them.* They, and their pupils and honest seekers after truth in general, will benefit from reading Professor Shanks's admirable book.

---

* A wonderful example of such support now comes from the Teacher Institute for Evolutionary Science (TIES). See footnote on p. 51.

'NOTHING WILL COME OF NOTHING': WHY LEAR WAS WRONG

Afterword to Lawrence M. Krauss's *A Universe from Nothing* (2012). I am pleased to note this book had its genesis in a 2009 symposium that I organized in Los Angeles on behalf of the Richard Dawkins Foundation for Reason and Science.

Nothing expands the mind like the expanding universe. The music of the spheres is a nursery rhyme, a jingle to set against the majestic chords of the Symphonie Galactica. Changing the metaphor and the dimension, the dusts of centuries, the mists of what we presume to call 'ancient' history, are soon blown off by the steady, eroding winds of geological ages. Even the age of the universe, accurate – so Lawrence Krauss assures us – to the fourth significant figure at 13.72 billion years, is dwarfed by the trillennia that are to come.

But Krauss's vision of the cosmology of the remote future is paradoxical and frightening. Scientific progress is likely to go into reverse. We naturally think that, if there are cosmologists in the year 2 trillion AD, their vision of the universe will be expanded over ours. Not so – and this is one of the many shattering conclusions I take away on closing this book. Give or take a few billion years, ours is a very propitious time to be a cosmologist. Two trillion years hence, the universe will have expanded so far that all galaxies but the cosmologist's own (whichever one it happens to be) will have receded behind an Einsteinian horizon so absolute, so inviolable, that they are not only invisible but beyond all possibility of leaving a trace, however indirect. They might as well never have existed. Every trace of the Big Bang will most likely have gone, for ever and beyond recovery. The cosmologists of the future will be cut off from their past, and from their situation, in a way that we are not.

We know we are situated in the midst of 100 billion galaxies, and we know about the Big Bang because the evidence is all around us: the red-shifted radiation from distant galaxies tells us of the Hubble expansion and we extrapolate it backward. We are privileged to see the evidence because we look out on an infant universe, basking in that dawn age when light can still travel from galaxy to galaxy. As Krauss and a colleague wittily put it, 'We live at a very special time . . . the only time when we can observationally verify that we live at a very special time!' The cosmologists of the third trillennium will be forced back to the stunted vision of our early twentieth century, locked as we were in a single galaxy which, for all that we knew or could imagine, was synonymous with the universe.

Finally, and inevitably, the flat universe will further flatten into a nothingness that mirrors its beginning. Not only will there be no cosmologists to look out on the universe, there will be nothing for them to see even if they could. Nothing at all. Not even atoms. Nothing.

If you think that's bleak and cheerless, too bad. Reality doesn't owe us comfort. When Margaret Fuller remarked, with what I imagine to have been a sigh of satisfaction, 'I accept the universe,' Thomas Carlyle's reply was withering: 'Gad, she'd better!' Personally, I think the eternal quietus of an infinitely flat nothingness has a grandeur that is, to say the least, worth facing off with courage.

But if something can flatten into nothing, can nothing spring into action and give birth to something? Or why, to quote a theological chestnut, is there something rather than nothing? Here we come to perhaps the most remarkable lesson that we are left with on closing Lawrence Krauss's book. Not only does physics tell us how something could have come from nothing, it goes further, by Krauss's account, and shows us that nothingness is unstable: something was almost bound to spring into existence from it. If I understand Krauss aright, it happens all the time: the principle sounds like a sort of physicist's version of two wrongs making a right. Particles and antiparticles wink in and out of existence like subatomic fireflies, annihilating each other, and then re-creating themselves by the reverse process, out of nothingness.

The spontaneous genesis of something out of nothing happened in a big way at the beginning of space and time, in the singularity known as the Big Bang followed by the inflationary period, when the universe, and everything in it, took a fraction of a second to grow through twenty-eight orders of magnitude (that's a 1 with twenty-eight zeroes after it – think about it).

What a bizarre, ridiculous notion! Really, these scientists! They're as bad as medieval Schoolmen counting angels on pinheads or debating the 'mystery' of the transubstantiation.

No, not so, not so with a vengeance and in spades. There is much that science still doesn't know (and it is working on it with rolled-up sleeves). But some of what we do know, we know not just approximately (the universe is not mere thousands but billions of years old): we know it with confidence and with stupefying accuracy. I've already mentioned that the age of the universe is measured to four significant figures. That's impressive enough, but it is nothing compared to the accuracy of some of the predictions with which Lawrence Krauss and his colleagues can amaze us. Krauss's hero Richard Feynman pointed out that some of the predictions of quantum theory – again based on assumptions that seem more bizarre than anything dreamed up by even the most obscurantist of theologians – have been verified with such accuracy that they are equivalent to predicting the distance between New York and Los Angeles to within one hairsbreadth.

Theologians may speculate about angels on pinheads or whatever is the current equivalent. Physicists might seem to have their own angels and their own pinheads: quanta and quarks, 'charm', 'strangeness' and 'spin'. But physicists can count their angels and can get it right to the nearest angel in a total of 10 billion: not an angel more, not an angel less. Science may be weird and incomprehensible – more weird and less comprehensible than any theology – but science works. It gets results. It can fly you to Saturn, slingshotting you around Venus and Jupiter on the way. We may not understand quantum theory (heaven knows, I don't), but a theory that predicts the real world to ten decimal places cannot in any straightforward sense be wrong. Theology not only lacks decimal places: it lacks even the smallest hint of a connection with the real world. As Thomas Jefferson said, when founding his University of Virginia: 'A professorship of Theology should have no place in our institution.'

If you ask religious believers why they believe, you may find a few 'sophisticated' theologians who will talk about God as the 'Ground of all Isness' or as 'a metaphor for interpersonal fellowship' or some such evasion. But the majority of believers leap, more honestly and vulnerably, to a version of the argument from design or the argument from first cause. Philosophers of the calibre of David Hume didn't need to rise from their armchairs to demonstrate the fatal weakness of all such arguments: they beg the question of the Creator's origin. But it took Charles Darwin, out in the real world on HMS *Beagle*, to discover the brilliantly simple – and non-question-begging – alternative to design. In the field of biology, that is. Biology was always the favourite hunting ground for natural theologians until Darwin – not deliberately, for he was the kindest and gentlest of men – chased them off. They fled to the rarefied pastures of physics and the origins of the universe, only to find Lawrence Krauss and his predecessors waiting for them.

Do the laws and constants of physics look like a finely tuned put-up job, designed to bring us into existence? Do you think some agent must have caused everything to start? Read Victor Stenger if you can't see what's wrong with arguments like that. Read Steven Weinberg, Peter Atkins, Martin Rees, Stephen Hawking. And now we can read Lawrence Krauss for what looks to me like the knock-out blow. Even the last remaining trump card of the theologian, 'Why is there something rather than nothing?' shrivels up before your eyes as you read these pages. If *On the Origin of Species* was biology's deadliest blow to supernaturalism, we may come to see *A Universe from Nothing* as the equivalent from cosmology. The title means exactly what it says. And what it says is devastating.

THE FAST FOOD THESIS:
RELIGION AS EVOLUTIONARY
BY-PRODUCT

Foreword to J. Anderson Thomson's *Why We Believe in Gods* (2011).
Andy Thomson was a founder member of the board of the Richard
Dawkins Foundation for Reason and Science (RDFRS) before we
merged with the Center for Inquiry (CFI); he is now a sagacious
member of the amalgamated board and our only medical doctor
and psychiatrist. In this book he trains his psychologist's eye on the
question of what makes people religious.

In one of the great understatements of history, *The Origin of Species*
confines its discussion of human evolution to a laconic prophecy:
'Light will be thrown on the origin of man and his history.' Less
often quoted is the beginning of the same paragraph: 'In the distant future I see open fields for far more important researches.
Psychology will be based on a new foundation.' Dr Thomson is one
of the evolutionary psychologists now making Darwin's forecast
come true, and this book about the evolutionary drivers of religiosity would have delighted the old man.

Darwin, though not religious in his maturity, understood the
religious impulse. He was a benefactor of Down church and he regularly walked his family there on Sundays (then continued his walk

while they went inside). He had been trained to the life of a clergyman, and William Paley's *Natural Theology* was his favoured undergraduate reading. Darwin killed natural theology's *answer* stone dead, but he never lost his preoccupation with its *question*: the question of function. It is no surprise that he was intrigued by the functional question of religiosity. Why do most people, and all peoples, harbour religious beliefs? 'Why' is to be understood in the special functional sense that we today, though not Darwin himself, would call 'Darwinian'.

How, to put the Darwinian question in modern terms, does religiosity contribute to the survival of genes promoting it? Thomson is a leading proponent of the 'by-product' school of thought: religion itself need have no survival value; it is a *by-product* of psychological predispositions that have.

'Fast food' is a *leitmotif* of the book: 'If you understand the psychology of fast food, you understand the psychology of religion.' Sugar is another good example. It was impossible for our wild ancestors to get enough of it, so we have inherited an open-ended craving that, now that it is easily met, damages our health. 'These fast-food cravings are a by-product. And now they become dangerous, because, uncontrolled, they can lead to health problems our ancestors likely never faced . . . Which brings us to religion.'

Another leading evolutionary psychologist, Steven Pinker, explains our love of music in a similar 'by-product' way, as 'auditory cheesecake, an exquisite confection crafted to tickle the sensitive spots of at least six of our mental faculties'. For Pinker, the mental faculties supernormally tickled as a by-product by music are mostly concerned with the sophisticated brain software required to disentangle meaningful sounds (for example, language) from background bedlam.* Thomson's fast-food theory of religion emphasizes, rather,

---

* For more on the human taste for music, see the passage of my conversation with Steven Pinker reproduced above, pp. 166–7.

those psychological predispositions that can be called *social*: 'adaptive psychological mechanisms that evolved to help us negotiate our relationships with other people, to detect agency and intent, and to generate a sense of safety. These mechanisms were forged in the not-so-distant world of our African homeland.'

Thomson's chapters identify a series of evolved mental faculties exploited by religion, each one beguilingly labelled with a line familiar from scripture or liturgy: 'Our Daily Bread', 'Deliver Us from Evil', 'Thy Will Be Done', 'Lest Ye Be Judged'. There are some compelling images:

> Think of a two-year-old child reaching out to be picked up and cuddled. He extends his hands above his head and beseeches you. Think now of the Pentecostal worshipper who speaks in tongues. He stretches out his hands above his head, beseeching god in the same 'pick-me-up-and-hold-me' gesture. We may lose human attachment figures through death, through misunderstandings, through distance, but a god is always there for us.

To most of us, that arms-extended gesture of the worshipper looks merely foolish. After reading Thomson we shall see it through more penetrating eyes: it is not just foolish, it is infantile.

Then there is our eagerness to detect the deliberate hand of agency.

> Why is it you mistake a shadow for a burglar but never a burglar for a shadow? If you hear a door slam, why do you wonder who did it before you consider the wind as the culprit? Why might a child who sees blowing tree limbs through a window fear that it's the bogeyman come to get him?

The hyperactive agency detection device evolved in the brains of our wild ancestors because of a risk asymmetry. A rustle in the long grass is statistically more likely to be the wind than a leopard. But the cost of a mistake is higher one way than the other. Agents, like leopards and burglars, can kill. Best go with the statistically unlikely guess. (Darwin himself made the point, in an anecdote about his dog's response to a wind-blown parasol.) Thomson pursues the thought – oversensitivity to *agents* where there are none – and gives us his elegant explanation of another of the psychological biases upon which religiosity is founded.

Our Darwinian preoccupation with kinship is yet another. For example, in Roman Catholic lore, 'the nuns are "sisters" or even "mother superiors," the priests are "fathers," the monks are "brothers," the Pope is the "Holy Father," and the religion itself is referred to as the "Holy Mother Church"'.

Dr Thomson has made a special study of suicide bombers, and he notes how kin-based psychology is exploited in their recruitment and training:

> Charismatic recruiters and trainers create cells of fictive kin, pseudobrothers outraged at the treatment of their Muslim brothers and sisters and separated from actual kin. The appeal of such martyrdom is not just the sexual fantasy of multiple heavenly virgins, but the chance to give chosen kin punched tickets to paradise.

One by one, the other components of religion – community worship, obedience to priestly authority, ritual – receive the Thomson treatment. Every point he makes has the ring of truth, abetted by a crisp style and vivid imagery. Andy Thomson is an outstandingly persuasive lecturer, and it shines through his writing. This short, punchy book will be swiftly read – and long remembered.

AN AMBITIOUS BANANA SKIN

This review of Richard Swinburne's *Is There a God?* was published
in the *Sunday Times* on 4 February 1996. Swinburne is a
distinguished theologian and philosopher of religion. Enough said.

It is a virtue of clear writing that you can see what is wrong with a
book as well as what is right. Richard Swinburne is clear. You can
see where he is coming from. You can also see where he is going to,
and there is something almost endearing in the way he lovingly
stakes out his own banana skin and rings it about with converging
arrows boldly labelled 'Step here'.

It is surprising that a writer as clear as Swinburne has risen to
the top of his profession as Nolloth Professor of Philosophy of Religion at Oxford. Theology is a field in which obscurantism is the
normal path to success, and a favourite trick is to insist that religion has its own 'dimension(s)', completely separate from those of
science; science and religion are about different kinds of truth and
you cannot use the criteria of one to judge the other; religion
answers those questions which are outside the territory of science.

Richard Swinburne will have none of these flabby evasions. His
opening chapter expounds what he is going to mean by the God
whose existence he plans to demonstrate, and it is very much not a
vague synonym for The Ground of All Isness or Caring in the Community, but a spirited, supernatural intelligence whose existence,
if demonstrated, would actually make a difference to something.

Swinburne returns to an earlier, braver and more intellectually honest – some might say foolhardy – theology.

Swinburne is ambitious. He will not shrink into those few remaining backwaters which scientific explanation has so far failed to reach. He offers a theistic explanation for those very aspects of the world where science claims to have succeeded, and he insists that his explanation is better. Better, moreover, by a criterion likely to appeal to a scientist: simplicity. He shows that his heart is in the right place by convincingly demonstrating why we should always prefer the simplest hypothesis that fits the facts. But then comes the great banana skin experience. By an amazing exploit of double-think, Swinburne manages to convince himself that theistic explanations are simple explanations.

Science explains complex things in terms of the interactions of simpler things, ultimately the interactions of fundamental particles. I (and I dare say you) think it a beautifully simple idea that all things are made of different combinations of fundamental particles which, although exceedingly numerous, are drawn from a small, finite set. If we are sceptical, it is likely to be because we think the idea too simple. But for Swinburne it is not simple at all, quite the reverse.

His reasoning is very odd indeed. Given that the number of particles of any one type, say electrons, is large, Swinburne thinks it too much of a coincidence for so many to have the same properties. One electron, he could stomach. But billions and billions of electrons, *all with the same properties*, that is what really excites his incredulity. For him it would be simpler, more natural, less demanding of explanation, if all electrons were different from each other. Worse, no one electron should naturally retain its properties for more than an instant at a time, but would be expected to change capriciously, haphazardly and fleetingly from moment to moment. That is Swinburne's view of the simple, native state of affairs. Anything more uniform (what you or I would call more simple) requires a special explanation. 'It is only because electrons and bits of copper and all other material objects have the same powers in the twentieth century as they did in the nineteenth century that things are as they are now' (p. 42).

Enter God. God comes to the rescue by deliberately and continuously sustaining the properties of all those billions of electrons and bits of copper, and neutralizing their otherwise ingrained inclination to wild and erratic fluctuation. That is why when you've seen one electron you've seen them all, that is why bits of copper all behave like bits of copper, and that is why each electron and each bit of copper stays the same as itself from microsecond to microsecond. It is because God is constantly hanging on to each and every particle, curbing its reckless excesses and whipping it into line with its colleagues to keep them all the same.

Oh, and in case you wondered how the hypothesis that God is simultaneously keeping a billion fingers on a billion electrons can be a *simple* hypothesis, the reason is this. God is only a *single* substance. What brilliant economy of explanatory causes compared with all those billions of independent electrons all just happening to be the same!

> Theism claims that every other object which exists is caused to exist and kept in existence by just one substance, God. And it claims that every property which every substance has is due to God causing or permitting it to exist. It is a hallmark of a simple explanation to postulate few causes. There could in this respect be no simpler explanation than one which postulated only one cause. Theism is simpler than polytheism. And theism postulates for its one cause, a person [with] infinite power (God can do anything logically possible), infinite knowledge (God knows everything logically possible to know), and infinite freedom. (p. 43)

Swinburne generously concedes that God cannot accomplish feats that are *logically* impossible, and one feels grateful for this forbearance. That said, there is no limit to the explanatory purposes to which God's infinite power is put. Is science having a little difficulty explaining X? No problem. Don't give X another glance. God's infinite power is effortlessly wheeled in to explain X (along with everything else), and it is always a supremely *simple* explanation because, after all, there is only one God. What could be simpler than that?

Well, actually, almost everything. A God capable of continuously monitoring and controlling the individual status of every particle in the universe is *not* going to be simple. His existence is therefore going to need a modicum of explaining in its own right (it is often considered bad taste to bring that up, but Swinburne does rather ask for it by pinning his hopes on the virtues of simplicity). Worse (from the point of view of simplicity), other corners of God's giant consciousness are simultaneously preoccupied with the doings and emotions and prayers of every single human being. He even, according to Swinburne, has to decide continuously *not* to intervene miraculously to save us when we get cancer. That would never do, for, 'if God answered most prayers for a relative to recover from cancer, then cancer would no longer be a problem for humans to solve'. And *then* where would we be?

If this is theology, perhaps Professor Swinburne's colleagues are wise to be less lucid.

HEAVENLY TWINS

Foreword to *Jesus and Mo: Folie à Dieu* (2013).

Where shall we look for the shrewdest, wittiest, most critically pen-etrating running commentary on the absurdities of contemporary religion – and even some absurdities of the organized opposition to today's religion? To books? To blogs? To print journalism? Radio? TV? Websites? You'll find good things, as well as plenty of bad, in all those media. But if I had to award the Palme d'Or for the most original and wittiest of all (amid stiff competition from such gems as Brian Dalton's Mr Deity and the songs of Roy Zimmerman), I would have to nominate an unassuming strip cartoon from my home country: Jesus and Mo.

Folie à Dieu is the latest in a marvellous series of collections of Jesus and Mo cartoons. Every intelligent observer of contemporary disputation will enjoy it. The protagonists, Jesus and Mo them-selves, are drawn with such disarming affection, it would be hard to take offence – even given the voracious appetite for offence that the faithful uniquely indulge.* Smile your way through this book, and you end up with a real liking for Jesus and Mo, a sympathy for their touchingly insecure tussles with each other, an empathy

---

* Actually, I'm no longer so sure it is unique. Other candidates today might include the common student appetite for 'safe spaces' protected from 'hurt' and 'offence'.

with their endearingly naive struggle to justify their respective faiths in the teeth of harsh reality: the reality of science and critical reason, often given voice by the never-seen character of the friendly but no-nonsense barmaid.

There is something pleasingly formulaic about the drawings. The gentle expressions of Jesus and Mo, and of Moses who occasionally enters to represent Judaism, never change. The backdrop to our heroes' conversation shifts between only about four scenes: the park bench; the stage, where they introduce their double act with Mo improbably on guitar; the bar where they improbably drink Guinness together; the double bed which they even more improbably share (with not the slightest hint of sexual innuendo). The scenes recycle with a comforting familiarity, which gives added punch to the satire.

The satire's breadth of coverage runs the gamut of contemporary controversy. I can think of no major issue in the whole absurd panoply of religious discourse which has escaped the penetrating eye of the anonymous Author of Jesus and Mo. Turning even-handedly to those who oppose contemporary religion, he accurately lampoons the bitchy infighting to which right-on progressives are unfortunately prone – shades of Monty Python and the Judean People's Front. There's topicality too: the presidential candidacy of a Mormon serves as the pretext for a cameo walk-on by the nineteenth-century charlatan Joseph Smith, his face completely covered by his preposterous magic hat.

But of all the victims of this splendid mockery, perhaps the most deeply wounded will be 'sophisticated theologians', those paragons of puffed-up vacuity, puffing out their soggy, infinitely yielding clouds of self-deceiving, apophatic obscurity. 'Sophisticated theology' is oxymoronic because, in truth, there is nothing in theology to be sophisticated about, but it has pretensions that are interminably spun out in verbiage whose very length contrasts with the devastating economy with which the Jesus and Mo author slices it up. To do this so effectively requires a firm grasp not just of 'theology' but of philosophy too. The laconic elegance with which our Author takes out the 'theologians' could only be achieved by somebody who has taken the trouble to immerse himself thoroughly in their self-deluding claptrap. Where a professional philosopher might take a thousand words to puncture the balloon of apophatuous obscurantism, the J & M strip achieves the same result at a fraction of the length and no diminution of critical effect.

Gentle these cartoons may be, but their gentleness belies a satirical bite which is all the more effective in consequence. *Folie à Dieu* would make an ideal Christmas present. Especially for our religious friends.

A TALE OF HORROR
AND HEROISM

In 2016 Farida Khalaf sent me her remarkable book *The Girl Who Beat ISIS*. I read it with growing horror and was so moved when I finished it that I immediately wrote this review and published it on my own website.

This book is the ghostwritten memoir of an almost superhumanly brave and heroic young woman, captured and sold into sexual slavery by ISIS. Farida Khalaf (not her real name, for obvious reasons) is a Yazidi teenager from the Kurdish region of northern Iraq. A high-flying mathematical scholar, she dreamed of becoming a teacher and won a coveted scholarship to Germany. But her dream and her happy family life were shattered when jihadist scum invaded her home village. The men of the village were lined up and shot for the crime of not being Muslims (the Yazidis are monotheists, but their God is evidently distinguishable enough from Allah to justify murder) and the women were taken away and sold as slaves: sex slaves in the case of the young women and children, virgins being especially prized. In the slave market, customers would come to inspect the merchandise before haggling over the price in full hearing of the goods themselves. One prospective buyer put his finger in Farida's mouth to check her teeth, as one might when buying a horse. She bit him, and I'm sorry she failed to bite his finger right off. Farida and her dear friend Evin deliberately tried to make themselves as unattractive as possible, in the hope of postponing the moment of purchase, although the conditions in which they were kept while imprisoned and awaiting sale were beyond appalling.

Farida was bought, sold on, bought again, raped repeatedly by her 'owners', starved, and beaten to the point of serious injury. Mercifully, we are spared the details of the rapes, but one horrific scene sticks in the memory:

'I've waited long enough', he said. 'God is my witness that this is so. I have a right to you.'

His long 'wait' had been while Farida was incapacitated by her attempted suicide after he bought her, cutting her wrists with a broken bottle, the only weapon she could procure. She had barely recovered from the massive loss of blood before this rape scene.

He rolled out his mat and got ready to kneel down and pray. I'd heard from my friends that the particularly religious ones commonly did this before taking a woman, thereby celebrating their rape as a form of worship.

Farida desperately tried to jump out of the window while he was distracted at prayer but he caught her.

I tried biting his arm. But nothing helped. I could not prevent Amjed from doing what he'd planned. When he finally got off me, I curled up into a ball and stayed on the bed, crying.

That night, Farida had an epileptic fit.

Evin was bought too and the two friends saw each other only intermittently after that, which added to their distress, for they found great solace in each other, and their mutually supportive comradeship in unspeakable adversity is among the most moving features of the book. Farida pretended not to speak Arabic and Evin posed as her sister who had to stick with her to translate into Kurdish.

With nauseating regularity, the girls' 'owners' would explain that their conduct was sanctified by the Qur'an: infidel women taken in war are your property to do what you like with. Obviously, everyone knows that, just ask the nearest 'scholar'!* And of course, infidel men are to be killed unless they convert to Islam. It must be great to have such confidence in your religion that you find it necessary to kill people who don't follow it. Their captors made repeated attempts to convert the girls to Islam, and made them learn the Qur'an by heart, on penalty of being caned if they failed.

After several brave but unsuccessful escape attempts, Farida and Evin eventually led a group of six girls in a hazardous breakout. Evin had an uncle living in Germany, whom she had managed to telephone with a mobile phone that they stole from their guards. He made contact with a Scarlet Pimpernel-style underground which, for a price, would smuggle them to safety if only they could escape. In a terrifying feat of daring, they managed it. The escaping party included twelve-year-old Besma (twelve is not too young to be raped; holy scripture sanctions it). Their epic trudge through hostile ISIS-held territory has the reader's heart in the mouth; the danger of recapture was ever-present, and the fear of what would happen to them if they were recaptured was inerasable. Their escape was worthy of a Colditz thriller, and Farida and Evin covered themselves with glory, shepherding the younger girls safely through, including little Besma who was so ill with starvation they feared for her life. Their odyssey ended with a boat crossing of the Euphrates, where Farida's uncle, and other relatives of the girls, had been briefed to meet them. Farida was later tearfully

---

* The quotation marks are deliberate. As I first remarked in a public speech in London, real scholars have read more than one book.

reunited with her mother, who had also managed to escape slavery but who was almost unrecognizable, so vicious had been her treatment. Farida's younger brother was the only survivor when the males of the village were shot because they weren't Muslims. He was wounded but feigned death and got away with it.

Even after her escape, a shadow hung over Farida. By the lights of the culture in which she was reared, the fact that she had been raped was seen as dishonouring her family, almost as though it was her fault. It was only once spoken out loud, but she and Evin and their comrades could sense it. She finally realized her ambition of going to Germany, where she now is. She is making good progress in learning German, and has revived her hopes of becoming a mathematics teacher. All credit to Germany. Would Britain have accepted her? Brexit Britain? Farage's shameful Britain? I hate to say it but I think I know the answer.

What a wonderfully gallant young woman, what a shining example to all of us spoiled brats fretting about our first world problems. Read the book, although I must warn you it's highly distressing. But also uplifting. Never to be forgotten.

VI

TENDING THE FLAME:
EVANGELIZING EVOLUTION

IN CONVERSATION
WITH MATT RIDLEY

FROM DARWIN TO DNA – AND BEYOND

This conversation, which took place in Oxford on 6 February 2020, was conducted and recorded specifically for inclusion in this volume. A fuller version appears in the audiobook. Matt Ridley is a dear friend, though we have our disagreements. He is among the best writers of science books that I know. And not just science books. Though originally educated in zoology, he is well read in economics and well deserves – though he might modestly disclaim it – the title of public intellectual.

Was Darwin a man of his time or a timeless genius? Why did no one think of evolution by natural selection before the nineteenth century? Can fashion drive evolution? Did humans create morality or did it evolve? What about technology? Will Darwin ever be superseded?

RD: Darwin was a Victorian, and it's often said that we have to read him in the context of his age. At least one of the biographies of Darwin has emphasized the economic things that were going on in Victorian times, and suggested that some of this was conducive to the idea of the survival of the fittest. On the other hand, I feel it rather lets Darwin down suggesting that all he cared about was politics.

MR: Well, I think many of us feel that he's somewhat timeless, more so than most people. He rose above his period rather better than most, and you see that in the way some of his ideas weren't appreciated in their time but are appreciated now. The classic example for me is sexual selection by female choice, which is a very specific idea that he comes up with, which is not fashionable, and rather gets buried – because in the nineteenth century you're not supposed to give females agency. And of course he comes out of the anti-slavery movement and yet what he's saying ends up being a little handy to racists. So there are paradoxes, and he's clearly uncomfortable with some of the ways people in his age are interpreting and using his stuff, but he's no saint; he gets a few things wrong; he's a creature of his age at the same time.

RD: Of course. I mean, you can't blame people for having opinions that come out of the climate of their time; obviously he was a racist by modern standards—

MR: —but much less so than most of his contemporaries. You can't help but admire him when you see him in the context of Victorian times, I think, and that's rather an interesting point.

RD: Exactly. Well, he was on non-speaking terms with Fitzroy because of a dispute about slavery. And I think Wallace was even more un-racist, so to speak, than Darwin.

MR: I think that's right. I think Wallace is, in a sense, a true democrat. I mean he's more radical than Darwin but he gets more wrong because he goes down the path of spiritualism – in a really weird way. And he refuses to follow the logic that if the body is produced by natural selection then the mind must be, too. Darwin never flinches at stuff like that, and that's where I think he scores above Wallace, who really did believe that natural selection couldn't explain the human

mind. It took another century for us to come round to that with evolutionary psychology, to start saying, 'Look, come on, we've got to
apply these principles to the human mind as well as the human body.'

RD: Yes, but funnily enough, their disagreement about sexual selection was almost the other way, because Wallace was the one who
was, as he himself said, 'more Darwinian than Darwin'; he hated the
idea of female choice being, somehow, almost mystical, whereas
Darwin was prepared to countenance the idea that there was such a
thing as female preference, female taste, female whim, the aesthetic
thing. And Wallace wanted that not to be the case – so that's almost
the opposite of Wallace's later flirtation with spiritualism.

MR: Yes. And recent scholarship on female choice theory and sexual selection theory seems to me to have unearthed Darwin from
under a lot of stuff, with the help of Fisher in this idea that it isn't
necessarily about good genes. So the peacock with the longest tail
may not necessarily have the best genes, or that may not be the
reason why it's being selected by the female. It's just a despotic fashion. If everyone else is choosing long-tailed males, then you've got
to choose long-tailed males too otherwise you won't have sexy sons.
And that's such a simple idea, and Darwin almost talks about
female birds being hypnotized by male displays. Which is an idea
that's come up recently – 'sensory drive' I think is the phrase used
now. And so he does feel ahead of his time on some of this stuff.

RD: I agree with that, and I love that idea of fashion as being something that has momentum in its own right . . . I've even gone so far
as to suggest that the bipedality of humans might have been a fashion. Many apes do rise up on their hind legs, and I could imagine
it being a kind of fashionable trend. The trendy thing to do is to
increase the amount of time you spend on your hind legs.

MR: I think you're dead right there, and something I've been fascinated by is the idea that culture precedes biology. You see this very
clearly in the case of lactose tolerance, which is something that some

human beings have because they also breed cattle and drink milk. And if you're drinking milk as an adult it helps to have your lactase gene switched on so you can digest the sugar as well as the protein in the milk. Now, isn't it lucky that the people whose lactase gene stayed switched on happen to then domesticate cattle? No, of course not, it's the other way round. It's because they domesticated cattle that they put selective pressure on themselves. And I feel the same is true of language, for example. I think probably, if we find the genes that are associated with linguistic ability – and we've found a few already – they aren't the cause of language, they're the consequence of language. In other words, it's because we were starting to use acoustic communication that we put those genes under selective pressure to get better at making our lips move or whatever it was.

RD: Why do you think – I've often wondered this – why do you think it took so long, up until the nineteenth century, before we got Darwin and Wallace? It seems, with hindsight, such an obvious thing to think of. In a way, I feel that what Isaac Newton did was so much cleverer, but it was two hundred years earlier.

MR: It's a very good point. I've had a similar thought, and I make the case in one of my books that Lucretius was jolly close to getting there two thousand years ago. He was saying some remarkably evolutionary things at the time of Cicero, and he gets buried by Christian scholarship, which doesn't like him, and he has to be rediscovered – and he then infects the Enlightenment with this idea of emergent properties, of spontaneous order, and not needing to think in terms of skyhooks, of top-down organization. And that then led me to the rather glib suggestion that there can be a special theory of evolution, which is the one that Darwin came up with, and a general theory of evolution, which arguably – exactly a hundred years before *The Origin of Species* – Adam Smith touches on in his book *The Theory of Moral Sentiments*. Because he's essentially arguing there that morality might have evolved, might be something that emerged spontaneously in human beings and wasn't taught to us by priests.

And I think Smith is hinting at the same thing, these emergent orders in human society. And then we forget about that and go round for a hundred years saying 'Darwinism applies to biology but it doesn't apply to culture', and I think we have to rediscover that. And it is true, of course, that Darwin read Adam Smith; he almost certainly read *The Theory of Moral Sentiments*, so he was influenced by this. So, I like to think we needed the eighteenth-century Enlightenment to get to the Darwinian ideas – but I agree with you, it ought to have been obvious before that.

RD: It's interesting that both Darwin and Wallace were travelling naturalists. You wouldn't think you'd need that – you'd think the idea could occur to anybody in an armchair – but apparently not. I was interested in what you were saying about Adam Smith and the idea – which I think is absolutely crucial for understanding evolution – that the whole of life is about small, local interactions between things rather than top-down control from above. And you make a big thing of this throughout your book *The Evolution of Everything*. I think one of the main problems with getting people to understand evolution is that it's very hard to resist the top-down idea which they've absorbed; everything about our life seems to be top-down.

MR: I completely agree, and I think there seems to be a human instinct here – what Daniel Dennett calls 'the intentional stance'. When a thunderstorm ruins your party plans, you cannot stop yourself – even you and me – saying, 'That was really unfair of someone to start that thunderstorm.' We walk around assuming that there is a sort of top-down cause for almost everything, and a top-down solution as well. And I suppose that made sense in terms of stone age psychology; if a rock hit you on the back of the head, you wouldn't say, 'Aren't I unlucky?', you'd turn around and say, 'Who threw that?'

RD: One of the things that you notice when reading Darwin is that he was constantly trying to cut down the idea of human exceptionalism. He's constantly trying – in *The Expression of the Emotions*, for example, but most of his books – to remove, or at least narrow, the barrier between humans and the rest of the animal kingdom. Is that still a problem, do you think?

MR: Yes, I do. I think we are dualists about the animal kingdom, still. It's a long battle, it's not won. Even today, when we've discovered that all living things on the planet use the same genetic material, DNA – that might not have been the case – we've discovered that the cleverest animal on the planet, or the one that thinks it's the cleverest, doesn't have the biggest number of genes, doesn't have the biggest quantity of DNA, doesn't even have any special genes at all: none of that was expected. Even twenty years ago people were saying, 'Yes, but there's going to be a whole bunch of genes that are human-specific' – and it turns out that no, we've got all the same genes as a mouse, we just use them in a different pattern and a different order, in the same way that a book is written with the same words but in a different order. So those molecular discoveries, cracking the genetic code in the 1960s, ought to have had an enormous impact in dethroning us from being special. In fact, the whole of the twentieth century is a sort of Copernican dethroning of human beings from being the centre of the universe, and yet I don't feel it's quite sunk in.

RD: Matt, occasionally there's the dark night of the soul . . . Can it really be the case that all this complexity and all this wonderful, beautiful, elegant life, and the universe and everything, can it really be that it all came about by bottom-up, non-top-down selection . . . ? I believe it could, but nevertheless it's a hard thing to swallow. Do you ever get that feeling of doubt?

MR: Well, I don't think I do, and I'm slightly alarmed to find that you do.*

---

* No need to be alarmed, Matt. The weakness soon passes. It's just that if I'm ever pressed hard to nominate the nearest approach to a

RD: I don't succumb to it, but I sympathize with others who have it.

MR: All sorts of highly intelligent people get the point, but then forget it again as soon as they look at some other aspect of life. For example: we are very creationist about technology. We assume that brilliant people sit in ivory towers and come up with bright ideas and force them on the world. Well, actually, the more closely you look at technology, the more evolutionary it appears. My favourite example is the fact that twenty-one people came up with the idea of the light bulb, independently, in the same decade – it was inevitable that light bulbs would be invented in the 1870s, just as it was inevitable that search engines would be invented in the 1990s. There's an evolutionary momentum there – Kevin Kelly writes about this very beautifully in his book *What Technology Wants*. And, of course, I can't go too far down that route because then you have to puzzle with the question, 'Well, in that case, why didn't it start two thousand years ago or five hundred years ago? Why did it start two hundred years ago?' Do you see what I mean?

RD: Well, that is a problem, but then you answer that by saying, 'Look, there had to be the previous cultural evolution in order to get up to that stage, and obviously you can't have search engines until you've got computers' . . . There's certainly a kind of software–hardware coevolution going on.

---

fundamental doubt of my world-view, I have to say that it lies somewhere hereabouts. I hasten to add that, even in the darkest dark night of the soul I soon reflect that, however hard it may occasionally be to credit bottom-up explanations for the full panoply of complexity in the universe, it is immediately clear to me that postulating even more complex top-down explanations is a non-starter. Even if we live in a top-down simulation programmed by a higher civilization, as in Daniel Galouye's science-fiction novel *Counterfeit World* (see the note on p. 78 above), that higher civilization must itself ultimately have a bottom-up explanation for its own existence.

MR: I think the key phrase here – and I think it's one of Stuart Kauffman's phrases – is 'the adjacent possible'. In other words, evolution can't make jumps. It has to get to A in order to get to B to get to C; it can't just go A to C. And that tends to be true of technology; as you say, in order to invent the search engine, you had to first invent the internet and so on. So I think there are similarities there; but that brings up – and I hope you don't mind me introducing a slightly new wrinkle here – a new idea in evolution that is floating around that just might be revolutionary enough to qualify – not as a superseding of Darwin, but as a big idea on its own. Andreas Wagner has written a book called *The Arrival of the Fittest*, which is a nice pun on 'survival of the fittest', and which makes this argument about genes in teams through hybridization events; getting over valleys in the adaptive landscape. It's the only one that I've read about in recent years that has made me think, 'Yes, maybe Darwinism needs reforming to understand this.' It's not group selection – which I think on the whole is a blind end – it's not epigenetics – which on the whole I think is a blind alley. But how much do you feel it is possible for Darwin, like Newton, to turn out to need to be overturned by an Einstein?

RD: I wouldn't call that an overturning. It seems to me that the key to that is a kind of game theory answer, really. You certainly need genes to form kind of clubs which work together. The way that happens is they're not selected as a group; each one is selected against the background of the others. It's like a kind of ecological context; there's an ecology of genes, where the gene pool of a species represents a population of genes that have to form teams with each other. They have to get on well with each other. They have to work collaboratively with each other. I think Ford showed this with moths. On his view, dominance was formed by cartels of genes being selected to modify each other's effects, so they were cooperating with each other. If you hybridize a pair of species of moths, which you can sometimes do, then dominance breaks down. But that's because, on his view, genes which had got used to working with each other in this climate of other genes or that climate of other genes could no longer do so. It's a bit like – one extreme example – if you could place herbivore guts together with carnivore teeth.

MR: I think what Andreas Wagner is saying is that you get hybridization events that give you whole new combinations, that can be useful, which you couldn't have got step-wise. You need to get them in chunks.

RD: That's right, but you mustn't fall into the trap of thinking this is a kind of group effect. It's an ecology of genes where each gene gets selected for its capacity to get on well with the other genes it's ever likely to meet in a body, which means the other genes in a gene pool of the species.

MR: Well, back to what we were saying right at the start. When Mendel gets rediscovered, it's thought to be a bit of a death knell to Darwin.

RD: It is very odd that when Mendel was rediscovered, the leading geneticists of the time – people like William Bateson – were anti-Darwinian. Or they thought they were. How on earth they thought the magnificent, beautiful sculpting of nature for function could come about without natural selection . . . but some of them thought mutation on its own would do the trick.

MR: But isn't that one of the interesting things about science: how obvious things look in retrospect, and not at the time? You can say similar things, I think, about the discovery of the double helix and the genetic code in the 1950s. It wasn't interpreted as an anti-Darwinian discovery in the same way that Mendel was, but in the long run it became an enormous vindication of it. Partly because it showed the unity of all life, and partly because it got rid of this final pocket of mysticism that there might be something very, very odd about life that was different from non-life in terms of physics or chemistry. The book that Schrödinger wrote in 1944 based on lectures he gave in Dublin in 1944 called 'What Is Life?' infected a lot of physicists with the idea that this problem was soluble, that life was going to turn out to be visible at a molecular level. But it also introduced the idea that there was some weird piece of quantum physics that we hadn't yet thought of that might be the answer here. And instead, what emerges with blinding clarity on 28 February 1953, thanks to Watson and Crick but also thanks to Rosalind Franklin and other predecessors, is the idea that life is just a four-letter word. It's a digital, linear code, and there's nothing else to it.

RD: As time goes by, ideas, theories, which have what it takes are robust, and the more the evidence comes in, the more confident we can be. I feel this is true of Darwinism; that in Darwin's own time it needed very passionate advocacy and was not accepted by many people, and yet the more we go on, the more the evidence in its favour comes in from all sort of different places – from molecular evidence and so on – until it becomes so robust that there's not the slightest chance that it will ever be proved wrong. That's not to say it won't ever be modified, of course – what do you feel about that?

MR: Well, I think the modification is in the detail. And I think if you look at the detail of evolutionary theory now, there's quite a lot that's different from Darwin; but if you look at the top of the tree it's absolutely spot-on. But of course you don't expect him to fill in the details of how evolution happened or why it happened; the simple, beautiful idea of natural selection leading to greater and greater complexity between form and function is so robust and so instantiated, as you say, in every other discovery we've made about life since, that there is not a chance in a million of it being overturned. I completely agree with that. And I think if we were sitting here a hundred years from now and arguing about evolution, we might still have political battles to fight. We might still have people saying, 'Oh, it's a heartless creed, it leads to survival of the fittest and

devil take the hindmost,' and we might have people saying, 'It con-
flicts with our religion, so we don't want to believe it,' but I don't
think we would have to be starting all over again and saying, 'Can
we please go back three hundred years to Charles Darwin?' That
idea is not going away. It's in the culture, it's here to stay, it's a beautiful idea, it's a much more general idea. I think we're only going to
deepen our understanding of it.

THE 'LITTLE PENGUIN' RELAUNCHED

Foreword to the new edition of John Maynard Smith's *The Theory of Evolution*, published by Cambridge University Press in 1993.

This book, in its original shorter edition, was my first introduction to John Maynard Smith and one of my first introductions to evolution. I bought it as a schoolboy, instantly captivated by the jacket blurb and author's photograph. The wild, nutty-professor hair, aslant like the pipe in the cheerfully smiling mouth; even the obviously intelligent eyes seemed somehow askew as they laughed their way through thick, round glasses (this was before John Lennon made them fashionable) badly in need of a clean. The picture perfectly complemented the quirky biographical note: 'Deciding that aeroplanes were noisy and old-fashioned, he entered University College, London, to study zoology.' I kept peeping at the back cover as I read, then returned to the text with a smile and renewed confidence that this was a man whose views I wanted to hear. I have known him personally now for twenty-six years and my initial impression has only deepened. This is a man whose views I want to hear, and so says everyone who knows him or reads his books, or even casually encounters him. At a conference for example.

Readers of 'campus novels' know that a conference is where you can catch academics at their worst.* The conference bar, in particular, is the academy in microcosm. Professors huddle together in exclusive, conspiratorial corners, talking not about science or scholarship but about 'tenure-track hiring' (their word for jobs) and 'funding' (their word for money). If they do talk shop, too often it will be to make an impression rather than to enlighten. John Maynard Smith is a splendid, triumphant, lovable exception. He values creative ideas above money, plain language above jargon. He is always the centre of a lively, laughing crowd of students and young research workers of both sexes. Never mind the lectures or the 'workshops'; be blowed to the motor-coach excursions to local beauty spots; forget your fancy visual aids and radio microphones; the only thing that really matters at a conference is that John Maynard Smith must be in residence and there must be a spacious, convivial bar. If he can't manage the dates you have in mind, you must just reschedule the conference. He doesn't have to give a formal talk (although he is a riveting speaker) and he doesn't have to chair a formal session (although he is a wise, sympathetic and witty chairman). He has only to turn up and your conference will succeed. He will charm and amuse the young research workers, listen

---

* I used this entire paragraph in the dedication to *The Ancestor's Tale*. John died before that book was published but after agreeing to accept the dedication. He was a major influence on me and I miss him terribly. One of the more worthwhile things I have done is interview him at the suggestion of Graham Massey, formerly Head of BBC Science, for a series now available on YouTube called *Web of Stories*. Massey's idea was to preserve for posterity a record of the spoken words of distinguished elder statesmen of science by getting younger colleagues to interview them. I was chosen to interview John Maynard Smith. I hardly needed to say anything. I simply prompted him to tell stories, or asked him questions, over a delightful two days at his Sussex home, and the result is a priceless archive of his wisdom, knowledge and humour.

to their stories, inspire them, rekindle enthusiasms that might be flagging, and send them back to their laboratories or their muddy fields, enlivened and invigorated, eager to try out the new ideas he has generously shared with them.

Not just ideas but knowledge, too. He sometimes quaintly poses as a workaday engineer who doesn't know anything about animals and plants. He *was* originally trained as an engineer, and the mathematical outlook and skills of his old vocation invigorate his present one. But he has been a professional biologist for a good forty years and a naturalist since childhood. He is leagues away from that familiar menace: the brash physical scientist who thinks he can wade in and clean up biology because, no matter how poorly he shows up against his fellow physicists, he at least knows more mathematics than the average biologist. John does know more mathematics, more physics and more engineering than the average biologist. But he also knows more biology than the average biologist. And he is incomparably more gifted in the arts of clear thinking and communicating than most physicists or biologists or anybody else. More, like a finely tuned antenna, he has the rare gift of biological intuition. Walk through wild country with him as I am privileged to have done, and you learn not just facts about natural history but the right way to ask questions about those facts. Better still, unlike some theorists, he has deep respect for good naturalists and experimentalists, even if they lack his own theoretical clout. He and I were once being shown around the Panama jungle by a young man, one of the staff of the Smithsonian tropical research station, and John whispered to me: 'What a privilege to listen to a man who really loves his animals.' I agreed, though the young man in this case was a forester and his 'animals' were various species of palm tree.

He is generous and tolerant of the young and aspiring, but a merciless adversary when he detects a dominating, powerful academic figure in pomposity or imposture. I have seen him turn red with anger when confronted with a piece of rhetorical duplicity from a senior scientist before a young audience.* If you ask him to name his own greatest virtue I suspect that, though he would be modest about nearly all his many skills and accomplishments, he would make one claim for himself: that he cares passionately about the truth. He is one of the few opponents who is seriously feared by creationist debaters. The slickest of these, like glib lawyers paid to advocate a poor case, are accustomed to bamboozling innocent audiences. They are eager to take on respectable scientists in debate, partly because they gain kudos and credibility from sharing a platform, on apparently equal terms, with a legitimate scholar. But they fear John Maynard Smith because, though he doesn't enjoy it, he always trounces them. Only a few weeks ago an anti-evolutionist author, basking in the short-term publicity that grows out of publishers' buying journalists lunch, was booked to have a debate in Oxford. Press and television interest had been easily whipped up, and the author's publishers must have been rubbing their hands with glee. Then the unfortunate fellow discovered who his opponent was to be: John Maynard Smith! He instantly backed out, and his supporters could do nothing to change his mind. If the debate had taken place John would indeed have routed him. But he'd have done it without rancour, and afterwards he'd have bought the wretched man a drink and even got him laughing.

I suppose some successful scientists make their careers by hammering away at one experimental technique that they are good at, and by gathering a gang of co-workers to do the donkey work. Their continued success rests primarily on their ability to coax a steady supply of money out of the government.† John Maynard Smith, by contrast, makes his way almost entirely by original thought, needing to spend very little money, and there is scarcely a branch of evolutionary or population genetic theory that has not been illuminated by his vivid and versatile inventiveness. He is one of that rare

---

* A professor of physics, if you please, trying to preach creationism at the Oxford Union.

† Or other granting agency.

company of scientists that changes the way people think. Together with only a handful of others, including W. D. Hamilton and G. C. Williams,* Maynard Smith is one of today's leading Darwinians. Perennially versatile, he has also made important contributions to the theory of biomechanics, of ecology, and of animal behaviour, in which he was largely responsible for promoting the persistently fashionable methods of game theory. He is in the forefront of the study of sex, probably the most baffling topic in modern evolutionary theory. Indeed, he was largely responsible for recognizing that sex constituted a problem in the first place, the problem now universally known by his phrase, 'the twofold cost of sex'.

He is an infectiously felicitous phrasemaker. His coinings have become a prevailing shorthand among the cognoscenti – 'Genetic Hitch-hiking', 'the Sir Philip Sidney Game', 'Partridge's Fallacy', the 'Haystack Model', 'chaps' as an abbreviation for *Homo sapiens* – you could fill a small dictionary with words and phrases that he introduced and which are now understood and daily used by evolutionary biologists the world over. He is also responsible for reviving and promulgating the earlier coinings of his mentor, the formidable J. B. S. Haldane: 'Pangloss's Theorem', 'The Bellman's Theorem' (What I tell you three times is true) and 'Aunt Jobiska's Theorem' (It's a fact the whole world knows). In turn, new generations of biologists are inspired to create their own Maynard Smithian phrases – 'the Beau Geste Effect', 'the Vicar of Bray Theory' – to lighten and refresh the pages of normally staid and rather dull academic journals. The pompous high priests of 'political correctness' don't like this kind of verbal informality. Maynard Smith, like Haldane before him, is too big a man to go along with their puritanical emasculation of language (and if my use of 'emasculation' gives offence to somebody, what a pity).

The qualities that make John Maynard Smith the life and soul of a

---

* And, I should have added, Robert Trivers.

good conference, the nemesis of creationists and charlatans, and the inspiration of so much youthful research, are also the qualities that make him the ideal author of a book for intelligent, critical lay people. This book – which, thanks to Cambridge University Press, he will now have to call something other than 'my little Penguin' – never had the flavour of ephemerality. Publishers never needed to buy lunches in order to get this book noticed. Through three editions and numerous reprintings, it has simply won its own place on the shelves of students and the generally literate; a staple that has seen silly fads and frothy fancies come and go. Few people in the world are better qualified than John Maynard Smith to explain evolution to us, and no subject more than evolution deserves such a talented teacher. You can hear his clear, logical, patient tones on every page. Not least, there is a total absence of pretentious languaging-up. Like Darwin himself, Maynard Smith knows that his story is intrinsically interesting enough and important enough to need no more than clear, patient, honest exposition.

It is a measure of both the brilliance of the book and the endurance of the neo-Darwinian synthesis itself that the 1975 text can stand its ground without revision today. There have, of course, been exciting new developments in the field. It would be worrying if there had not, and they are discussed in his new introduction. But the fundamental ideas and the great bulk of the detailed assertions of the original book remain as important and as true as ever. The new introduction itself is an elegant essay which can be recommended in its own right as a summary of important recent developments in evolutionary theory.

Darwin's theory of evolution by natural selection is the only workable explanation that has ever been proposed for the remarkable fact of our own existence, indeed the existence of all life wherever it may turn up in the universe. It is the only known explanation for the rich diversity of animals, plants, fungi and bacteria; not just the leopards, kangaroos, Komodo dragons, dragonflies, corncrakes, Coast redwood trees, whales, bats, albatrosses, mushrooms and bacilli that share our time, but the countless others – tyrannosaurs, ichthyosaurs, pterodactyls, armour-plated fishes, trilobites and giant sea scorpions – that we know only from fossils but which, in their own eons, filled every cranny of the land and sea. Natural selection is the only workable explanation for the beautiful and compelling illusion of 'design' that pervades every living body and every organ. Knowledge of evolution may not be strictly useful in everyday commerce. You can live some sort of life and die without ever hearing the name of Darwin. But if, before you die, you want to understand why you lived in the first place, Darwinism is the one subject that you must study. This book is the best general introduction to the subject now available.

FOXES IN THE SNOW

Foreword to a new edition of George Williams' *Adaptation and Natural Selection* (2018).

The neo-Darwinian synthesis of the 1930s and 1940s was a collective Anglo-American achievement, defined by a recognizable 'canon' of seminal books, those of Fisher, Haldane, Mayr, Dobzhansky, Simpson and others. Julian Huxley's *Evolution: the modern synthesis* bequeathed its title to the whole movement although for its theoretical content it doesn't stand out. If I were asked to nominate one book from the second half of the twentieth century that deserves to take an honoured place alongside the 'canon' of the 1930s and 1940s, I would choose *Adaptation and Natural Selection* by George C. Williams. On opening it I have the feeling of being ushered into the presence of a penetrating and outstanding mind, the same feeling I get, indeed, from reading *The Genetical Theory of Natural Selection*, although Williams, unlike Fisher, was no mathematician. In George Williams we have an author of immense learning and incisive critical intelligence, who thought deeply about every aspect of evolution and ecology. Williams not only enlarged the synthesis, he exposed with great clarity where many of its followers had gone astray, even in some cases the original authors themselves. This is a book that every serious student of biology must read, a book that irrevocably changes the way we look at life. Throughout my career as an Oxford tutor, I obviously recommended many books to my students. But I think this was the only one I insisted that all should read.

Here's a list of major mistakes a student is likely to make before reading this book, but will not make afterwards. 'Mutations are an adaptation to speed up evolution.' 'Dominance hierarchies are an adaptation to make sure the strongest individuals reproduce.' 'Territoriality is an adaptation to space the species out and beneficially limit the population.' 'Sex ratios are optimized to make the best use of species resources.' 'Death from old age is an adaptation to clear superannuated individuals out of the way and make room for the young.' 'Natural selection favours species that resist extinction.' 'Species parcel out niches for the benefit of a balanced ecosystem.' 'Predators hunt "prudently," taking care not to deplete prey that they are going to need in the future.' 'Individuals limit their reproduction to avoid overpopulation.'

'Adaptation' is the first word in the title and the book is largely a plea for a proper, scientific study of adaptation – a scientific *teleonomy*, to adopt Pittendrigh's term as advocated by Williams. But Williams is the last person who could justly be tarred as a naive 'adaptationist'. This pejorative was given wide currency by Gould and Lewontin in their overrated 'spandrels paper' of 1979. It denotes those who assume without evidence that everything an animal is or does must be an adaptation. Unfortunately, their critique of adaptationism has been misunderstood, not least by some philosophers such as the late Jerry Fodor,* as a critique of the very idea of adaptation itself.

A 'spandrel' is a non-adaptive by-product. The name comes from the gaps between gothic arches which are a necessary but non-functional by-product of the functionally important arches themselves. Long before the word was introduced into biology, Williams, a leading advocate of adaptation as a proper subject for

---

* Daniel Dennett, personal communication.

scientific study, gave an incisive critique of what would later be called spandrels. His vivid example, which regularly grabbed the attention of my Oxford students, was a fox repeatedly running along its own tracks in the snow. Its paws increasingly flattened the snow, which made each successive journey easier and faster. But it would be wrong to say the fox's paws were adapted to flatten snow. They can't help flattening snow. This particular beneficial effect is a by-product. Williams summed up the message pithily: adaptation is an 'onerous concept'.

If I might paraphrase the Anglican marriage service in a way that Williams might not, any attribution of adaptation should not be entered into unadvisedly or lightly; but reverently, discreetly, advisedly, soberly and in the fear of Occam's Razor. You must first assure yourself that you could, if called upon to do so, translate your adaptation theory back into the rigorous terms of neo-Darwinism. The 'adaptation' you postulate must not just be 'beneficial' in some vague, Panglossian sense. You must clearly set out, and be prepared to defend, a strictly Darwinian pathway to the evolution of the alleged adaptation. The 'benefit' must accrue at the proper level in the hierarchy of life, which is the unit of Darwinian natural selection. And the proper level, for Williams as for me, is that of the individual genes responsible for the putative adaptation.

The term 'Panglossian' was introduced into biology by J. B. S. Haldane, one of the architects of the synthesis. His star pupil John Maynard Smith reported that Haldane proposed three 'theorems' to satirize errors in scientific thinking.

Aunt Jobiska's Theorem (from Edward Lear): 'It's a fact the whole world knows.'

The Bellman's Theorem (from Lewis Carroll): 'What I tell you three times is true.'

Pangloss's Theorem (from Voltaire and applying especially to biology): 'All is for the best in the best of all possible worlds.'

My second paragraph above was a list of Panglossian errors frequently perpetrated by professors and students alike (including my own undergraduate self). Adaptations cannot be just 'good'. It is not enough that they convey 'benefit'. They have to be good for some entity that has been naturally selected precisely because of benefit to itself. And that entity, as Williams powerfully argues, will normally be the gene. I'm fond of a Williams *bon mot* from his later book, *Natural Selection*: 'A gene pool is an imperfect record of a running average of selection pressures over a long period of time in an area often much larger than individual dispersal distances.' But why the gene? And 'gene' in what sense? Williams' rationale was so clear and irrefutable, I'm inclined to quote it in full, but you only have to turn to page 23, the 'Socrates paragraph', which also grabbed my Oxford students by the collar when they read it. Here's the central point.

> With Socrates' death, not only did his phenotype disappear, but also his genotype ... Socrates' genes may be with us yet, but not his genotype because meiosis and recombination destroy genotypes as surely as death ... It is only the meiotically dissociated fragments of the genotype that are transmitted in sexual reproduction, and those fragments are further fragmented by meiosis in the next generation. If there is an ultimately indivisible fragment it is, by definition, 'the gene' that is treated in the abstract discussions of population geneticists.

That last sentence is the answer to my second question, 'Gene, in what sense?' I summed up the Williams answer a decade later when I jokingly wrote that *The Selfish Gene* might better have been called *The slightly selfish big bit of chromosome and the even more selfish little bit of chromosome*. It could also have been called *The Cooperative Gene*, and here lies the answer to perhaps the commonest criticism of the 'gene's-eye view' of natural selection. There is no simple, atomistic, one-to-one mapping between single genes and units of phenotype. Most genes have effects in many parts of the body, and most phenotypic features are influenced by many genes: how then, the critics bleat, can 'the gene' be the unit of natural selection? The objection is easily answered and Williams dispatches it with characteristic aplomb:

> No matter how functionally dependent a gene may be, and no matter how complicated its interactions with other genes and environmental factors, it must always be true that a given gene substitution will have an arithmetic mean effect on fitness in any population. (p. 57)

Williams is eloquent on the idea that the other genes in the genome (which in the long run means in the population gene pool) constitute the main environment in which a gene operates – the 'background' against which it is naturally selected. The fallacy (a sadly common one) is to assume that a coadapted gene complex is necessarily selected as a unit. Rather, each gene in the complex is selected individually for its compatibility with the other genes in the complex, which are in turn being selected for the very same compatibility.

Return for a moment to Williams' picturesque example of the fox in the snow. I think he'd have accepted the following reservation to his 'spandrel' or by-product lesson. Natural selection actually could favour an adaptive broadening of fox paws for the function of flattening snow. But *only* if the resulting path benefited the fox itself (and its family) alone, rather than foxes in general. It might, for example, be confined to the individual fox's own territory. This brings me to the central core of the book, which is Williams' critique of 'group selection'. This is as needed today as it was in 1966, for group selectionism won't lie down. With its magnetic allure, perhaps politically or even aesthetically motivated, group selectionism keeps coming back for more, in ways that, I can't resist confessing, remind me of Monty Python's Black Knight.*

Williams admits that natural selection could theoretically choose among groups. He just doesn't think it's important in practice. His meaning of group selection includes what he later called 'clade selection'. A hypothetical example might be a tendency for within-species natural selection (which Williams calls 'organic selection') to favour larger-sized individuals while at the same time whole *species* of smaller individuals are less likely to go extinct ('biotic selection'). Some authors espouse a different form of group selection in which altruistic or cooperative behaviour of individuals, or indeed a tendency to live in groups, is thought to be favoured because it benefits the group. Williams declines to invoke group selection where the phenomena are more parsimoniously explained by kin selection (Hamilton's seminal papers had just appeared) or reciprocation (Trivers' clever theorizing lay in the future, but Williams anticipates the basic idea). As for living in groups, there are, of course, numerous ways in which individuals benefit: huddling for warmth, safety in numbers when predators strike, the 'many eyes effect' when spotting opportunities, aerodynamic or hydrodynamic facilitation in flocking birds or schooling fish, 'non-zero-sum games' in bringing down large prey, etc. Indeed, all these examples are nowadays often handled by game theory models in which individuals maximize their own benefit in the context of other individuals maximizing *their* own benefit. Group benefit plays absolutely no part in such models. Incidentally, Williams has a prescient anticipation of evolutionary game theory, in a slightly different context. Williams revisited and updated his critique of group selection in his 1992 book *Natural Selection*, but I'll say no more about it here.

---

* Played by John Cleese, wearing full armour, in their film *Monty Python and the Holy Grail.*

In the final chapter of *Adaptation and Natural Selection*, where he lays out his programme for a scientific teleonomy, Williams quotes William Paley's *Natural Theology* on the vertebrate eye. His purpose is to illustrate the self-evident 'design' of living creatures, an immensely powerful illusion of design, which pervades some (though not all – that would be adaptationism) biological entities such as the eye. The complex, statistically improbable juxtaposition of mutually suited functionally cooperating parts – precision-focusing lens, precision-adjusting iris diaphragm, retina with millions of colour-coding photocells, optic nerve trunk cable to the brain: such phenomena (and they are legion in all parts of all animals and plants) can only be explained if the principles of chemistry and physics are supplemented by 'the one additional postulate of natural selection and its consequence, adaptation'. Philosophers and others who don't see the glaring need for natural selection (or divine creation as Paley would have it) must simply be ignorant of the relevant beautiful facts. Have they never seen a David Attenborough film? Or looked down a microscope at a cell? Or contemplated their own hand?

Williams urges us to take seriously the need for a special kind of explanation of adaptation, but to pay cautious attention to the precise mechanism of natural selection and the level in the hierarchy of life where it acts. It is his contention that genic selection occupies the appropriate level. Group selection is a theoretical possibility, but it lacks the power to build up Paleyesque complexity: Darwin's 'organs of extreme perfection and complexity', organs which, for Hume, 'ravish into admiration all men who have ever contemplated them'. We marvel at complex organs that give individuals the power to see, birds to fly, bats to echolocate, dogs to smell, cheetahs to sprint. There are no complex organs that give species, or groups, or ecosystems, the power to do anything. Those larger groupings of individuals are just not the kind of entity that has complex 'organs' or, indeed, adaptations of any kind. What groups do is a consequence, a by-product indeed, of what their component individuals do.

George Williams, a tall, imposing, Abe Lincolnesque figure, quiet, kind, thoughtful, modest, made major research contributions to solving outstanding problems in evolutionary biology – really big problems like the evolution of sex, of senescence, of life-history strategies. He was a pioneer of the up-and-coming but still under-valued subject of Darwinian medicine. His *Natural Selection: domains, levels and challenges* was an important successor to this book. But I think *Adaptation and Natural Selection* is his outstanding achievement. When I re-read it before writing this foreword I expected to find passages that needed critical updating or even deleting. I failed. It can still be recommended to today's students without reserve. Not for its historical interest like some books of the synthesis, but because this fifty-year-old book is still biologically illuminating, wise and – as far as I can judge – correct.

TELLING TRUTH
IN A DARK TIME

This review of the paperback edition of *Blueprints: solving the mystery of evolution* by Maitland Edey and Donald Johanson appeared in the *New York Times* on 9 April 1989. Weirdly, my review provoked a lawsuit from a man in Texas who spotted it 26 years later and thought he recognized himself in one of its phrases. He sued me – unsuccessfully – for $58 million.

'Do you realize,' said Don, 'that nearly half the people in the United States don't believe in evolution?'

This sentence epitomizes both the provocation for and the odd provenance of the book under review. To take the latter first, *Blueprints* purports to be the joint work of a distinguished scientist and a journalist, Donald C. Johanson and Maitland A. Edey. It is their second collaboration; the first was *Lucy: the beginnings of humankind*. Such a combination is bound to arouse suspicions of ghostwriting by the journalist, cashing in on the name of the scientist. The difference here is that the ghost manifests himself with unusual frankness. Mr Johanson* enters the book only as Don, a third-person character

---

* I would normally have said 'Professor Johanson', but it is the *New York Times'* (actually rather endearing) house style to call all (male) authors 'Mr'.

who occasionally drops in, looks over the author's shoulder and comments on whatever he happens to be working on at the moment. '"Those things are called Punnett squares," said Don, watching as I laboriously completed the large square on the preceding page. "Boy, are they dull."'

In other places, especially in the sections on molecular genetics and bacterial evolution, there is an odd role reversal: 'Don' comes off as pupil, his colleague as master. 'Mait' indulges in pedagogical questions like 'Does that suggest anything to you?' and Don's answer is rewarded with a magisterial 'Right'. Mr Johanson, the director of the Institute of Human Origins in Berkeley, California, is a fine paleontologist and anthropologist. He has many achievements to his name, but writing this book is not one of them, and I shall henceforth refer to the author in the singular. But it is a shame to carp, for this book should be welcomed by anyone with a love of truth in a dark time. It has an important and true story to tell – the story of evolution. As far as I am able to judge, the science in the book is accurate and up-to-date. On the whole it is pleasantly written, in spite of the reservations entered above (and a few others: I had earlier promised myself that if I had to endure the silly story about Thomas Henry Huxley's schoolboy triumph over Bishop Wilberforce one more time, I'd scream; and I duly did so).*

Following a history of Darwin and his predecessors, the large middle section of the book covers the important science of genetics, from Gregor Mendel through the American geneticist

---

* Professor Richard Wrangham, the eminent Harvard authority on apes, has uncovered an unpublished poem written by Bishop Wilberforce entitled 'Lines written on hearing that Professor Huxley said that "he did not care whether his grand-father was an Ape"'. The poem – published in *Nature*, vol. 287, 18 Sept. 1980 – demonstrates that the bishop had a delightful sense of humour and took the Huxley encounter in good part. The poem itself is reproduced at the end of this piece.

T. H. Morgan to Francis Crick – giving too little credit, for my money, to the English geneticist R. A. Fisher and his colleagues in the 1930s. The section called 'The origin of life' is notable for its courageous attempt (which I have shirked in my own writings) to explain the difficult ideas of the German chemist Manfred Eigen. For me, the most interesting chapter is the one devoted to the work of the American bacteriologist Carl R. Woese because it deals with the earliest phases of evolution, the split between our remotest cousins, the archaebacteria, and all the rest of us.

The chapters on human evolution display predictable expertise on fossils, but it is also good to see Mr Johanson's arid home ground irrigated by a refreshing trickle of molecular evidence, and par- ticularly gratifying to find at last proper recognition of the enormously important work of the American biochemist Vincent Sarich. Contrary to the erstwhile conclusions of all paleontologists, we now know from the work of Mr Sarich and his colleague, the molecular biologist Allan Wilson, that our common ancestor with chimpanzees lived astonishingly recently. Moreover, we are closer cousins to African apes (chimpanzees and gorillas) than those apes are to other apes (orang-utans and gibbons). We are not, then, merely like apes or descended from apes; we are apes, and African apes at that. The final chapter, a reflection on extinction and the dangers of being too smart, moves towards being noticeably well written. Mr Edey may call himself a journalist, but he evidently is a pretty high-class journalist.

So to the book's provocation, the statement that nearly half the people in the United States don't believe in evolution. Not just any people but powerful people, people who should know better, people with too much influence over educational policy. We are not talk- ing about Darwin's particular theory of natural selection. It is still (just) possible for a biologist to doubt its importance, and a few claim to. No, we are here talking about the fact of evolution itself, a fact that is proved utterly beyond reasonable doubt. To claim equal time for creation science in biology classes is about as sensible as to claim equal time for the flat-earth theory in astronomy classes. Or, as someone has pointed out, you might as well claim equal time in sex education classes for the stork theory. It is absolutely safe* to say that if you meet somebody who claims not to believe in evolution, that person is ignorant, stupid or insane (or wicked, but I'd rather not consider that).

If that gives you offence, I'm sorry. You are probably not stupid, insane or wicked; and ignorance is no crime in a country with strong local traditions of interference in the freedom of biology educators to teach the central theorem of their subject. I recently toured East Coast radio stations, doing phone-ins. I came away optimistic. I had expected hostile barracking from creationists with closed minds. Instead, what I found was genuine curiosity and honest interest. I got sincere questions from intelligent people who really wanted to know because they had had literally no education in evolution.

I don't think it is too melodramatic to say that civilization is at war. It is a war against religious bigotry. In Britain recently our newspapers have shown crowds of fundamentalists (they happen to be Muslim rather than Christian, but in this context the distinction is of no importance) baying for the death of the distinguished novelist Salman Rushdie, displaying his effigy with its eyes put out and publicly burning his books. The truly appalling thing all such people have in common, whether they

---

* So it is, but unsafe in another sense. This was the phrase that provoked the lawsuit mentioned in the opening note to this piece. The litigious Texan claimed that he was a creationist, therefore I was saying that he personally was ignorant, stupid or insane, and that this lowered him in the eyes of the community and deterred third persons from associating with him. He evidently rated his prestige in the community so highly that it was worth $58 million to him. Perhaps the most amazing aspect of his case is that he managed to find a law firm willing to take it. They must have known full well that it would lose as, not surprisingly, it did – but not before I was obliged to engage a Texas lawyer to defend me.

are incited to murder by ayatollahs or to less violent observances by television evangelists, is that they know, for certain,* that their particular brand of revealed truth is absolute and needs no reasoned defence. In Iran I don't suppose evolution is even an issue, but in the United States a case can be made that it is right there on the front line.

If you feel even vaguely in the mood to stand up and be counted, evolution is a pretty good issue on which to take your stand. It is an excellent standard-bearer for reason and the gentle virtues of civilization. This is because the more you read, quietly and soberly, the evidence for evolution, the more powerful will you discover that evidence to be. You are as safe taking your stand on the fact of evolution as you would be on the fact that the Earth goes round the sun. But the latter is not – any longer – at stake in the war against fundamentalism. Evolution is on the front line because it is an important issue disputed by fundamentalists, and you can be completely confident that you can easily prove them wrong.

Blueprints is not the only book, and probably not the best book, in which you may locate the ammunition. Even in time of war one should not suppress criticism of one's own side, and I haven't done so. But this is an honest book, telling the truth in an area where half the authors' country claims to believe an absurd and palpable falsehood. I say 'claims' because a belief that is held in carefully nurtured ignorance of the alternative is hardly a belief to be taken seriously. For all its faults, *Blueprints* is about more important matters than many a book you will find displayed in your bookshop or, I dare say, reviewed in these pages.

*

---

* Demonstrating the chilling power of childhood indoctrination.

'Lines written on hearing that Professor Huxley had said that "he did not care whether his grand-father was an Ape"'

Oft had I heard, but deemed the tale untrue,
That man was cousin to the Kangaroo;
That he before whose face all nature quailed,
Was but the monkey's heir, though unentailed;
And that the limber Ape, whose knavish ways
And tricks fantastic oft our laughter raise,
Was just what *we* were in some previous state,
Ages ere Noah shipped his living freight.
But now a learn'd Professor, grave and wise,
Stoutly maintains what I supposed were lies;
And, while each listening sage in wonder gapes,
Claims a proud lineage of ancestral Apes.
Alas! cried I, if such the sage's dreams,
Save me, ye powers, from these unhallowed themes;
From self-degrading science keep me free,
And from the pride that apes humility!
But O should fate bring back these dreams accursed,
And shuddering Nature find her laws reversed;
Should this, the age of wonders, see again
Men sunk to monkeys, monkeys raised to men:
Be mine the lot, on some far-distant shore,
Where Science wearies not nor savants bore—
Where no learn'd Apes our fallen race may scorn,
Nor point the moral which our tails adorn—
To shun the sight of metamorphosed friends,
Till time again shall shape their altered ends,
To soothe each fond regret, howe'er I can;
And, at the least, to dream myself a Man!

S. Wilberforce

IRRESPONSIBLE PUBLISHING?

This review of Richard Milton's *The Facts of Life: shattering the myth of Darwinism*, appeared in *New Statesman*, 28 Aug. 1992.

Every day I get letters, in capitals and obsessively underlined if not actually in green ink, from flat-earthers, young-earthers, perpetual-motion merchants, astrologers and other harmless fruitcakes. The only difference here is that Richard Milton managed to get his stuff published. The publisher – we don't know how many decent publishers turned it down first – is called 'Fourth Estate'. Not a house that I had heard of,* but apparently neither a vanity press nor a fundamentalist front. So, what are 'Fourth Estate' playing at? Would they publish – for this book is approximately as silly – a claim that the Romans never existed and the Latin language is a cunning Victorian fabrication to keep schoolmasters employed?

A cynic might note that there is a paying public out there, hungry for simple religious certitude, who will lap up anything with a subtitle like 'shattering the myth of Darwinism'. If the author pretends not to be religious himself, so much the better, for he can then be exhibited as an unbiased witness. There is – no doubt about it – a fast buck to be made by any publishers unscrupulous enough

---

* I've heard of them now, but in 1992 they were comparatively unknown. They have now become quite a respectable publisher, having learned not to publish books like Richard Milton's.

to print pseudoscience that they know is rubbish but for which there is a market.

But let's not be so cynical. Mightn't the publishers have an honourable defence? Perhaps this unqualified hack is a solitary genius, the only soldier in the entire platoon – nay, regiment – who is in step. Perhaps the world really did bounce into existence in 8000 BC. Perhaps the whole vast edifice of orthodox science really is totally and utterly off its trolley. (In the present case, it would have to be not just orthodox biology but physics, geology and cosmology too.) How do we poor publishers know until we have printed the book and seen it panned?

If you find that plea persuasive, think again. It could be used to justify publishing literally anything; flat-earthism, fairies, astrology, werewolves and all. It is true that an occasional lonely figure, originally written off as loony or at least wrong, has eventually been triumphantly vindicated (though not often a journalist like Richard Milton, it has to be said). But it is also true that a much larger number of people originally regarded as wrong really were wrong. To be worth publishing, a book must do a little more than *just* be out of step with the rest of the world.

But, the wretched publisher might plead, how are we, in our ignorance, to decide? Well, the first thing you might do – it might even pay you, given the current runaway success of some science books – is employ an editor with a smattering of scientific education. It needn't be much: A-level biology would have been ample to see off Richard Milton. At a more serious level, there are lots of smart young science graduates who would love a career in publishing (and their jacket blurbs would avoid egregious howlers like calling Darwinism the 'idea that chance is the mechanism of evolution'). As a last resort you could even do what proper publishers do and send the stuff out to referees. After all, if you were offered a manuscript claiming that Tennyson wrote *The Iliad*, wouldn't you consult somebody, say with an O-level in history, before rushing into print?

You might also glance for a second at the credentials of the author.

If he is an unknown journalist, innocent of qualifications to write his book, you don't have to reject it out of hand but you might be more than usually anxious to show it to referees who do have some credentials. Acceptance need not, of course, depend on the referees' endorsing the author's thesis: a serious dissenting opinion can deserve to be heard. But referees will save you the embarrassment of putting your imprint on twaddle that betrays, on almost every page, complete and total pig-ignorance of the subject at hand.

All qualified physicists, biologists, cosmologists and geologists agree, on the basis of massive, mutually corroborating evidence, that the Earth's age is at least four billion years. Richard Milton thinks it is only a few thousand years old, on the authority of various creation 'science' sources including the notorious Henry Morris (Milton himself claims not to be religious, and he affects not to recognize the company he is keeping). The great Francis Crick (himself not averse to rocking boats) recently remarked that 'anyone who believes that the Earth is less than ten thousand years old needs psychiatric help'. Yes yes, maybe Crick and the rest of us are all wrong and Milton, an untrained amateur with a 'background' as an engineer, will one day have the last laugh. Want a bet?

Milton misunderstands the first thing about natural selection. He thinks the phrase refers to selection among species. In fact, modern Darwinians agree with Darwin himself that natural selection chooses among individuals *within* species. Such a fundamental misunderstanding would be bound to have far-reaching consequences; and they duly make nonsense of several sections of the book.

In genetics, the word 'recessive' has a precise meaning, known to every school biologist. It means a gene whose effect is masked by another (dominant) gene at the same locus. Now it also happens that large stretches of chromosomes are inert – untranslated. This kind of inertness has not the smallest connection with the 'recessive' kind. Yet Milton manages the feat of confusing the two. Any slightly qualified referee would have picked up this clanger.

There are other errors from which any reader capable of thought would have saved this book. Stating correctly that Immanuel Velikovsky was ridiculed in his own time, Milton goes on to say: 'Today, only forty years later, a concept closely similar to Velikovsky's is widely accepted by many geologists – that the major extinction at the end of the Cretaceous . . . was caused by collision with a giant meteor or even asteroid.' But the whole point of Velikovsky (indeed, the whole reason why Milton, with his eccentric views on the age of the Earth, champions him) is that his collision was supposed to have happened recently; recently enough to explain biblical catastrophes like Moses' parting of the Red Sea. The geologists' meteorite, on the other hand, is supposed to have impacted sixty-five million years ago! There is a difference – approximately sixty-five million years' difference. If Velikovsky had placed his collision tens of millions of years ago he would not have been ridiculed. To represent him as a misjudged wilderness figure who has finally come into his own is either disingenuous or – more charitably and plausibly – stupid.

In these post-Leakey, post-Johanson days, creationist preachers are having to learn that there is no mileage in 'missing links'. Far from being missing, the fossil links between modern humans and our ape ancestors now constitute an elegantly continuous series. Richard Milton, however, still hasn't got the message. For him, 'the only "missing link" so far discovered remains the bogus Piltdown Man'. *Australopithecus*, correctly described as a human body with an ape's head, doesn't qualify because it is 'really' an ape. And *Homo habilis* – 'handy man' – which has a brain 'perhaps only half the size of the average modern human's' is ruled out from the other side: 'The fact remains that handy man is a human – not a missing link.' One is left wondering what a fossil has to do – what more *could* a fossil do – to qualify as a 'missing link'?

No matter how continuous a fossil series may be, the conventions of zoological nomenclature will always impose discontinuous names. At present, there are only two generic names to spread over all the hominids. The more ape-like ones are shoved into the genus *Australopithecus*; the more human ones into the genus *Homo*. Intermediates are saddled with one name or the other. This would still be true if the series were as smoothly continuous as you can possibly imagine. So, when Milton says, of Johanson's 'Lucy' and associated fossils, 'the finds have been referred to either *Australopithecus* and hence are apes, or *Homo* and hence are human,' he is saying something (rather dull) about naming conventions, nothing at all about the real world.

But this is a more sophisticated criticism than Milton's book deserves. The only serious question raised by its publication is why. As for would-be purchasers, if you want this sort of silly-season drivel you'd be better off with a couple of Jehovah's Witness tracts. They are more amusing to read, they have rather sweet pictures, and they put their religious cards on the table.

*

The above review was followed by a pair of letters to *New Statesman* including one by Christopher Potter, Milton's publisher (which I'm afraid I don't have), to which I replied as follows:

September 1992

The Editor
New Statesman

Two of your correspondents spring gallantly to the defence of Richard Milton, following my review of his book. One turns out to be Milton himself. The other is his publisher. Why no letter from his mother?

'[Dawkins] implies that I am a Creationist, whereas I make it clear in my book that I have no religious views . . .' Exactly. That is why I wrote, 'Milton himself states that he is not religious . . .'

'Milton is not a crypto-Creationist as Dawkins implies.' Quite. That is why I wrote of Milton that he 'appears not to be religious himself...'

'Even the lightest of research would have revealed that Fourth Estate is not a fundamentalist front ...' Ah yes, that must have been why I wrote that Fourth Estate is 'neither a vanity press, nor a fundamentalist front'. My whole *point* was that Fourth Estate is not a fundamentalist front. It is precisely this that makes their gullibility and lack of discrimination so remarkable.

'Dawkins says that I think the earth is only a few thousand years old. I say no such thing.' Oh *of course*, how silly of me. I must have been misled by the publisher's words on the jacket: 'The rocks of the earth's crust have been formed in thousands rather than millions of years. *The Facts of Life* is the first book to present these recent findings as part of a coherent and devastating case against Darwinism.' Where did the publisher get this from? I can only think that he, like me, was foolish enough to read the book. We must hope that others will be wiser.

As for the 'recent findings', the 'professional scientists from many countries of the world' who are quoted as authority for them are mainly two: Melvin Cook and Henry Morris. Cook, whose three quoted works were published as recently as 1957, 1966 and 1968, hails from Salt Lake City and patronizes a journal called *Creation Research Science Quarterly*. Morris is the aged director of the notorious Institute for Creation Research of San Diego, California. These two constitute what Milton's publisher Christopher Potter, who tells us that he is a science graduate, calls 'the scientific community'.

Perhaps Mr Potter would have made a better publishing decision if he had been a history graduate instead. He might then have woken up to the ludicrous implications of Milton's thesis. Here we have an author capable of taking seriously the idea that the Earth sprang into existence in 8000 BC. Presumably the dinosaurs came and went just before the Bronze Age? Were iguanadons trained to haul rocks to Stonehenge?

Richard Milton may not be a creationist, but that doesn't stop his anti-evolution arguments being pure, bog-standard, creationist propaganda. I am a connoisseur of this kind of drivel and I can assure you that there is absolutely nothing new in his book. Richard Milton himself is no more worth arguing with than a Jehovah's Witness doorstepper. The only argument I would have is with his publishers. And the argument would not be about evolution (since they are not a fundamentalist front) but about the responsibilities of a decent publisher.

INFERIOR DESIGN

This review of Michael Behe's *The Edge of Evolution* was published in the *New York Times* on 1 July 2007.

I had expected to be as irritated by Michael Behe's second book as by his first. I had not expected to feel sorry for him. The first – *Darwin's Black Box* (1996), which purported to make the scientific case for 'intelligent design' – was enlivened by a spark of conviction, however misguided. The second is the book of a man who has given up. Trapped along a false path of his own rather unintelligent design, Behe has left himself no escape. Poster boy of creationists everywhere, he has cut himself adrift from the world of real science. And real science, in the shape of his own department of biological sciences at Lehigh University, has publicly disowned him, via a remarkable disclaimer on its website: 'While we respect Prof. Behe's right to express his views, they are his alone and are in no way endorsed by the department. It is our collective position that intelligent design has no basis in science, has not been tested experimentally and should not be regarded as scientific.' As the Chicago geneticist Jerry Coyne wrote recently, in a devastating review of Behe's work in *The New Republic*, it would be hard to find a precedent.

For a while, Behe built a nice little career on being a maverick. His colleagues might have disowned him, but they didn't receive flattering invitations to speak all over the country and to write for the *New York Times*. Behe's name, and not theirs, crackled triumphantly around the memosphere. But things went wrong, especially at the famous 2005

414

trial where Judge John E. Jones III immortally summed up as 'breath-taking inanity' the effort to introduce intelligent design into the school curriculum in Dover, Pa. After his humiliation in court, Behe – the star witness for the creationist side – might have wished to re-establish his scientific credentials and start over. Unfortunately, he had dug himself in too deep. He had to soldier on. *The Edge of Evolution* is the messy result, and it doesn't make for attractive reading.

We now hear less about 'irreducible complexity', with good reason. In *Darwin's Black Box*, Behe simply asserted without justification that particular biological structures (like the bacterial flagellum, the tiny propeller by which bacteria swim) needed all their parts to be in place before they would work, and therefore could not have evolved incrementally. This style of argument remains as unconvincing as when Darwin himself anticipated it. It commits the logical error of arguing by default. Two rival theories, A and B, are set up. Theory A explains loads of facts and is supported by mountains of evidence. Theory B has no supporting evidence, nor is any attempt made to find any. Now a single little fact is discovered, which A allegedly can't explain. Without even asking whether B can explain it, the default conclusion is fallaciously drawn: B must be correct. Incidentally, further research usually reveals that A can explain the phenomenon after all: thus the biologist Kenneth R. Miller (a believing Christian who testified for the other side in the Dover trial) beautifully showed how the bacterial flagellar motor could evolve via known functional intermediates.

Behe correctly dissects the Darwinian theory into three parts: descent with modification, natural selection and mutation. Descent with modification gives him no problems, nor does natural selection. They are 'trivial' and 'modest' notions, respectively. Do his creationist fans know that Behe accepts as 'trivial' the fact that we are African apes, cousins of monkeys, descended from fish?

The crucial passage in *The Edge of Evolution* is this: 'By far the most critical aspect of Darwin's multifaceted theory is the role of random mutation. Almost all of what is novel and important in Darwinian thought is concentrated in this third concept.'

What a bizarre thing to say! Leave aside the history: unacquainted with genetics, Darwin set no store by randomness. New variants might arise at random, or they might be acquired characteristics induced by food, for all Darwin knew. Far more important for Darwin was the non-random process whereby some survived but others perished. Natural selection is arguably the most momentous idea ever to occur to a human mind, because it – alone as far as we know – explains the elegant illusion of design that pervades the living kingdoms and explains, in passing, us. Whatever else it is, natural selection is not a 'modest' idea, nor is descent with modification.

But let's follow Behe down his solitary garden path and see where his overrating of random mutation leads him. He thinks there are not enough mutations to allow the full range of evolution we observe. There is an 'edge', beyond which God must step in to help. Selection of random mutation may explain the malarial parasite's resistance to chloroquine, but only because such micro-organisms have huge populations and short life-cycles. *A fortiori*, for Behe, evolution of large, complex creatures with smaller populations and longer generations will fail, starved of mutational raw materials.

If mutation, rather than selection, really limited evolutionary change, this should be true for artificial no less than natural selection. Domestic breeding relies upon exactly the same pool of mutational variation as natural selection. Now, if you sought an experimental test of Behe's theory, what would you do? You'd take a wild species, say a wolf that hunts caribou by long pursuit, and apply selection experimentally to see if you could breed, say, a dogged little wolf that chivvies rabbits underground: let's call it a Jack Russell terrier. Or how about an adorable, fluffy pet wolf called, for the sake of argument, a Pekinese? Or a heavyset, thick-coated wolf, strong enough to carry a cask of brandy, that thrives in Alpine passes and might be named after one of them, the St Bernard? Behe has to predict that you'd wait till hell freezes over, but the necessary mutations would not be forthcoming. Your wolves would stubbornly remain unchanged. Dogs are a mathematical impossibility.

Don't evade the point by protesting that dog breeding is a form of intelligent design. It is (kind of), but Behe, having lost the argument over irreducible complexity, is now in his desperation making a completely different claim: that mutations are too rare to permit significant evolutionary change anyway. From Newfies to Yorkies, from Weimaraners to water spaniels, from Dalmatians to dachshunds, as I incredulously close this book I seem to hear mocking barks and deep, baying howls of derision from five hundred breeds of dogs – every one descended from a timber wolf within a time frame so short as to seem, by geological standards, instantaneous.

If correct, Behe's calculations would at a stroke confound generations of mathematical geneticists, who have repeatedly shown that evolutionary rates are not limited by mutation. Single-handedly, Behe is taking on Ronald Fisher, Sewall Wright, J. B. S. Haldane, Theodosius Dobzhansky, Richard Lewontin, John Maynard Smith and hundreds of their talented co-workers and intellectual descendants. Notwithstanding the inconvenient existence of dogs, cabbages and pouter pigeons, the entire corpus of mathematical genetics, from 1930 to today, is flat wrong. Michael Behe, the disowned biochemist of Lehigh University, is the only one who has done his sums right. You think?

The best way to find out is for Behe to submit a mathematical paper to the *Journal of Theoretical Biology*, say, or the *American Naturalist*, whose editors would send it to qualified referees. They might liken Behe's error to the belief that you can't win a game of cards unless you have a perfect hand. But, not to second-guess the referees, my point is that Behe, as is normal at the grotesquely ill-named Discovery Institute (a tax-free charity, would you believe?), where he is a senior fellow, has bypassed the peer-review procedure altogether, gone over the heads of the scientists he once aspired to number among his peers, and appealed directly to a public that – as he and his publisher know – is not qualified to rumble him.

THE ONLY KIND OF
TRUTH THAT WORKS

This is an abbreviated version of my review of Jerry Coyne's *Why Evolution is True*, published in the *Times Literary Supplement* in 2009.

How can you say that evolution is 'true'? Isn't that just your opinion, of no more value than anybody else's? Isn't every view entitled to equal 'respect'? Maybe so where the issue is one of, say, musical taste or political judgement. But when it is a matter of scientific fact? Unfortunately, scientists do receive such relativistic protests when they dare to claim that something is factually true in the real world. Given the title of Jerry Coyne's book, this is a distraction that I must deal with.

A scientist arrogantly asserts that thunder is not Thor's hammer, nor is it the sound of God's balls triumphantly banging together as he impregnates the Earth Goddess. It is, instead, the reverberating echoes from the electrical discharges that we see as lightning. Poetic (or at least stirring) as those tribal myths may be, they are false.

But now a certain kind of anthropologist can be relied upon to jump up and say something like the following. Who are you to elevate scientific 'truth' so? The tribal beliefs are true in the sense that they hang together in a meshwork of consistency with the rest of the tribe's world-view. Scientific 'truth' is only one kind ('Western' truth, the anthropologist may add, or even 'patriarchal'). Like tribal truths, yours merely hang together with the world-view that you happen to hold, which you call scientific. An extreme version of this viewpoint (no joke – I have actually encountered it) goes so far as to say that logic and evidence themselves are nothing more than instruments of masculine oppression over the 'intuitive mind'.

Listen, anthropologist.* Just as you entrust your travel to a Boeing 747 rather than a magic carpet or a broomstick; just as you take your tumour to the best surgeon available, rather than a shaman or a mundu mugu, so you'll find that the scientific version of truth works. You can use it to navigate through the real world. Science predicts, with complete certainty unless the end of the world intervenes, that the city of Shanghai will experience a total eclipse of the sun on 22 July 2009. Theories about the moon god devouring the sun god may be poetic, and they may cohere with other aspects of a tribe's world-view, but they won't predict the date, time and place of an eclipse. Science will, and with an accuracy you could set your watch by. Science gets you to the moon and back. Even if we bend over backwards to concede that scientific truth is no more than that which enables you to pilot your way reliably, safely and predictably around the real universe, it is in exactly this sense that – at very least – evolution is true. Evolutionary theory pilots us around biology reliably and predictively, with a detailed and unblemished success that rivals anything in science. The least you can say about evolutionary theory is that it works. All but pedants would go further and assert that it is true.

Whence, then, comes the oft-parroted canard, 'Evolution is *only a theory*'? Perhaps from a misunderstanding of philosophers who assert that science can never demonstrate truth. All it can do is fail to disprove a hypothesis. The more strenuously you work to

---

* It isn't only anthropologists (and, I'm glad to say, it's far from being all anthropologists). Similar nonsense can be heard from an all-too-powerful species of literary 'theorist'.

disprove it, and the more persistently you fail, the more inclined you feel to ditch 'hypothesis' in favour of 'fact', but you never get there. Evolution is an unfalsified hypothesis – one that was vulnerable to falsification but has so far survived.

Scientists generally don't mind this kind of philosopher and even thank him for taking care of such stuff, thereby freeing them to get on with advancing knowledge. They might, however, venture that what is sauce for the goose of science is sauce for the gander of everyday experience. If evolution is an unfalsified hypothesis, then so is every fact about the real world; so is the very existence of a real world.

This kind of argument is swiftly and rightly sidelined. Evolution is true in whatever sense you accept it as true that New Zealand is in the Southern Hemisphere. If we refused ever to use a word like 'true', how could we conduct our day-to-day conversations? Or fill in a census form: 'What is your sex?' 'The hypothesis that I am male has not so far been falsified, but let me just check again.' As Douglas Adams might have said, it doesn't read well. Yet the philosophy that imposes such scruples on science has no basis for absolving everyday facts from the same circumlocution. It is in this sense that evolution is true – provided, of course, that the scientific evidence for it is strong. It is very strong, and Jerry Coyne displays it for us in a way that no objective reader could fail to find compelling.

Here I must anticipate another favourite accusation that will, as I know from personal experience, be plonkingly levelled against Dr Coyne and his book: 'Why bother? You are tilting at a dead horse, flogging windmills. Nobody takes creationism seriously, nowadays.' Translation: 'The Regius Professor of Theology at my university is no creationist, the Archbishop of Canterbury accepts evolution, therefore you are wasting your time setting out the evidence.' The melancholy facts are these. Polls in both Britain and America show a majority wanting 'intelligent design' to be taught in science classes. In Britain, only 69 per cent want evolution to be taught at all (MORI). In America, more than 40 per cent* believe that 'Life on Earth has existed in its present form since the beginning of time' (Pew) and that 'God created human beings pretty much in their present form at one time within the last 10,000 years or so' (Gallup).

Science teachers, especially in America but increasingly in Britain, feel beleaguered, and it is small comfort to them if a handful of theologians and bishops occasionally murmur a word of support for evolutionary science. Occasional murmurs are not enough. In October 2008, a group of about sixty American science teachers met to compare notes, at the Center for Science Education at Emory University in Atlanta, and they had some revealing experiences to relate. One teacher reported that students 'burst into tears' when told they would be studying evolution. Another teacher described how students repeatedly screamed, 'No!' when he began talking about evolution in class.[†]

Such experiences are common throughout the United States, but also, once again I am loath to admit, in Britain. The *Guardian* reported that, in February 2006, 'Muslim medical students in London distributed leaflets that dismissed Darwin's theories as false'. The Muslim leaflets were produced by the Al-Nasr Trust, a registered charity with tax-free status. The British taxpayer, that is to say, is subsidizing the systematic distribution of scientific falsehood to educational institutions. Science teachers across Britain will

---

* Now slightly less than 40 per cent – a small step in the right direction.
[†] I've already mentioned (see footnote on p. 51 above) TIES, the Teacher Institute for Evolutionary Science, which my foundation set up under Bertha Vasquez, herself an outstanding teacher, to help American science teachers deal with this kind of thing. Nothing else in the science curriculum is met with outright hostility (which is bizarre, because nothing else in the science curriculum is more securely established than the fact of evolution) and the teachers need all the help they can get. Bertha and TIES are giving it to them.

confirm that they are coming under slight, but growing, pressure from creationist lobbies, usually inspired by American or Islamic sources.

So, let nobody have the gall to deny that Professor Coyne's book is necessary. Not just his book, and here I must declare an interest. February 12th 2009 is Darwin's 200th birthday, and the 150th anniversary of *The Origin of Species* falls this autumn. Publishers being as anniversary-minded as they are, Darwin-related books were obviously to be expected this Darwin year. Nevertheless, it is true to say that neither Jerry Coyne nor I was aware of the other's book on the evidence for evolution when we began our own – his published now, mine* in the autumn. And our two books may not be the only ones. Bring them on, I say. The more the merrier. The evidence is massive, the modern version of the story would surprise and inspire even Darwin, and it cannot be told too often. Evolution is, after all, the true story of why we all exist, and an exhilaratingly powerful and satisfying explanation. It supersedes – and devastates – all predecessors, no matter how devoutly and sincerely believed.

Jerry Coyne's book is outstandingly good. His knowledge of evolutionary biology is prodigious, his deployment of it as masterful as his touch is light. His coverage is enviably comprehensive, yet he simultaneously manages to keep the book compact and readable. His nine chapters include 'Written in the rocks', laced with examples that make short work of the most popular of all creationist lies, the one about unbridgeable 'gaps' in the fossil record. 'Show me your intermediates!' Jerry Coyne shows you – and very numerous and convincing they are. Not just fossils of large charismatic animals like whales and birds, and the coelacanth-cousins that made the transition from water to land, but also microfossils. These have the advantage of sheer numbers: some kinds of sedimentary rock

---

* *The Greatest Show on Earth.*

are almost entirely made of the tiny fossilized skeletons of foraminiferans, radiolarians and other calcareous or siliceous protozoa. This means you can plot a sensitive graph of some chosen measurement, as a continuous function of geological time, while you systematically work your way through a core of sediments. One of Coyne's graphs shows a genus of radiolarians (beautiful protozoans with minute, lantern-like shells) caught in the act, two million years ago, of 'speciating' – splitting into two species.

Such splitting of one species into two is what Darwin's title actually means, and it is one of the few weak areas in that great book. Jerry Coyne is probably today's leading authority on speciation, and it is not surprising that his chapter called 'The origin of species' is so good. So also is 'The geography of life'. Possibly the most immediately convincing evidence against creationism is to be found in the geographical distribution of animals and plants, on continents and islands (in the broad sense, 'islands' include lakes, mountaintops, oases – from an animal's point of view any small area where it can live, surrounded by a larger area where it can't). After setting out the voluminous evidence on the subject, Coyne concludes:

> Now try to think of a theory that explains the patterns we've discussed by invoking the special creation of species on oceanic islands and continents . . . There are no good answers – unless, of course, you presume that the goal of a creator was to make species *look* as though they evolved on islands. Nobody is keen to embrace that answer, which explains why creationists simply shy away from island biogeography.

Such dishonesty by omission is lamentably characteristic of creationists. They love fossils because they have been schooled, wrongly as Coyne shows, to believe that 'gaps' in the fossil record are an embarrassment to evolutionary theorists. The geographical distribution of species really is an embarrassment to creationists – and they conspicuously ignore it.

The book includes a lucid exposition of natural selection at the level of the gene – Darwin expressed it at the level of the individual organism. Coyne describes how a parasitic worm changes the appearance and behaviour of its ant host, turning the ant's abdomen into a simulacrum of a red berry, angled temptingly up in the air with carefully weakened stalk joining it to the thorax. You've guessed the sequel. The 'berry', full of worm eggs, is eaten by a bird, which is the definitive host of the worm. In Coyne's own words,

> All of these changes are caused by the genes of the parasitic worm as an ingenious ploy to reproduce themselves . . . It is staggering adaptations like this – the many ways that parasites control their carriers, just to pass on the parasites' genes – that get an evolutionist's juices running.

Very true. That kind of gene-centred 'adaptationist' language has become all but universal among evolutionary biologists working in the field. It is amusing, therefore, to recall the overbearing hostility with which it was attacked thirty years ago by the dedicatee of Coyne's book, his old teacher, the distinguished geneticist Richard Lewontin. It is not irrelevant that Coyne also has a very necessary clarification of the idea of the 'selfish gene', in which he correctly explains that it has no connection with spurious claims that we are deterministically hard-wired to be selfish. Thirty years on, how things have changed.

Coyne's chapter on 'The engine of evolution' begins with a splendidly macabre example. Giant Japanese hornets raid the nests of honey bees to feed their larvae. A single hornet scout discovers a beehive and marks it 'for doom' with a sort of chemical black spot.

> Alerted by the mark, the scout's nestmates descend on the spot, a group of twenty or thirty hornets arrayed against a colony of up to thirty thousand honeybees.

But it's no contest. Wading into the hive with jaws slashing, the hornets decapitate the bees one by one. With each hornet making heads roll at a rate of forty per minute, the battle is over in a few hours: every bee is dead, and body parts litter the hive. Then the hornets stock their larder.

Coyne's purpose in telling the story is to contrast the terrible fate of European bees, introduced into Japan, with native Japanese bees that have had time to evolve a defence.

And their defense is stunning – another marvel of adaptive behavior. When the hornet scout first arrives at the hive, the honeybees near the entrance rush into the hive, calling nestmates to arms while luring the hornet inside. In the meantime, hundreds of worker bees assemble inside the entrance. Once the hornet is inside, it is mobbed and covered by a tight ball of bees. Vibrating their abdomens, the bees quickly raise the temperature inside the ball to about 117 degrees Fahrenheit ... In twenty minutes the hornet scout is *cooked to death*, and – usually – the nest is saved.

Coyne adds that the bees can survive the high temperature, but it's another insight of the 'gene's-eye view' that this would not be necessary in order for natural selection to favour the adaptation. Worker bees are sterile: their genes survive, not in the workers themselves but *as copies* in the bodies of the minority of hive members destined for reproduction. If the workers in the centre of the ball were cooked alongside the hornet, it would be well worth the sacrifice. Copies of their genes 'for cooking' live on.

There's a good chapter on 'Remnants, vestiges, embryos and bad design', topics that Darwin himself treated well, and also on 'How sex drives evolution', and 'What about us?' on human evolution. But Coyne really comes into his own with another strand of powerful evidence that was not available to Darwin. The molecular genetics revolution, which began in 1953, would have taken Darwin's breath away and filled him with exultation. Every living creature carries within each of its cells a voluminous textual recipe for making itself. Nowadays, we can read these messages, accurately and with a completeness that is limited only by (rapidly shrinking) money and time. Because the DNA texts of all animals and plants use the identical four-letter code, we have a goldmine of opportunity for comparison. In his own time, Darwin could compare, say, the wing of a bat, the flipper of a whale and the spade of a mole, and spot the relationships among a handful of bones. Today – and more cheaply in the near tomorrows – we can do it on an altogether grander scale, lining up billion-letter DNA texts from bat, whale and mole, and literally counting the single-letter discrepancies and resemblances. Moreover, we don't have to limit our comparisons to one group, such as the mammals. The universal genetic code allows us to make letter-for-letter textual comparisons across plants, snails and bacteria, as well as vertebrates. This not only provides evidence for the fact of evolution that is orders of magnitude more solid even than the powerful evidence Darwin could muster. We can also construct, finally and definitively, the complete tree of all life, the universal pedigree. And we can find, in huge numbers, the molecular equivalents of vestigial evolutionary relics like the human appendix and the kiwi's wings.

For the genome is littered with dead genes. Huge wastes of DNA territory comprise a graveyard of discarded, superseded old genes (plus meaningless sequences of nonsense DNA that never functioned) with occasional islands of current, extant genes that are actually read by the translating machinery and turned into action. Dead, untranslated genes are called pseudogenes. The reason our sense of smell is poor, compared with, say, dogs', is that most of our ancestral genes for smelling have been rendered inactive. We still have them but they are dead. Molecular biologists can still read them – serried ranks of molecular 'fossils' – but the body does not.

It is wonderful enough that we can construct a tree of life based upon active genes, and find that different genes agree on the same pedigree. It is even more convincing that we get the same pedigree with dead genes, whose DNA sequences represent nothing, and must be regarded only as the inert legacy of history. How would a creationist explain that? How would he explain the very existence of pseudogenes? Why would the creator litter the genome with useless, untranslated variants of genes, and locate them, moreover, in exactly the right pattern around the animal and plant kingdoms to give the impression – the deceptive impression, as he would presumably have to admit – that they evolved and were not created?

Coyne is right to identify the most widespread misunderstanding about Darwinism as 'the idea that, in evolution, "everything happens by chance" . . . This common claim is flatly wrong.' Not only is it flatly wrong, it is obviously wrong, transparently wrong, even to the meanest intelligence (a phrase that has me actively restraining myself). If evolution worked by chance, it obviously couldn't work at all. Unfortunately, instead of working out that they must therefore have misunderstood evolution, creationists conclude, instead, that evolution must be false. This one misunderstanding, single-handed, accounts for much of the uncomprehending opposition to evolution that made it necessary for Jerry Coyne to write his book in the first place. The need was great; the execution is superb. Please read it.

EPILOGUE:
TO BE READ AT MY FUNERAL

The title of this epilogue is my only justification for including it. It is extracted and edited from the opening chapter of *Unweaving the Rainbow*.

We are going to die, and that makes us the lucky ones. Most people are never going to die because they are never going to be born. The potential people who could have been here in my place but who will in fact never see the light of day outnumber the sand grains of Sahara. Certainly those unborn ghosts include greater poets than Keats, scientists greater than Newton. We know this because the set of possible people allowed by our DNA so massively outnumbers the set of actual people. In the teeth of these stupefying odds it is you and I, in our ordinariness, that are here.

We live on a planet that is all but perfect for our kind of life: not too warm and not too cold, basking in kindly sunshine, softly watered; a gently spinning, green and gold harvest festival of a planet. What are the odds that a planet picked at random would have these complaisant properties? Imagine a spaceship full of sleeping explorers, deep-frozen would-be colonists of some distant world. Perhaps the ship is on a forlorn mission to save the species before an unstoppable comet, like the one that killed the dinosaurs, hits the home planet. The voyagers go into the deep-freeze soberly

429

reckoning the odds against their spaceship's ever chancing upon a planet friendly to life. If one in a million planets is suitable at best, and it takes centuries to travel from each star to the next, the spaceship is pathetically unlikely to find a tolerable, let alone safe, haven for its sleeping cargo.

But imagine that the ship's robot pilot turns out to be unthinkably lucky. After millions of years the ship has the extraordinary luck to happen upon a planet capable of sustaining life: a planet of equable temperature, bathed in warm starshine, refreshed by oxygen and water. The passengers, Rip van Winkles, wake stumbling into the light. After a million years of sleep, here is a whole new fertile globe, a lush planet of warm pastures, sparkling streams and waterfalls, a world bountiful with creatures, darting through alien green felicity. Our travellers walk entranced, stupefied, unable to believe their unaccustomed senses or their luck.

I am lucky to be alive and so are you. Privileged, and not just privileged to enjoy our planet. More, we are granted the opportunity to understand why our eyes are open, and why they see what they do, in the short time before they close for ever.

SOURCES AND ACKNOWLEDGEMENTS

The author, editor and publishers gratefully acknowledge the permission of copyright holders to reproduce material in this volume.

I. TOOLS OF TWO TRADES: WRITING SCIENCE

'In conversation with Neil deGrasse Tyson: On science and scientists, in public and private': first broadcast on *StarTalk*, Sept. 2015.

'The uncommon sense of science': Review of Lewis Wolpert, *The Unnatural Nature of Science*, first published in the *Sunday Times*, 1992.

'Are we all related?': First published in Gemma Elwin Harris, *Big Questions from Little People, answered by some very big people* (Faber/ NSPCC, 2012).

'The timeless and the topical': First published in Tim Folger, ed., *The Best American Science and Nature Writing* (Boston: Houghton Mifflin, 2003).

'Fighting on two fronts': First published as the introduction to the 'Edge' event 'Napoleon Chagnon: blood is their argument' (see https:// www.edge.org/conversation/ napoleon-chagnon-blood-is-their-argument).

'Pornophilosophy': Review of Lynn Margulis and Dorion Sagan, *Mystery Dance: on the evolution of human sexuality*, first published in *Nature*, vol. 354, 12 Dec. 1991.

'Determinism and dialectics: a tale of sound and fury': Review of Steven Rose, Leon J. Kamin and Richard Lewontin, *Not in Our Genes: biology, ideology, and human nature*, first published in *New Scientist*, 24 Jan. 1985.

'Tutorial-driven teaching': First published in David Palfreyman, ed., *The Oxford Tutorial: 'Thanks, you thought me how to think'*, 2nd edn (Oxford, Oxford Centre for Higher Education Policy Studies, 2008).

'Life after light': First published as introduction to the audio edition of *Dark Universe* by Daniel F. Galouye (London, Audible Audiobooks, 2009).

'A scientific education and the Deep Problems': First published as afterword to the paperback edition of Fred Hoyle, *The Black Cloud* (London, Penguin, 2010).

'Rationalist, iconoclast, Renaissance man': First published as foreword to Jacob Bronowski, *The Ascent of Man*, new edn (London, BBC Books, 2011).

'Revisiting *The Selfish Gene*': First published as introduction to *The Selfish Gene*, 30th-anniversary edition (Oxford, Oxford University Press, 2006).

II. WORLDS BEYOND WORDS: CELEBRATING NATURE

'In conversation with Adam Hart-Davis: Evolution and plain writing in science': First published in Adam Hart-Davis, *Talking Science* (Chichester, Wiley, 2004).

'Close encounters with the truth': Review of Carl Sagan's *The Demon-Haunted World: science as a candle in the dark* (New York, Random House, 1995), first published in *The Times*, February 1996.

'Conserving communities': First published in Art Wolfe, *The Living Wild*, ed. Michelle A. Gilders (Seattle, Wildlands, 2000).

'Darwin on the slab': First published as foreword to David Dugan, *Inside Nature's Giants* (London, Collins for Channel 4, 2011).

'Life within life': First published as foreword to David P. Hughes,

Jacques Brodeur and Frédéric Thomas, eds, *Host Manipulation by Parasites* (Oxford, Oxford University Press, 2012).

'Pure delight in a godless universe': Previously unpublished.

'Travelling with Darwin': Extracted from Everyman's introduction to Charles Darwin, *The Origin of Species* and *The Voyage of the Beagle* (New York and London, Random House, 2003). Reproduced by permission of Everyman's Library, an imprint of Alfred A. Knopf.

'Pictures of paradise': First published as foreword to Paul D. Stewart, *Galápagos: the islands that changed the world* (London, BBC, 2006).

III. INSIDE THE SURVIVAL MACHINE: EXPLORING HUMANITY

'In conversation with Steven Pinker: Language, learning and debugging the brain': first broadcast in *The Genius of Charles Darwin* (Channel 4 Television, 2008).

'Old brain, new brain': First published as foreword to Jeff Hawkins, *A Thousand Brains: a new theory of intelligence* (New York: Basic Books, forthcoming 2021).

'Breaking the species barrier': First published in John Brockman, ed., *This Will Change Everything: ideas that will shape the future* (New York, Harper, 2009).

'Branching out': Review of Ian Tattersall and Jeffrey H. Schwartz, *Extinct Humans* (Boulder, Colo., Westview, 2000), first published in *New York Times*, 6 Aug. 2000.

'Darwinism and human purpose': First published in John R. Durant, ed., *Human Origins* (Oxford, Clarendon Press, 1989).

'Worlds in microcosm': Extracted from the essay first published in Neil Spurway, ed., *Humanity, Environment and God: The Glasgow Centenary Gifford Lectures* (Oxford, Wiley-Blackwell, 1993).

'Real genes and virtual worlds': First published as afterword to David Buss, ed., *Handbook of Evolutionary Psychology* (Hoboken, NJ: Wiley, 2005).

'Nice guys (still) finish first': First published as foreword to Robert

Axelrod, *The Evolution of Cooperation*, new edn (New York, Basic Books, 2006).

'Art, advertisement and attraction': First published as foreword to Robin Wight, *The Peacock's Tail and the Reputation Reflex: the neuroscience of art sponsorship* (London, Engine/Arts & Business, 2007).

'From African Eve to the Banda strandlopers': Review of Jonathan Kingdon, *Self-Made Man and his Undoing* (London: Simon & Schuster, 1993), first published in the *Times Literary Supplement*, 26 March 1993.

'We are stardust': First published as foreword to Bailey Harris and Douglas Harris, *My Name is Stardust*, illus. Natalie Malan (Augusta, Mo., Storybook Genius, 2017).

'The descent of Edward Wilson': Review of Edward O. Wilson, *The Social Conquest of Earth* (New York, Norton, 2012), first published in *Prospect*, June 2012.

IV. THE MINER'S CANARY: SUPPORTING SCEPTICISM

'In conversation with Christopher Hitchens: Is America heading for theocracy?': Extracted from interview first published in *New Statesman*, Dec. 2011.

'Witness of internal delusion': First published as foreword to Dan Barker, *Godless: how an evangelical preacher became one of America's leading atheists*, pb (Berkeley, Ca., Ulysses, 2008)

'Kicking the habit': First published as foreword to Daniel C. Dennett and Linda LaScola, *Caught in the Pulpit* (Durham, NC, Pitchstone, 2015).

'The unburdening lightness of relief': First published as foreword to Catherine Dunphy, *Apostle to Apostate: the story of the Clergy Project* (Durham, NC, Pitchstone, 2015).

'A public and political atheist': First published as foreword to Herb Silverman, *Candidate without a Prayer: an autobiography of a Jewish atheist in the Bible Belt* (Durham, NC, Pitchstone, 2012).

'The great escape': First published as foreword to Seth Andrews,

Deconverted: a journey from religion to reason (Parker, Colo., Outskirts, 2019).

'In His own words: a portrait of God': First published as foreword to Dan Barker, *God: the most unpleasant character in all fiction* (New York, Sterling, 2016).

'Liberation from theology': First published as foreword to Tom Flynn, ed., *The New Encyclopedia of Unbelief* (Amherst, NY: Prometheus, 2007).

'The God Temptation': First published as introduction to the tenth-anniversary edition of *The God Delusion* (London, Bantam, 2016).

'The intellectual and moral courage of atheism': first published in *The Four Horsemen* (London, Bantam, 2019).

V. COUNSEL FOR THE PROSECUTION: INTERROGATING FAITH

'In conversation with Lawrence Krauss: Should science speak to faith?': Extracted from interview first published in *Scientific American*, 19 June 2007.

'Defending the wall of separation': First published as foreword to Sean Faircloth's *Attack of the Theocrats: how the religious right harms us all – and what we can do about it* (Durham, NC, Pitchstone, 2012).

'A moral and intellectual emergency': First published as foreword to Sam Harris, *Letter to a Christian Nation* (London, Bantam, 2007).

'Unmasking the design illusion': First published as foreword to Niall Shanks, *God, the Devil and Darwin: a critique of intelligent design theory* (New York, Oxford University Press, 2004).

'"Nothing will come of nothing": why Lear was wrong': First published as afterword to Lawrence M. Krauss's *A Universe from Nothing* (New York, Free Press, 2012).

'The fast food thesis: religion as evolutionary by-product': First published as foreword to J. Anderson Thomson with Clare Aukofer, *Why We Believe in Gods: a concise guide to the science of faith* (Charlottesville, Va., Pitchstone, 2011).

'An ambitious banana skin': Review of Richard Swinburne, *Is There a God?* (Oxford, Oxford University Press, 1996), first published in the *Sunday Times*, 4 Feb. 1996.

'Heavenly twins': First published as foreword to 'Mohammed Jones', *Jesus and Mo: folie à dieu* (2013).

'A tale of horror and heroism': Review of Farida Khalaf and Andrea C. Hoffman, *The Girl Who Beat ISIS: my story* (London, Vintage, 2016), published on www.richarddawkins.net.

VI. TENDING THE FLAME: EVANGELIZING EVOLUTION

'In conversation with Matt Ridley: From Darwin to DNA – and beyond': conversation recorded for this volume.

'The "little Penguin" relaunched': First published as foreword to John Maynard Smith, *The Theory of Evolution*, new edn (Cambridge, Cambridge University Press, 1993).

'Foxes in the snow': First published as foreword to George Williams, *Adaptation and Natural Selection: a critique of some current evolutionary thought*, new edn (Princeton, Princeton University Press, 2018).

'Telling truth in a dark time': Review of Maitland Edey and Donald Johanson, *Blueprints: solving the mystery of evolution*, pb edn (New York, Penguin, 1989), first published in *New York Times*, 9 April 1989.

'Irresponsible publishing?': Review of Richard Milton, *The Facts of Life: shattering the myth of Darwinism* (London, Fourth Estate, 1992), first published in *New Statesman*, 28 Aug. 1992.

'Inferior design': Review of Michael Behe, *The Edge of Evolution: the search for the limits of Darwinism* (New York, Free Press, 2007), first published in *New York Times*, 1 July 2007.

'The only kind of truth that works': Extracted from review of Jerry Coyne's *Why Evolution is True* (Oxford, Oxford University Press, 2009), first published in *Times Literary Supplement*, 13 Feb. 2009.

BIBLIOGRAPHY OF
WORKS CITED

The following list gives publication details of works mentioned in the text and footnotes.

Atkins, Peter W., *The Creation* (London, Freeman, 1981)

Atkins, Peter W., *Creation Revisited* (London, Freeman, 1993)

Axelrod, Robert, *The Evolution of Cooperation*, new edn (New York, Basic Books, 2006)

Axelrod, Robert, and Dion, Douglas, 'The further evolution of cooperation', *Science*, vol. 242, 1988, pp. 1385–90

Behe, Michael J., *Darwin's Black Box: the biochemical challenge to evolution* (New York, Free Press, 1996)

Burt, Austin, and Trivers, Robert, *Genes in Conflict: the biology of selfish genetic elements* (Cambridge, Mass., Harvard University Press, 2006)

Colman, Andrew M., and Woodhead, Peter, 'The origin of the juxtaposition of "nature" and "nurture": not Galton, Shakespeare, or Mulcaster, but Socrates', *British Psychological Society History and Philosophy of Psychology Newsletter*, vol. 8, 1989, pp. 35–7

Cosmides, Leda, Tooby, John, and Barkow, Jerome, eds, *The Adapted Mind* (Oxford, Oxford University Press, 1992)

Cronin, Helena, *The Ant and the Peacock: altruism and sexual selection from Darwin to today* (Cambridge, Cambridge University Press, 1991)

Dawkins, Richard, *The Ancestor's Tale: a pilgrimage to the dawn of life* (London, Weidenfeld & Nicolson, 2004; 2nd edn with Yan Wong, 2016)

Dawkins, Richard, *The Blind Watchmaker* (London, Norton, 1986)

Dawkins, Richard, *Climbing Mount Improbable* (London, Viking, 1996)

Dawkins, Richard, *A Devil's Chaplain* (London, Weidenfeld & Nicolson, 2003)

Dawkins, Richard, *The Extended Phenotype* (London, Oxford University Press, 1982)

Dawkins, Richard, *The God Delusion* (London, Bantam, 2006)

Dawkins, Richard, *The Greatest Show on Earth: the evidence for evolution* (London, Bantam, 2009)

Dawkins, Richard, *Science in the Soul: selected writings of a passionate rationalist* (London, Bantam, 2017)

Dawkins, Richard, *The Selfish Gene* (Oxford, Oxford University Press, 1976; 30th anniversary edn, 2006)

Dawkins, Richard, *Unweaving the Rainbow* (London, Allen Lane, 1998)

Dawkins, Richard, Dennett, Daniel C., Harris, Sam, and Hitchens, Christopher, *The Four Horsemen: the discussion that sparked an atheist revolution* (London, Bantam, 2019)

Dennett, Daniel C., *Breaking the Spell: religion as a natural phenomenon* (New York, Viking, 2006)

Eberhard, W. G., *Sexual Selection and Animal Genitalia* (Cambridge, Mass., Harvard University Press, 1860)

Edwards, A. W. F., 'Natural selection and the sex ratio: Fisher's sources', *American Naturalist*, vol. 151, 1998, pp. 564–69

Eiseley, Loren, *The Firmament of Time* (London, Gollancz, 1961)

Eldredge, Niles, *Unfinished Synthesis* (New York, Oxford University Press, 1985)

Gould, Stephen Jay, 'Evolution as fact and theory', in *Hen's Teeth and Horse's Toes* (New York, Norton, 1994)

Gould, Stephen J., 'The meaning of punctuated equilibrium and its role in validating a hierarchical approach to macroevolution', in R. Milkman, ed., *Perspectives on Evolution* (Sunderland, Mass., Sinauer, 1982)

Grafen, Alan, 'A geometric view of relatedness', in Richard Dawkins and Mark Ridley, eds, *Oxford Surveys in Evolutionary Biology*, Vol. 2 (Oxford, Oxford University Press, 1985), pp. 28–89

Harris, Sam, *The End of Faith: religion, terror, and the future of reason* (London, Simon & Schuster, 2005)

Jeans, James, *The Mysterious Universe* (Cambridge, Cambridge University Press, 1930)

Krauss, Lawrence M., *A Universe from Nothing: why there is something rather than nothing* (New York, Free Press, 2012)

Marchant, James, ed., *Alfred Russel Wallace: letters and reminiscences*, vol. 1 (London, Cassell, 1916)

Medawar, Peter B., *Pluto's Republic* (Oxford, Oxford University Press, 1982)

Medawar, Peter B., and Medawar, Jean S., *Aristotle to Zoos: a philosophical dictionary of biology* (Cambridge, Mass., Harvard University Press, 1983)

Miller, Geoffrey, *The Mating Mind: how sexual choice shaped the evolution of human nature* (New York, Random House, 2000)

Morris, Desmond, *The Secret Surrealist* (Oxford, Phaidon, 1987)

Nesse, Randolph M., and Williams, George C., *Why We Get Sick: the new science of Darwinian medicine* (New York, Random House, 1994)

Norris, Pippa, and Inglehart, Ronald, *Sacred and Secular: religion and politics worldwide* (New York, Cambridge University Press, 2004)

Pinker, Steven, *The Blank Slate* (London, Allen Lane, 2002)

Pinker, Steven, *How the Mind Works* (London, Allen Lane, 1997)

Pinker, Steven, *The Language Instinct* (London, Penguin, 1994)

Ridley, Matt, *Nature via Nurture: genes, experience and what makes us human* (London, Fourth Estate, 2003)

Richmond, M. H., and Smith, D. C., *The Cell as a Habitat* (London, Royal Society of London Publications, 1979), ch. 1

Sagan, Carl, *Pale Blue Dot: a vision of the human future in space* (New York, Random House, 1994)

Segerstråle, Ullica, *Defenders of the Truth: the sociobiology debate* (Oxford, Oxford University Press, 2000)

Smith, John Maynard, 'Current controversies in evolutionary biology', in M. Grene, ed., *Dimensions of Darwinism* (Cambridge, Cambridge University Press, 1983)

Thomas, Lewis, *The Lives of a Cell* (London, Futura, 1974)

Weismann, August, *The Germ Plasm: a theory of heredity*, trans. W. N. Parker and H. Rönnfeldt (London, W. Scott, 1893)

Wilson, Edward O., *Sociobiology: the new synthesis* (Cambridge, Mass., Harvard University Press, 1975)

Wolpert, Lewis, *Malignant Sadness: the anatomy of depression* (London, Faber, 2006)

INDEX

A for Andromeda (Hoyle and Elliott), 80

Adams, Douglas, 420

Adaptation and Natural Selection (Williams), 393–400

The Adapted Mind (Barkow, Cosmides and Tooby), 221

advertising, 93n, 234–9

'African Eve', 191, 243, 246–8

After Many a Summer (Huxley), 17

Al-Nasr Trust, 421

Alexander, Richard D., 54, 257

altruism, 92–4, 97, 169, 260

American Naturalist (journal), 417

The Ancestor's Tale (Dawkins and Yan Wong), 136n

Andersson, Malte, 256

Andrews, Seth, 296–7

Animal Behaviour (journal), 5

The Ant and the Peacock (Cronin), 237

Anticipations of the Reaction of Mechanical and Scientific Progress upon Human Life and Thought (Wells), 17

The Ants (Hölldobler and Wilson), 265

Apostle to Apostate (Dunphy), 287

An Appetite for Wonder (Dawkins), 26

The Arrival of the Fittest (Wagner), 382

The Ascent of Man (BBC TV), 85–9

Asimov, Isaac, 80

Atheist Alliance International, 269

Atiyya, Dr Izzat, 316

Atkins, Peter, 8–9, 96–7, 357

Attack of the Theocrats (Faircloth), 338

Attenborough, David, 85, 120, 231n

Attneave, Fred, 206

Australopithecus: Africa, 241; evolution, 188, 189; extinction, 184; fossil record, 410–11; reconstruction, 185; Taung Child, 86–7

Axelrod, Robert, 229–33

Ayer, A. J., 91

Bachmann, Michele, 340

'Banda strandlopers', 248–9

Barfoot, Rev. Milton, 281

Barker, Dan, 279, 280–3, 293n, 298–301

Barkow, Jerome, 221

Barlow, Horace, 205, 206, 209, 210

Barlow, Nora, 155

Barnes, Russell, 163

Barrow, John, 199

Barton, Nick, 256

Bateson, Patrick, 68

Bateson, William, 147, 383

beauty, 27, 115, 142, 171–3, 237

Beethoven, Ludwig van, 321
Behe, Michael, 414–17
Benedict XIV, Pope, 318
Benn, Tony, 336
Bentham, Jeremy, 271n
Benveniste, Jacques, 42n
Bergson, Henri, 6
The Best American Science and Nature Writing (anthology), 47
Bible: belief in, 293; character of God, 299–301; Gideon, 233, 298–9; Jefferson, 273; literal interpretation, 294, 300–1, 334; as literature, 276–7; translations, 277
Big Bang, 80, 324, 354–5
Big Questions from Little People (anthology), 44
The Black Cloud (Hoyle), 17, 79–84
Blair, Tony, 342
The Blank Slate (Pinker), 170n
The Blind Watchmaker (Dawkins), 349
Blueprints: solving the mystery of evolution (Edey and Johanson), 332n, 401, 405
Bodleian Library, 4
Boghossian, Peter, 13, 57
Born, Max, 87
Bostrom, Nick, 78n
bower birds, 113–14, 235–7
Box, Joan Fisher, 148
brain: ancestral, 228, 360; ant, 114, 137; bird, 236; 'constrained virtual reality', 226–8; constructing a model of reality, 177–8, 202; evolution, 39, 101, 322, 359; goal-seeking, 195–8; language learning, 171; logical thinking, 22; mind and, 168; models, 179–80, 210–12, 215, 216–20; music and speech, 166–7, 358; Necker Cube illusion,

219–20, 226; old reptilian and new mammalian, 178–81; on-board computer, 193–5; sensory recognition systems, 205–12; simulation of future events, 215–20; size, 241, 410
Brave New World (Huxley), 17, 185n
Breaking the Spell (Dennett), 137
Brecht, Bertolt, 87
Brief Candle in the Dark (Dawkins), 117n
Brockman, John, 54, 183
Brodeur, Jacques, 132, 133
Bronowski, Jacob, 5, 85–9
Brown, Derren, 311
Brown, Pastor, 294
Browne, Sir Thomas, 240
Brunet, Peter, 71
Burt, Austin, 96n
Bush, George W., 341–2
Buss, David, 221, 223
Buxton, John, 73n

Cain, Arthur, 71, 91
Candidate without a Prayer (Silverman), 290
Carlyle, Thomas, 353
Carroll, Lewis, 395
Catholic Encyclopedia, 318–19
Caught in the Pulpit (Dennett and LaScola), 284
The Cell as a Habitat (Smith), 11
Center for Inquiry (CFI), 176, 357
Chagall, Marc, 87
Chagnon, Napoleon, 54–6
Charnov, Eric, 256
Chirac, Jacques, 341–2
Christianity: beliefs, 35–6, 336; Bible, 299–301; Catholicism, 118, 271; Clergy Project, 286; decline, 271n; fundamentalist, 270, 274, 304; 'intelligent design', 348, 415; Islam and, 272,

Christianity, *cont.*
304; liberation from, 306–7;
optimism, 229; scholarship,
378; in Uganda, 339; in US, 274,
291–5, 339–41, 343, 344–5; *see
also* Jesus
Christmas, 37, 183, 269, 277–8, 367
Civilisation (BBC TV), 85
cladogram, 244–6
Clark, Kenneth, 85
Clarke, Arthur C., 16, 80, 82
Clergy Project, 282n, 285–6, 287–9
Climbing Mount Improbable
(Dawkins), 112
Clutton-Brock, Tim, 256
Colburn, Henry, 154
Coleridge, Samuel Taylor, 137
Concise Commandments of Islam
(Golpaygani), 314–15
Contact (Sagan), 80
Cook, Melvin, 412
Copernicus, Nicolaus, 314
Cosmides, Leda, 221, 222, 223, 256
Cosmos (Sagan), 116
Counterfeit World (Galouye),
78n, 381n
Covington, Syms, 160
Coyne, Jerry, 256, 414, 418–27
Crank, Jessica, 341
The Creation (Atkins), 8–9
*Creation Research Science
Quarterly*, 412
Creation Revisited (Atkins), 96–7
creationism: argument from
improbability, 347–8, 350;
Behe's position, 414–15; Dar-
win's battle with creationists,
164; debates, 331, 332–3, 389, 391,
404; evidence against, 423, 427;
'intelligent design', 348, 420;
Milton's position, 411–13;
'missing links', 410; Old Earth
Creationism, 330; teaching

issue, 270, 276, 420–2; technol-
ogy, 381; Young Earth
Creationism, 330, 331, 332–3
Crespi, Bernard, 256
Crichton, Michael, 79–80
Crick, Francis, 349, 384, 403, 409
Cromwell, Oliver, 297
Cronin, Helena, 237
Cupitt, Don, 199

Daly, Martin, 221, 223
'Dan Dare, Pilot of the Future', 16
Dark Universe (Galouye), 76–8
Darwin, Charles: battle with
creationists, 164; *Beagle*
voyage, 154–6, 158, 160, 356,
379; beetle expert, 284n;
birthday, 130, 422; Doctor
Dolittle comparison, 16;
experiments on sweet peas,
150; gentleman scientist, 176;
historical context, 375–6,
378–9, 425–6; ideas on heredity, 143, 144–5, 147, 149–50;
Kelvin's criticisms, 150–1; kin
selection, 263–4; on origin of
design illusion, 120–1, 142,
309–12, 347, 355; on origins of
human ancestor, 316; pub-
lisher, 154, 255; question of
agency detection, 360; relationship with Wallace, 176,
237–8, 376–7; response to idea
of human exceptionalism,
379–80; revisions of his work,
145–7; sexual selection theory,
234–5, 237–8, 377; 'survival of
the fittest', 153–4, 258–9, 376;
view of religiosity, 357–8;
world-view, 156–7; *see also*
evolution; natural selection
Darwin, George, 151
Darwin, Leonard, 148

Darwinism: adaptation, 113, 114,
120–1, 132, 157; altruism, 93, 97;
answer to 'intelligent design',
349; explanations for emotions,
169–70, 173; 'fitness', 153–4, 259;
kin selection, 265; Mendelian
genetics, 147–8, 188, 383; misunderstandings of, 92, 427; Modern
Synthesis, 188, 391; neo-
Darwinism, 107–8, 143–4, 153,
188, 391, 393; objections to, 150–3,
407–9, 415–16, 421; sociobiology,
65; status of, 143, 392; teaching
issue, 276; values, 101
Darwin's Black Box (Behe), 414, 415
David, King, 319
Davy, Humphry, 137
Day-Lewis, Cecil, 230n
de Tocqueville, Alexis, 274
de Vries, Hugo, 147
*Deconverted: a journey from religion
to reason* (Andrews), 296–7
Defenders of the Truth (Segerstråle), 222
The Demon-Haunted World (Sagan),
47, 116–19
Dennett, Daniel: on ant behaviour,
137; on 'belief in belief', 337;
Cartesian Theatre, 179; *Caught
in the Pulpit*, 284–6; Clergy
Project, 285–6; 'crane' explan-
ation, 324; *The Four Horsemen*,
313; on 'the intentional stance',
379; at meeting to honour
Chagnon, 54; responses to
religious people, 281n, 305n;
source, 394n
Derrida, Jacques, 58, 157
The Descent of Man (Darwin), 145,
154, 163, 258
A Devil's Chaplain (Dawkins), 57, 101,
150n
Dhuhulow, Aisho Ibrahim, 339
Dickens, Charles, 277

Dion, Douglas, 232–3
Discovery Institute, 417
DNA: constructing tree of life, 426;
dead genes, 426, 427; human,
380; information, 108, 125, 323;
mitochondrial, 191, 243–6;
'parasitic', 95–6; preservation,
126; sequencing, 321
Dobzhansky, Theodosius, 187, 188,
393, 417
Doctor Dolittle books, 16, 155
dog breeds, 416–17
The Double Helix (Watson), 15
dreams, 30, 220, 226
Dunphy, Catherine, 287–9
Durant, John R., 192

Eagle comic, 16
East African Mammals
(Kingdon), 241
Edey, Maitland, 332n, 401, 403
Edge website, 54, 183
The Edge of Evolution (Behe), 414–17
Edwards, A. W. F., 147
Eigen, Manfred, 403
Einstein, Albert: exile, 87; explaining
science, 48; religion, 35, 304, 305;
status, 157, 382; understanding,
174, 175, 322
Eiseley, Loren, 5, 9–10
Elliott, John, 80
Emlen, Stephen, 256
Encyclopedia of Unbelief, 304
The End of Faith (Harris), 344
evolution: adaptive, 157; atlas of, 241;
convergent, 126n; 'cooperative
gene', 95; cultural, 198, 381;
'group selection' theory, 256;
hominid fossil record, 189;
human sexuality, 57–60; ideas
before Darwin, 378–9; influence of technology, 248–51;
Maynard Smith's work, 386–92;

evolution, *cont.*
merging identities, 136; misunderstanding of, 427; natural laboratory (Galápagos), 159; natural selection theory, *see* natural selection; of nervous system, 82, 219; 'random mutation', 415–17; second law of thermodynamics objection to, 152–3; seeded by deliberate design, 349; of self-deception, 102; 'survival of the fittest', 153–4; teaching of, 294, 351, 403–4, 421–2; truth of, 418–22; understanding timescale, 39, 111; versus 'intelligent design', 50–1, 348–50

Evolution: the modern synthesis (Huxley), 393

'Evolution as fact and theory' (Gould), 319–20

The Evolution of Cooperation (Axelrod), 229–33

The Evolution of Everything (Ridley), 379

evolutionary psychology, 140, 165, 168, 174, 221–6, 377

The Expression of the Emotions in Man and Animals (Darwin), 163–4, 380

The Extended Phenotype (Dawkins), 112–13, 132–7

Extinct Humans (Tattersall and Schwartz), 187

The Facts of Life: shattering the myth of Darwinism (Milton), 407–13

Faircloth, Sean, 338, 340–3

Falwell, Jerry, 270

Feldman, Marc, 132

Feynman, Richard, 27, 355

The Firmament of Time (Eiseley), 9–10

Fisher, Ronald A.: Darwin's notion of particulate inheritance, 149–50; *The Genetical Theory of Natural Selection*, 144, 393; influence, 55, 417; neo-Darwinism, 144, 147–8, 403; sex ratio theory, 145–7; on sexual selection, 237–8, 377; on sibling care, 259

Fitzroy, Captain Robert, 154, 155–6, 160, 376

Fletcher, Rev. Valentine, 284n

Flynn, Tom, 302

Fodor, Jerry, 394

Folger, Tim, 47

Ford, E. B., 382

forecasting, 201, 203, 205, 212–14

Foucault, Michel, 157

The Four Horsemen (Dawkins, Dennett, Harris and Hitchens), 313

Fourth Estate, 407, 412

Fox, Robin, 54

Franklin, Rosalind, 384

Freud, Sigmund, 87, 157

Fuller, Margaret, 353

Galápagos: the islands that changed the world (Stewart), 158–60

Galápagos islands, 158–60

Galileo, 33–4, 87, 118, 314, 319, 335

Galouye, Daniel F., 76, 78n, 381n

Gaylor, Annie Laurie, 280

genes: birds' tails, 192–3; calculation for maximizing long term survival, 98–9; community of, 122–6; cooperation among self-interested, 95–6, 108–9, 133–4; differential survival, 258; extended phenotype of, 112–15; frequencies in populations, 148; hybridization events, 382–3; kin altruism, 92–3, 259–61; parasite, 133, 134–6, 424; personification of, 96–7; pseudogenes, 426–7;

genes, *cont.*
real genes in a virtual world,
226–8, 251–2; religiosity, 358;
selfish, 95–6, 101, 180–1, 197,
424; sexual selection, 377;
sociobiology controversy, 64–6;
sterility or fertility, 263–4, 425;
units of natural selection,
107–8, 397
Genes in Conflict (Burt and Trivers),
96n
*The Genetical Theory of Natural
Selection* (Fisher), 144, 393
The Genius of Charles Darwin
(Channel 4), 163
Germain, Karl, 311n
Getty, Victoria, 158
Gifford, Adam, Lord Gifford, 199
Gifford Lectures, 199, 335n
The Girl Who Beat ISIS (Farida
Khalaf), 368–71
goals, 181, 193–8, 261
God: character of, 296, 298–301;
existence of, 164, 304, 361–4;
'Ground of all Isness', 355, 361;
involvement in politics, 340;
liberation from belief in, 306–7;
obedience to, 293–4; as the
universe, 35
God, the Devil and Darwin
(Shanks), 348
*God: the most unpleasant character in
all fiction* (Barker), 293n,
298–301
The God Delusion (Dawkins), 32,
299–300, 309
God Temptation, 309–10, 312
Godless (Barker), 279
Golding, William, 17
Goldwater, Barry, 340
Golpaygani, Ayatollah Ozma Sayyed
Mohammad Reda Musavi, 314
Goodall, Jane, 120

Goodenough, Ursula, 305
Gould, Stephen Jay, 116, 257,
319–20, 394
Grafen, Alan, 93, 238–9, 256, 260n
Graham, Billy, 292
The Greatest Show on Earth (Dawk-
ins), 87n, 138, 142, 422n
Greer, Germaine, 89n
Gregory, Richard, 219
group selection, 101, 256, 261–2,
264–5, 382, 397–9
Guardian, 421

Habgood, John, 199
Haig, David, 54
Haldane, J. B. S.: character, 40;
influence, 417; neo-Darwinism,
144, 148, 393; 'theorems', 64n,
390, 395; view of Medawar, 15
Hamilton, W. D. (Bill): calculation of
'inclusive fitness', 98, 260–1, 264;
death, 230; influence, 55, 97, 102,
230, 259, 264, 265, 390; kin
selection theory, 259–60, 264,
265, 398; personification of a
gene, 97; sexual selection theory,
237–9; work on social insects, 262
Hamilton's Rule, 260, 263
Handbook of Evolutionary Psychology
(edited by David Buss), 221, 226n
Hansell, Michael, 132
Hardy, Sir Alister, 74, 248
Hardy, G. H., 148
Harris, Bailey, 253–4
Harris, Douglas, 253
Harris, Sam, 313, 344–6
Hart-Davis, Adam, 105–15
Harvey, Paul, 256
Hawking, Stephen, 356
Hawkins, Jeff, 176–82
Heidegger, Martin, 58–9
heredity, 143–5, 147
Hinduism, 141, 270, 306

Hirsi Ali, Ayaan, 280
History of Western Philosophy
    (Russell), 85
Hitchens, Christopher, 269–78, 313
Hitler, Adolf, 87
Hodgkin, Alan, 177, 351
Hölldobler, Bert, 265
Homo (genus), 189, 411
Homo erectus, 184, 187, 188, 190,
    241, 247
Homo ergaster, 190, 241n
Homo floresiensis, 184
Homo habilis, 188, 410
Homo sapiens, 187, 188, 224,
    321, 390
Host Manipulation by Parasites
    (Brodeur and Thomas), 132
How the Mind Works
    (Pinker), 165n
Howell, Kenneth, 317–18
Hoyle, Fred, 17, 79–84, 90, 151
Hughes, David, 132, 133
Human Origins (edited by John R.
    Durant), 192
Humanity, Environment and God
    (Centenary Year Gifford
    Lectures), 199
Hume, David, 311, 347, 355, 399
Husain, Ed, 280
Huxley, Aldous, 17, 177, 185n
Huxley, Andrew F., 177, 350
Huxley, Julian, 17, 177, 259, 350
Huxley, T. H., 39, 149, 177, 181,
    402, 406

Ibn Baaz, Sheik Abdel-Aziz,
    118–19
Ibrahim, Mustafa, 339
'inclusive fitness', 98n, 261, 264
Infidel (Hirsi Ali), 280
Ingersoll, Robert, 305
Inside Nature's Giants (Channel 4),
    129–31

Institute for Creation Research of San
    Diego, 412
'intelligent design' (ID), 51, 348–52,
    414–15, 417, 420
interbreeding, 184, 188–90
Iraq War, 271n, 341–2
Is There a God? (Swinburne), 362
Islam: apostasy, 306; Christianity and,
    272, 304; *Concise Commandments
    of Islam*, 314–16; converting to,
    371; *fatwa*s, 118–19, 316; jihadists,
    304; optimism, 229; radical
    Islamism, 280; science teaching
    issues, 422; *see also* Muslims
The Islamist (Husain), 280
Island Africa (Kingdon), 241

Jardine, Lisa, 85
Jeans, Sir James, 6
Jefferson, Thomas, 273, 274, 339, 343,
    346, 355
Jenkin, Fleeming, 144–5, 147,
    148–9, 150
Jenkin, Patrick, 65
Jesus: appearance, 291; Christian
    beliefs, 35–7, 274, 281, 293–4, 296,
    306; in Jefferson Bible, 273; Lewis
    on, 273–4; mother Mary, 317;
    personality, 272; political
    influence, 340
Jesus and Mo: folie à dieu (cartoons),
    366–8
Jews: beliefs, 35; Bible, 299–300; Chief
    Rabbi, 299; Clergy Project, 286;
    discrimination against, 304;
    religious revival, 272; secular,
    274–5; Silverman, 290, 292–3;
    traditions, 36; violence in Israel,
    339; *see also* Judaism
Jillette, Penn, 293n
Joad, C. E. M., 41n
Johanson, Donald, 332n, 401–3,
    410, 411

Johansson, Tove, 339
Johnson, Samuel, 14
Journal of Theoretical Biology, 417
Jowett, Benjamin, 75n
Judaism, 36, 292, 306, 364; *see also* Jews

Kacelnik, Alex, 256
Kamin, Leon, 62, 65–8
Kato, David, 339
Kauffman, Stuart, 382
Keats, John, 27, 49n, 115, 236
Keller, Helen, 279
Kelly, Kevin, 381
Kelvin, Sir William Thomson, Lord, 150–1
Kenny, Anthony, 199
Keynes, John Maynard, 295n
Khalaf, Farida, 368–71
Khayyam, Omar, 85
kin selection, 67, 101, 256, 257, 264–5, 398
King, Captain Phillip Parker, 154
Kingdon, Jonathan, 240–1, 243, 247–52
Kitcher, Philip, 223
The Knife Edge (Moore's sculpture), 88
Krauss, Lawrence, 9, 84n, 323–4, 329–37, 352–6
Küng, Hans, 222

Lacan, Jacques, 58–9
Lady Chatterley's Lover (Lawrence), 255
Lamarckian theory, 144, 145, 164
Lane Fox, Robin, 64n
Language, Truth and Logic (Ayer), 91
The Language Instinct (Pinker), 171n
LaScola, Linda, 284–6
Lawrence, D. H., 40
Lear, Edward, 395
Leonardo da Vinci, 86

Letter to a Christian Nation (Harris), 344–6
Lettvin, Jerome, 205–6, 209, 210
Levin, Bernard, 117
Lewis, C. S., 87n, 273–4
Lewontin, Richard, 62, 65–8, 394, 417, 424
Lindsay, James, 13
The Lives of a Cell (Thomas), 10
The Living Wild (Wolfe), 120
Lofting, Hugh, 16, 155
Lord of the Flies (Golding), 17
Losing Faith in Faith (Barker), 282
Lucretius, 378
Lucy (*Australopithecus*), 411
Lucy: the beginnings of humankind (Johanson and Edey), 401

Maddox, John, 42n, 116
Madison, James, 273, 339
Magrack *Maximum Science* series, 105
Malan, Natalie, 254
Malignant Sadness (Wolpert), 38
Mammals of Africa (Kingdon), 240
Mann, Thomas, 87
Margulis, Lynn, 57–61
Marx, Karl, 157
Mary, Assumption of, 317–18
Maschler, Tom, 91–2
Massey, Graham, 387n
The Matrix (film), 78n
Maturana, Humberto, 209
Maynard Smith, John: appearance and character, 386–9; career, 389–90, 417; on Haldane's 'theorems', 390, 395; kin selection theory, 264; on literature, 4; at meeting to honour Chagnon, 54; mitochondria claim, 191; phrasemaker, 390; relationship with Rose, 101; relationship with Zahavi, 94n;

The Theory of Evolution, 386, 391–2
Mayr, Ernst, 95, 187, 188, 189, 393
Medawar, Peter, 5, 11–16, 49, 153n
Mendel, Gregor, 147, 150, 383, 402
Mendelian genetics, 53, 147, 149, 188, 259, 264
Menon, Latha, 101–2
Michelangelo, 321
Michod, Richard, 256
Miller, Kenneth R., 415
Milton, Richard, 407–13
'missing link', 410
mitochondria, 10, 134, 136, 191, 243–7
Mittelstaedt, Horst, 209
models, 199–205; *see also* brain
'Modern Synthesis', 187–8, 259
Monod, Jacques, 96
Monty Python, 366, 398
Moore, Henry, 88
Morgan, Elaine, 248
Morgan, T. H., 403
Morris, Desmond, 59–60, 91
Morris, Henry, 409, 412
Moses, 319, 366, 410
Murray, John, 154, 255
music: appreciating and enjoying, 82, 105, 109–10, 163, 165–8, 181, 359; Bach's brain, 66; birdsong, 236; creativeness and logic, 26; information theory, 82; of the spheres, 352
Muslims: apostasy, 272, 306; Clergy Project, 286; cosmological beliefs, 118–19; creationist lobby, 422; fundamentalists, 404–5; leaflets against Darwin's theories, 421; suicide bombers, 360; treatment of Yazidis, 368, 370; *see also* Islam
My Name is Stardust (Bailey and Douglas Harris), 253–4

The Mysterious Universe (Jeans), 6
Mystery Dance (Margulis and Sagan), 57–61

The Naked Ape (Morris), 59–60
natural selection: bodies and minds, 224, 376–7; brain models, 212, 215–16; capacity to simulate, 219–20; Chagnon's work, 55; chance and, 152, 348; debate on unit selected, 92, 94–5; differential survival of genes, 113; domestic breeding, 416–17; gene pool record, 125, 396; goal-seeking capacity, 196–7; Jenkin's argument, 144–5, 147; selfishness, 229; 'survival of the fittest', 153, 258–9; units, 92, 94–5, 261, 395, 397; universal significance, 156–7; versus design, 349, 392, 399; within species, 409; *see also* group selection, kin selection, sexual selection
Natural Selection: domains, levels and challenges (Williams), 396, 398, 400
Natural Theology (Paley), 358, 399
Nature (journal), 42n, 57, 116, 256–7, 264, 402n
Necker Cube, 219, 226
Needham, Joseph, 86n
Nesse, Randolph M., 136
The New Encyclopedia of Unbelief (edited by Tom Flynn), 302
The New Republic, 414
New Scientist, 62, 101, 151n
New Statesman, 269, 407, 411
New York Times, 187, 331, 401, 414
Newton, Isaac: Keats's response to, 27, 49n, 115; laws of motion, 148, 174; religion, 33–4; research strategy, 50; status, 157, 378, 382, 429

Nice Guys Finish First (BBC TV), 231
Nobel Prizes: Literature, 5–6, 116;
     prizewinners, 11, 49, 71, 80, 350
North British Review, 144
Not in Our Genes (Rose, Kamin and
     Lewontin), 62–8, 222n
Nowak, Martin, 256

Orgel, Leslie, 349
Origin of Species (Darwin), 143, 145,
     152, 154, 177, 181, 255, 356, 357, 378,
     422–3
Orwell, George, 277
*Oxford Surveys in Evolutionary
     Biology*, 260n
Oxford University Gazette, 69

pagan religions, 141
pain, 173, 178–9, 216–17
Paine, Thomas, 275
Pale Blue Dot (Sagan), 6–7, 305–6,
     313, 335n
Paley, William, 111, 112, 312, 358, 399
Palfreyman, David, 69, 75n
parasites, 114, 123, 132–7, 238, 424
Parker, Dorothy, 265
Parker, Geoffrey, 256
Pavlov, Ivan, 214–15
peacocks, 93n, 234–5, 237, 310, 377
*The Peacock's Tail and the Reputation
     Reflex* (Wight), 234
Penguin Books, 79, 229, 232
Penn and Teller, 311
Perry, Rick, 274
The Phenomenon of Man (Teilhard de
     Chardin), 13–14
Pinker, Steven, 54, 163–76, 221,
     256, 358
Pittendrigh, Colin S., 394
Pius XII, Pope, 317–18
Planck, Max, 157
Pluckrose, Helen, 13
Pluto's Republic (Medawar), 14

Point Counterpoint (Huxley), 17
Popper, Karl, 50
Porco, Carolyn, 7–8
Potter, Christopher, 411–12
Private Eye, 64n
Prospect (magazine), 255
Purgatory, 318–19
Pythagoras, 157, 265

quantum theory, 110–11, 174–5, 322–4,
     355–6
Queller, David, 256
Qur'an, 277, 315, 370

Randi, James 'The Amazing', 42n, 311
Reagan, Ronald, 341
Rees, Martin, 357
Reidenberg, Joy, 129
Rescorla, R. A., 214
Richard Dawkins Award, 269
Richard Dawkins Foundation for
     Reason and Science (RDFRS),
     35, 51n, 352, 357
Ridley, Matt, 375–85
Rodgers, Michael, 102
Rose, Steven, 62, 65–8, 101
Rushdie, Salman, 404
Russell, Bertrand, 85, 305

Sagan, Carl: *Contact*, 80; *Cosmos*,
     116; *The Demon-Haunted World*,
     47–8, 116–19; eloquence, 116;
     Gifford Lectures, 335; *Pale Blue
     Dot*, 6–7, 305–6, 313–14; religious
     position, 335; writing style, 5,
     6–7, 116
Sagan, Dorion, 57–61
Samta, Saron, 341
Sarich, Vincent, 403
Sartre, Jean-Paul, 59
Schaller, George, 120
Schrödinger, Erwin, 87, 384
Schwartz, Jeffrey H., 187, 190, 191

Science (journal), 230
'Science and Literature' (Medawar lecture), 49–50
Science in the Soul (Dawkins), 16, 18, 156n, 265n, 278n
Scientific American, 329
Secular Coalition for America, 295, 343
Segerstråle, Ullica, 62, 222
Self-Made Man and his Undoing (Kingdon), 240
The Selfish Gene (Dawkins), 90–102, 106, 230, 231, 232, 238n, 396
sexual selection, 93n, 113, 234–9, 376–7
Shakespeare, William, 22
Shanks, Niall, 348–52
Sherman, Paul, 256
Shermer, Michael, 48–9
Shoemaker, Eugene, 8
sibling care, 259–60
Silverman, Herb, 290–5
Silverman, Sharon, 294
Simpson, George Gaylord, 393
Skeptic magazine, 48
Smith, Adam, 378–9
Smith, Sir David, 10–11
Smolin, Lee, 311–12
The Social Conquest of Earth (Wilson), 255
sociobiology, 62, 64–7, 221–3
Sociobiology (Wilson), 222, 264
Socrates, 396
Sokal, Alan, 12–13, 57
Somerscales, Gillian, 18
speech, 166–8, 170, 172
Spencer, Herbert, 153, 258
Spinoza, Baruch, 35
StarTalk (radio show), 21
Stenger, Victor, 355–6
Stewart, Paul, 158–60
Sunday Times, 38, 361
Sweeney, Julia, 304–5, 306
Swinburne, Richard, 361–4

Swiss, Jamy Ian, 311

Talking Science (2004), 105
Tarnita, Corina, 256
Tasmanian wolf, 126
Tattersall, Ian, 187, 190, 191
Taung Child, 86–7
Taylor, Jeremy, 231
Teacher Institute for Evolutionary Science (TIES), 51n, 351n, 421n
Teilhard de Chardin, Pierre, 13–14
Texas Freethought Convention, 269
Thackeray, Francis, 87n
Thatcher, Margaret, 101
The Theory of Evolution (Maynard Smith), 386–92
The Theory of Moral Sentiments (Smith), 378–9
This Will Change Everything (symposium), 183
Thomas, Frédéric, 132, 133
Thomas, Lewis, 5, 10
Thompson, D'Arcy, 15–16
Thomson, J. Anderson, 357–60
A Thousand Brains (Hawkins), 176–82
The Time Machine (Wells), 17
Timeline (Crichton), 79–80
The Times, 47n, 116, 151–2
Times Literary Supplement, 12, 240, 418
Tinbergen, Niko, 71
Tolkien, J. R. R., 87n
Tooby, John, 221, 222, 223, 256
Toscanini, Arturo, 87
'Transgressing the boundaries: towards a transformative hermeneutics of quantum gravity' (Sokal), 12–13, 57
Trivers, Robert, 54, 96n, 102, 256, 390n, 398
Twain, Mark, 73n
Tyson, Neil deGrasse, 7, 21

The Unbelievers (film), 329n
United States of America: belief in
    'intelligent design', 50–1; Bible Belt,
    292, 297; Founding Fathers, 338,
    343–4; religiosity, 33–4, 273, 304;
    theocracy and 'theocratic attack',
    270, 304, 338–9, 341, 342; views on
    evolution, 401, 403, 420–1
universe: age, 24, 279, 314, 320, 330,
    346, 352, 354, 429; Big Bang, 80,
    324, 353–4; brain's model of
    reality, 177; expanding, 5, 313,
    324, 352–3; God and, 35; laws of,
    138, 304; multiverse theory,
    311–12, 335; origin, 324–5;
    'Steady State' theory, 80, 83
A Universe from Nothing (Krauss), 9,
    84n, 323–4, 329n, 352–6
The Unnatural Nature of Science
    (Wolpert), 38
Unweaving the Rainbow (Dawkins),
    26–7, 49n, 100, 115, 226,
    227, 429
Ussher, James, 314

Van Gogh, Vincent, 22
Variation under Domestication
    (Darwin), 144
Vasquez, Bertha, 51n, 421n
Vaulting Ambition (Kitcher), 223
Velikovsky, Immanuel, 410
Victoria, Queen, 339n
Voltaire, 64n, 395
von Holst, Erich, 209
The Voyage of the Beagle (Darwin),
    143, 154–6, 158

Wagner, Andreas, 382, 383
Wallace, Alfred Russel: relationship
    with Darwin, 149–50, 176, 234,
    237–8, 376–7; 'survival of the
    fittest', 258; travelling naturalist,
    379; view of sexual selection
    theory, 234, 237–8, 377
Walter, Bruno, 87
The War of the Worlds (Wells), 79
Watson, James, 15, 384
Web of Stories (online collection),
    387n
Weinberg, Steven, 356
Weinberg, Wilhelm, 148
Weissman, August, 143–4
Wells, H. G., 17, 79
West, Stuart, 256
West-Eberhard, Mary Jane, 256
What Technology Wants (Kelly), 381
White, Amiyah, 341
Why Evolution is True (Coyne),
    418–27
Why We Believe in Gods (Thomson),
    357
Why We Get Sick (Nesse and
    Williams), 136n
Wickramasinghe, Chandra, 151
Wight, Robin, 234, 238
Wilberforce, Bishop Samuel, 402, 406
Williams, George C., 136, 390,
    393–400
Wilson, Allan, 244–6, 403
Wilson, Edward O., 222, 223, 255–7,
    260–2, 264–5
Wilson, Margot, 221
Woese, Carl R., 403
Wolfe, Art, 120
Wolpert, Lewis, 38
Wordsworth, William, 197, 198, 308
Wrangham, Richard, 54, 256, 402n
Wren, Christopher, 321
Wright, Sewall, 144, 148, 417
Wynne-Edwards, V. C., 262

Yazidi people, 368

Zahavi, Amotz, 93, 238–9

ABOUT THE AUTHOR

Richard Dawkins was first catapulted to fame with his iconic work of 1976, *The Selfish Gene*, which he followed with a string of prestigious bestselling books including *The Blind Watchmaker, Climbing Mount Improbable, The Ancestor's Tale* and *The God Delusion*. He is also the author of the anthologies *A Devil's Chaplain* and *Science in the Soul* and two volumes of autobiography, *An Appetite for Wonder* and *Brief Candle in the Dark*. He is a Fellow of both the Royal Society and the Royal Society of Literature, and the recipient of numerous honours and awards. He remains a fellow of New College, Oxford. In 2013, Dawkins was voted the world's top thinker in *Prospect* magazine's poll of 10,000 readers from over a hundred countries.

#booksdofurnishalife
@richarddawkins